AVIATION WEATHER

AVIATION WEATHER

Arco Editorial Board

ARCO PUBLISHING, INC.
NEW YORK

Published by Arco Publishing, Inc.
219 Park Avenue South, New York, N.Y. 10003

Copyright © 1979 by Arco Publishing, Inc.

All rights reserved. No part of this book may be
reproduced, by any means, without permission in
writing from the publisher.

Library of Congress Cataloging in Publication Data

Arco Publishing, Inc., New York.
  Aviation weather.

  1. Meteorology in aeronautics. I. Title.
TL556.A84   629.132'4   78-23251
  ISBN 0-668-04413-6 (Paper Edition)

Printed in the United States of America

CONTENTS

Part I: What You Should Know About Weather   1

  Chapter 1. The Earth's Atmosphere   3
    Composition   4
    Vertical Structure   4
    The Standard Atmosphere   4
    Density and Hypoxia   5

  Chapter 2. Temperature   7
    Temperature Scales   8
    Heat and Temperature   8
    Temperature Variations   9
    In Closing   12

  Chapter 3. Atmospheric Pressure and Altimetry   13
    Atmosphere Pressure   13
    Altimetry   19
    In Closing   23

  Chapter 4. Wind   24
    Convection   24
    Pressure Gradient Force   25
    Coriolis Force   26
    The General Circulation   27
    Friction   31
    The Jet Stream   32
    Local and Small Scale Winds   32
    Wind Shear   35
    Wind, Pressure Systems, and Weather   36

  Chapter 5. Moisture, Cloud Formation, and Precipitation   38
    Water Vapor   38
    Change of State   40
    Cloud Formation   43
    Precipitation   43
    Land and Water Effects   44
    In Closing   46

  Chapter 6. Stable and Unstable Air   47
    Changes Within Upward and Downward Moving Air   47
    Stability and Instability   49
    What Does It All Mean?   52

  Chapter 7. Clouds   53
    Identification   53
    Signposts in the Sky   62

  Chapter 8. Air Masses and Fronts   63
    Air Masses   63
    Fronts   64
    Fronts and Flight Planning   78

Chapter 9. Turbulence ... 79
   Convective Currents ... 80
   Obstructions to Wind Flow ... 82
   Wind Shear ... 86
   Wake Turbulence ... 88
   In Closing ... 90

Chapter 10. Icing ... 91
   Structural Icing ... 92
   Induction System Icing ... 97
   Instrument Icing ... 98
   Icing and Cloud Types ... 99
   Other Factors in Icing ... 100
   Ground Icing ... 102
   Frost ... 102
   In Closing ... 102

Chapter 11. Thunderstorms ... 105
   Where and When? ... 105
   They Don't Just Happen ... 111
   The Inside Story ... 111
   Rough and Rougher ... 112
   Hazards ... 113
   Thunderstorms and Radar ... 120
   Do's and Don'ts of Thunderstorm Flying ... 121

Chapter 12. Common IFR Producers ... 124
   Fog ... 125
   Low Stratus Clouds ... 127
   Haze and Smoke ... 128
   Blowing Restrictions to Visibility ... 128
   Precipitation ... 129
   Obscured or Partially Obscured Sky ... 129
   In Closing ... 129

Part II: Aviation Weather Services ... **133**

Chapter 13. The Aviation Weather Service Program ... 135
   Data Flow ... 135
   Observations ... 135
   Meteorological Centers and Forecast Offices ... 136
   Service Outlets ... 136
   Users ... 137

Chapter 14. Surface Aviation Weather Reports ... 149
   Station Designator ... 149
   Type and Time of Report ... 150
   Sky Condition and Ceiling ... 151
   Visibilty ... 153
   Weather and Obstructions to Vision ... 154
   Sea Level Pressure ... 155
   Temperature and Dew Point ... 155
   Wind ... 156

Altimeter Setting	156
Remarks	156
Report Identifiers	160
Reading the Surface Aviation Weather Report	161

Chapter 15. Pilot and Radar Reports — 162
- Pilot Weather Reports (PIREPS) — 162
- Radar Weather Reports (RAREPS) — 163

Chapter 16. Aviation Weather Forecasts — 166
- Terminal Forecasts — 166
- Area Forecast (FA) — 173
- TWEB Route Forecasts and Synopsis — 175
- Inflight Advisories (WS, WA, WAC) — 176
- Winds and Temperatures Aloft Forecast (FD) — 177
- Special Flight Forecast — 178
- Hurricane Advisory (WH) — 179
- Convective Outlook (AC) — 179
- Severe Weather Watch Bulletin (WW) — 179

Chapter 17. Surface Analysis — 182
- Valid Time — 182
- Isobars — 182
- Pressure Systems — 182
- Fronts — 182
- Other Information — 184
- Using the Chart — 184

Chapter 18. Weather Depiction Chart — 186
- Plotted Data — 186
- Analysis — 188
- Using the Chart — 188

Chapter 19. Radar Summary Chart — 189
- Echo Pattern and Coverage — 189
- Weather Associated with Echoes — 189
- Intensity and Trend of Precipitation — 189
- Heights of Echo Bases and Tops — 190
- Movement of Echoes — 190
- Additional Information — 190
- Using the Chart — 190

Chapter 20. Significant Weather Prognostics — 193
- Domestic Flights — 193
- International Flights — 195
- Using Significant Weather Progs — 197

Chapter 21. Winds and Temperatures Aloft — 201
- Forecast Winds and Temperatures Aloft (FD) — 201
- Observed Winds Aloft — 201
- Using the Charts — 201

Chapter 22. Freezing Level Chart ... 204
    Plotted Data ... 204
    Analysis ... 204
    Using the Chart ... 204

Chapter 23. Stability Chart ... 206
    Lifted Index ... 206
    K Index ... 206
    Stability Analysis ... 206
    Using the Chart ... 208

Chapter 24. Severe Weather Outlook Chart ... 209
    General Thunderstorms ... 209
    Severe Thunderstorms ... 209
    Tornadoes ... 209
    Using the Chart ... 209

Chapter 25. Constant Pressure Charts ... 211
    Plotted Data ... 211
    Analysis ... 212
    Three Dimensional Aspects ... 213
    Using the Charts ... 213

Chapter 26. Constant Pressure Prognostics ... 220
    Height Contours/Streamlines ... 220
    Temperature ... 220
    Windspeed ... 220
    Formats ... 220
    Using the Charts ... 222

Chapter 27. Tropopause, Max Wind, and Wind Shear Charts ... 226
    Observed Tropopause Chart ... 226
    Domestic Tropopause Wind and Wind Shear Progs ... 226
    International Tropopause and Wind Shear Progs ... 230

Chapter 28. Tables and Conversion Graphs ... 235
    Icing Intensities ... 235
    Turbulence Intensities ... 236
    Locations of Probable Turbulence by Intensities versus
        Weather and Terrain Features ... 237
    Standard Conversions ... 238
    Density Altitude Computation ... 239
    Selected Contractions ... 240
    Acronyms ... 242

Chapter 29. Weather Information
    FSS-CS/T and National Weather Service Telephone Numbers ... 243

Part III: Aviation Weather—Review Questions for FAA Written Test ... **251**

**Part IV: NTSB Special Study—Nonfatal, Weather-Involved
General Aviation Accidents** ... **273**

Part V: Glossary of Weather Terms ... **293**

AVIATION WEATHER

Part I
What You Should Know About Weather

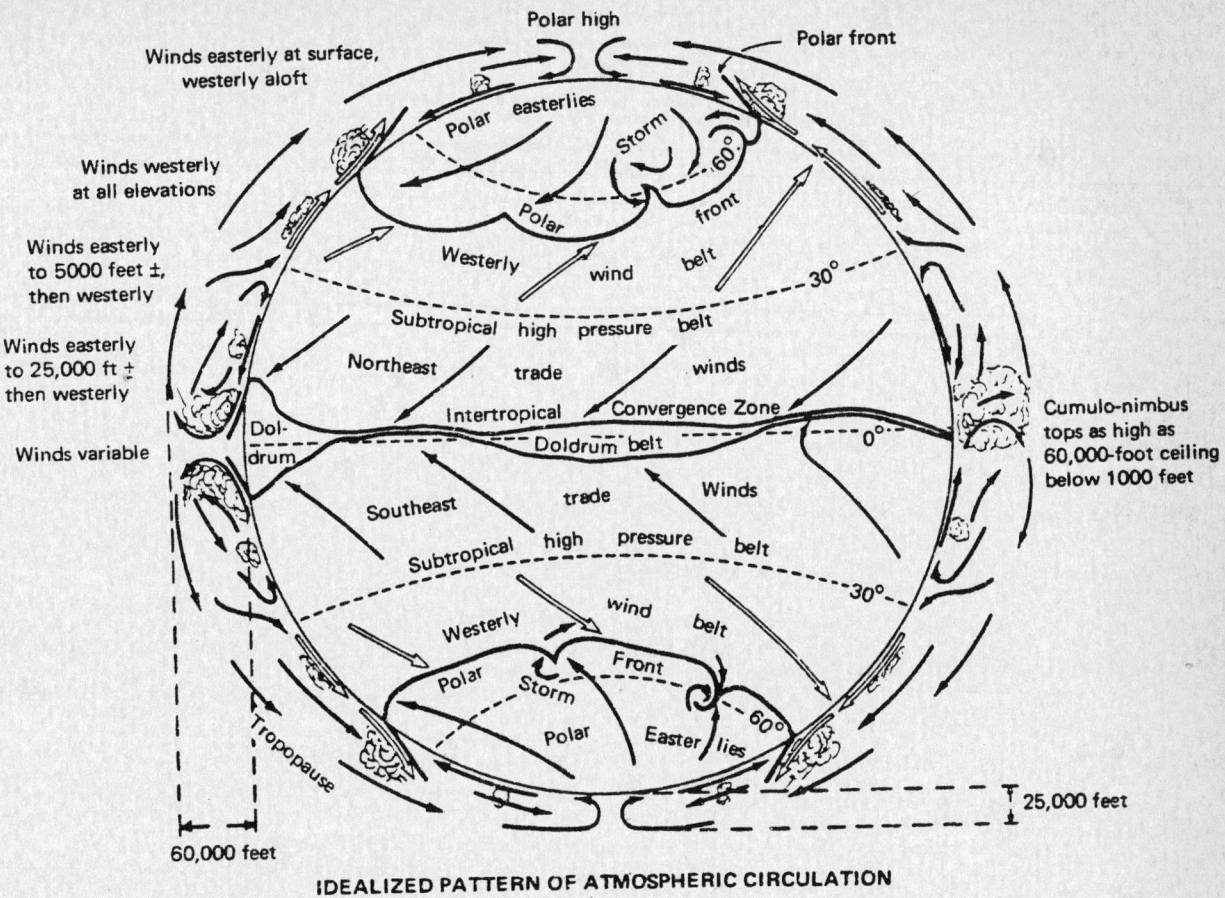

IDEALIZED PATTERN OF ATMOSPHERIC CIRCULATION

Chapter 1
THE EARTH'S ATMOSPHERE

Planet Earth is unique in that its atmosphere sustains life as we know it. Weather—the state of the atmosphere—at any given time and place strongly influences our daily routine as well as our general life patterns. Virtually all of our activities are affected by weather, but of all man's endeavors, none is influenced more intimately by weather than aviation.

Weather is complex and at times difficult to understand. Our restless atmosphere is almost constantly in motion as it strives to reach equilibrium. These never-ending air movements set up chain reactions which culminate in a continuing variety of weather. Later chapters in this book delve into the atmosphere in motion. This chapter looks briefly at our atmosphere in terms of its composition; vertical structure; the standard atmosphere; and of special concern to you, the pilot, density and hypoxia.

COMPOSITION

Air is a mixture of several gases. When completely dry, it is about 78% nitrogen and 21% oxygen. The remaining 1% is other gases such as Argon, Carbon Dioxide, Neon, Helium, and others. Figure 1 graphs these proportions. However, in nature, air is never completely dry. It always contains some water vapor in amounts varying from *almost* zero to about 5% by volume. As water vapor content increases, the other gases decrease proportionately.

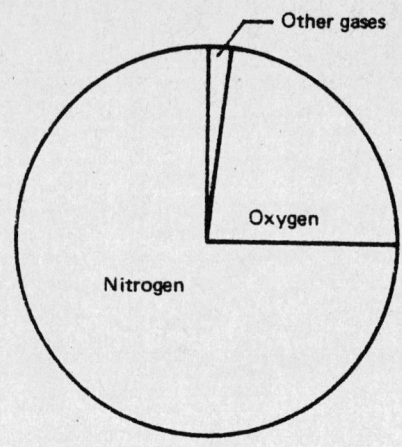

FIGURE 1. Composition of a dry atmosphere. Nitrogen comprises about 78%; oxygen, about 21%; and other gases, about 1%. When water vapor is added, the percentages decrease proportionately. Water vapor varies from almost none to about 5% by volume.

VERTICAL STRUCTURE

We classify the atmosphere into layers, or spheres, by characteristics exhibited in these layers. Figure 2 shows one division which we use in this book. Since most weather occurs in the troposphere and since most flying is in the troposphere and stratosphere, we restrict our discussions mostly to these two layers.

The TROPOSPHERE is the layer from the surface to an average altitude of about 7 miles. It is characterized by an overall decrease of temperature with increasing altitude. The height of the troposphere varies with latitude and seasons. It slopes from about 20,000 feet over the poles to about 65,000 feet over the Equator; and it is higher in summer than in winter.

At the top of the troposphere is the TROPOPAUSE, a very thin layer marking the boundary between the troposphere and the layer above. The height of the tropopause and certain weather phenomena are related. Chapter 13 discusses in detail the significance of the tropopause to flight.

Above the tropopause is the STRATOSPHERE. This layer is typified by relatively small changes in temperature with height except for a warming trend near the top.

THE STANDARD ATMOSPHERE

Continual fluctuations of temperature and pressure in our restless atmosphere create some problems for engineers and meteorologists who require a fixed standard of reference. To arrive at a standard, they averaged conditions throughout the atmosphere for all latitudes, seasons, and altitudes. The result is a STANDARD ATMOSPHERE with specified sea-level temperature and pressure and specific rates of change of temperature and pressure with height. It is the standard for calibrating the pressure altimeter and developing aircraft performance data. We refer to it often throughout this book.

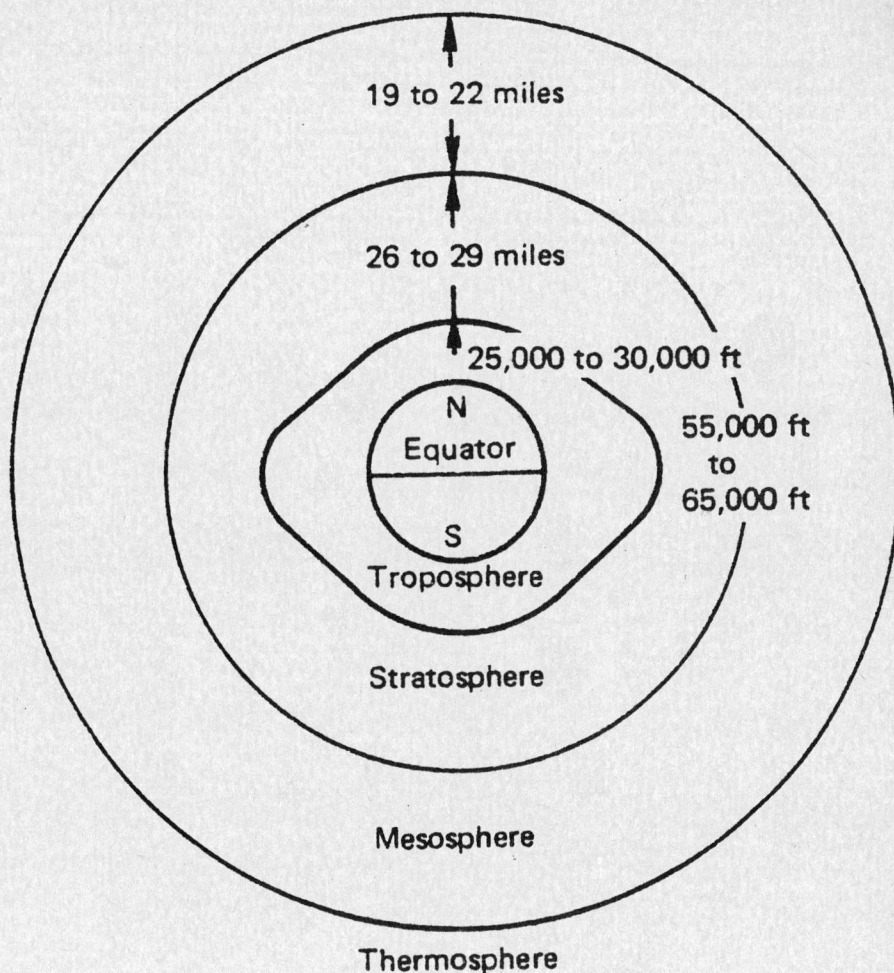

FIGURE 2. The atmosphere divided into layers based on temperature. This book concentrates on the lower two layers, the troposphere and the stratosphere.

DENSITY AND HYPOXIA

Air is matter and has weight. Since it is gaseous, it is compressible. Pressure the atmosphere exerts on the surface is the result of the weight of the air above. Thus, air near the surface is much more dense than air at high altitudes. This decrease of density and pressure with height enters frequently into our discussions in later chapters.

The decrease in air density with increasing height has a physiological effect which we cannot ignore. The rate at which the lungs absorb oxygen depends on the partial pressure exerted by oxygen in the air. The atmosphere is about one-fifth oxygen, so the oxygen pressure is about one-fifth the total pressure at any given altitude. Normally, our lungs are accustomed to an oxygen pressure of about 3 pounds per square inch. But, since air pressure decreases as altitude increases, the oxygen pressure also decreases. A pilot continuously gaining altitude or making a prolonged flight at high altitude without supplemental oxygen will likely suffer from HYPOXIA—a deficiency of oxygen. The effects are a feeling of exhaustion; an impairment of vision and judgment; and finally, unconsciousness. Cases are known where a person lapsed into unconsciousness without realizing he was suffering the effects.

When flying at or above 10,000 feet, force yourself to remain alert. Any feeling of drowsiness or undue fatigue may be from hypoxia. If you do not have oxygen, descend to a lower altitude. If fatigue or drowsiness continues after descent, it is caused by something other than hypoxia.

A safe procedure is to use auxiliary oxygen during prolonged flights above 10,000 feet and for even short flights above 12,000 feet. Above about 40,000 feet, pressurization becomes essential.

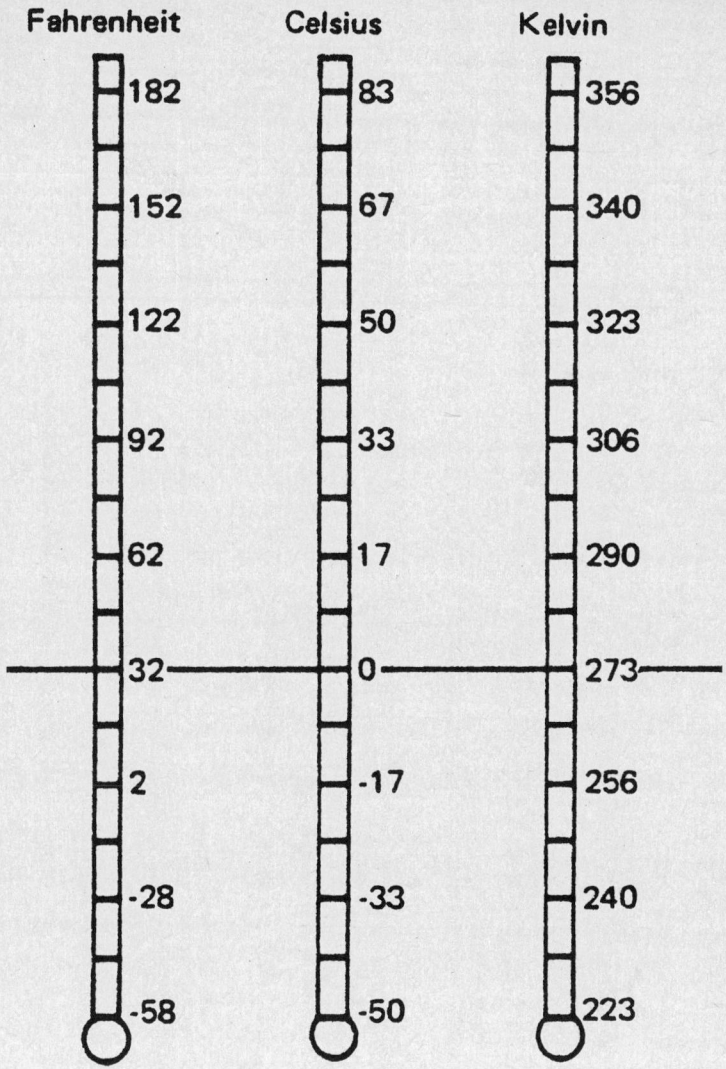

Chapter 2
TEMPERATURE

Since early childhood, you have expressed the comfort of weather in degrees of temperature. Why, then, do we stress temperature in aviation weather? Look at your flight computer; temperature enters into the computation of most parameters on the computer. In fact, temperature can be critical to some flight operations. As a foundation for the study of temperature effects on aviation and weather, this chapter describes commonly used temperature scales, relates heat and temperature, and surveys temperature variations both at the surface and aloft.

Temperature

TEMPERATURE SCALES

Two commonly used temperature scales are Celsius (Centigrade) and Fahrenheit. The Celsius scale is used exclusively for upper air temperatures and is rapidly becoming the world standard for surface temperatures also.

Traditionally, two common temperature references are the melting point of pure ice and the boiling point of pure water at sea level. The melting point of ice is 0° C or 32° F; the boiling point of water is 100° C or 212° F. Thus, the difference between melting and boiling is 100 degrees Celsius or 180 degrees Fahrenheit; the ratio between degrees Celsius and Fahrenheit is 100/180 or 5/9. Since 0° F is 32 Fahrenheit degrees colder than 0° C, you must apply this difference when comparing temperatures on the two scales. You can convert from one scale to the other using one of the following formulae:

$$C = \frac{5}{9}(F - 32)$$

$$F = \frac{9}{5}C + 32$$

where C is degrees Celsius and F is degrees Fahrenheit. Figure 3 compares the two scales. Many flight computers provide for direct conversion of temperature from one scale to the other. Section 16, AVIATION WEATHER SERVICES has a graph for temperature conversion.

Temperature we measure with a thermometer. But what makes a thermometer work? Simply the addition or removal of heat. Heat and temperature are not the same; how are they related?

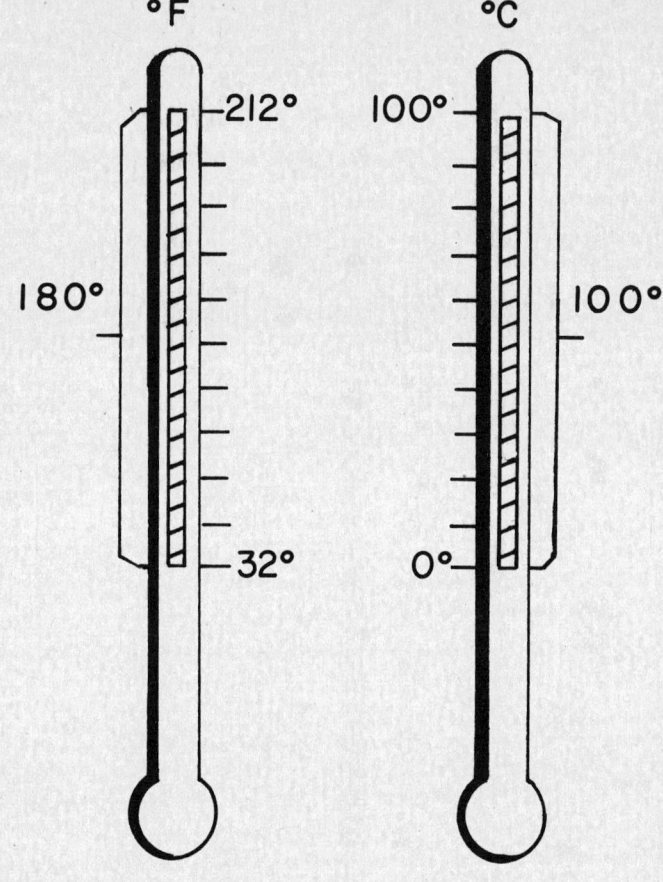

FIGURE 3. The two temperature scales in common use are the Fahrenheit and the Celsius. 9 degrees on the Fahrenheit scale equal 5 degrees on the Celsius.

HEAT AND TEMPERATURE

Heat is a form of energy. When a substance contains heat, it exhibits the property we measure as temperature—the degree of "hotness" or "coldness." A specific amount of heat absorbed by or removed from a substance raises or lowers its temperature a definite amount. However, the amount of temperature change depends on characteristics of the substance. Each substance has its unique temperature change for the specific change in heat. For example, if a land surface and a water surface have the same temperature and an equal amount of heat is added, the land surface becomes hotter than the water surface. Conversely, with equal heat loss, the land becomes colder than the water.

The Earth receives energy from the sun in the form of solar radiation. The Earth and its atmosphere reflect about 55 percent of the radiation and absorb the remaining 45 percent converting it to heat. The Earth, in turn, radiates energy, and this outgoing radiation is "terrestrial radiation." It is evident that the average heat gained from incoming solar radiation must equal heat lost through terrestrial radiation in order to keep the earth from getting progressively hotter or colder. However, this balance is world-wide; we must consider regional and local imbalances which create temperature variations.

TEMPERATURE VARIATIONS

The amount of solar energy received by any region varies with time of day, with seasons, and with latitude. These differences in solar energy create temperature variations. Temperatures also vary with differences in topographical surface and with altitude. These temperature variations create forces that drive the atmosphere in its endless motions.

DIURNAL VARIATION

Diurnal variation is the change in temperature from day to night brought about by the daily rotation of the Earth. The Earth receives heat during the day by solar radiation but continually loses heat by terrestrial radiation. Warming and cooling depend on an imbalance of solar and terrestrial radiation. During the day, solar radiation exceeds terrestrial radiation and the surface becomes warmer. At night, solar radiation ceases, but terrestrial radiation continues and cools the surface. Cooling continues after sunrise until solar radiation again exceeds terrestrial radiation. Minimum temperature usually occurs after sunrise, sometimes as much as one hour after. The continued cooling after sunrise is one reason that fog sometimes forms shortly after the sun is above the horizon. We will have more to say about diurnal variation and topographic surfaces.

SEASONAL VARIATION

In addition to its daily rotation, the Earth revolves in a complete orbit around the sun once each year. Since the axis of the Earth tilts to the plane of orbit, the angle of incident solar radiation varies seasonally between hemispheres. The Northern Hemisphere is warmer in June, July, and August because it receives more solar energy than does the Southern Hemisphere. During December, January, and February, the opposite is true; the Southern Hemisphere receives more solar energy and is warmer. Figures 4 and 5 show these seasonal surface temperature variations.

VARIATION WITH LATITUDE

The shape of the Earth causes a geographical variation in the angle of incident solar radiation. Since the Earth is essentially spherical, the sun is more nearly overhead in equatorial regions than at higher latitudes. Equatorial regions, therefore, receive the most radiant energy and are warmest. Slanting rays of the sun at higher latitudes deliver less energy over a given area with the least being received at the poles. Thus, temperature varies with latitude from the warm Equator to the cold poles. You can see this average temperature gradient in figures 4 and 5.

VARIATIONS WITH TOPOGRAPHY

Not related to movement or shape of the earth are temperature variations induced by water and terrain. As stated earlier, water absorbs and radiates energy with less temperature change than does land. Large, deep water bodies tend to minimize temperature changes, while continents favor large changes. Wet soil such as in swamps and marshes is almost as effective as water in suppressing temperature changes. Thick vegetation tends to control temperature changes since it contains some water and also insulates against heat transfer between the ground and the atmosphere. Arid, barren surfaces permit the greatest temperature changes.

These topographical influences are both diurnal and seasonal. For example, the difference between a daily maximum and minimum may be 10° or less over water, near a shore line, or over a swamp or marsh, while a difference of 50° or more is common over rocky or sandy deserts. Figures 4 and 5 show the seasonal topographical variation. Note that in the Northern Hemisphere in July, temperatures are warmer over continents than over oceans; in January they are colder over continents than over oceans. The opposite is true in the Southern Hemisphere, but not as pronounced because of more water surface in the Southern Hemisphere.

To compare land and water effect on seasonal temperature variation, look at northern Asia and at southern California near San Diego. In the deep continental interior of northern Asia, July average temperature is about 50° F; and January average, about −30° F. Seasonal range is about 80° F. Near San Diego, due to the proximity of the Pacific

Temperature

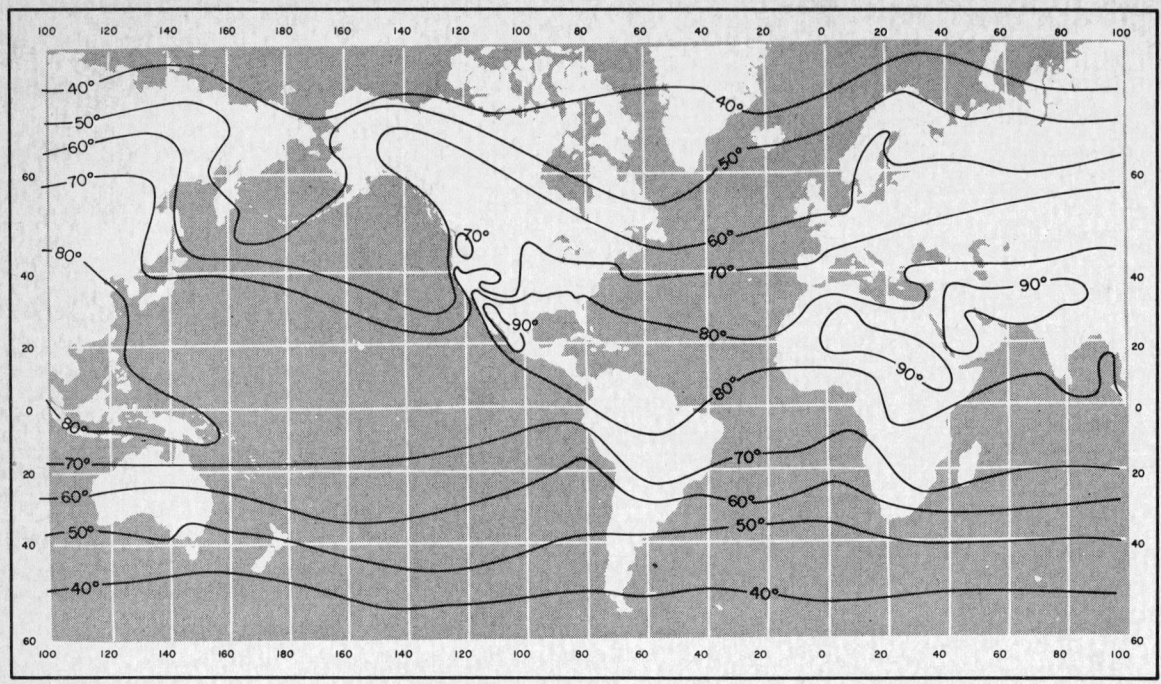

FIGURE 4. World-wide average surface temperatures in July. In the Northern Hemisphere, continents generally are warmer than oceanic areas at corresponding latitudes. The reverse is true in the Southern Hemisphere, but the contrast is not so evident because of the sparcity of land surfaces.

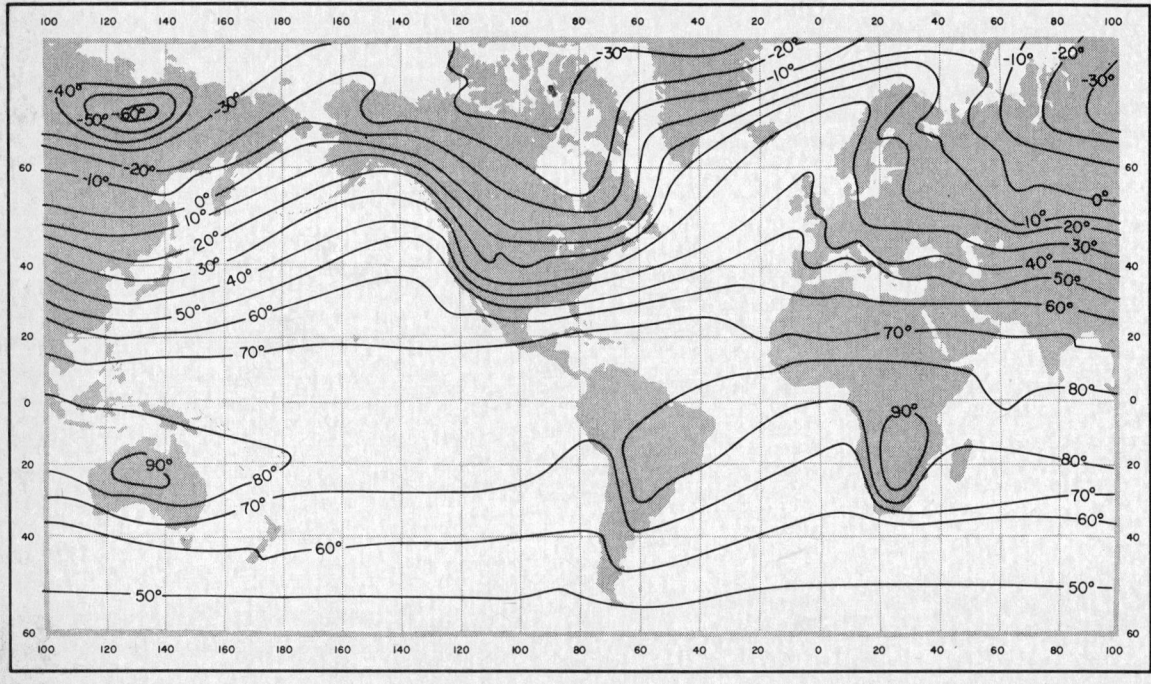

FIGURE 5. World-wide average surface temperatures in January when the Northern Hemisphere is in the cold season and the Southern Hemisphere is warm. Note that in the Northern Hemisphere, continents are colder than oceanic areas at corresponding latitudes, and in the Southern Hemisphere continents are warmer than oceans.

Ocean, July average is about 70° F and January average, 50° F. Seasonal variation is only about 20° F.

Abrupt temperature differences develop along lake and ocean shores. These variations generate pressure differences and local winds which we will study in later chapters. Figure 6 illustrates a possible effect.

Prevailing wind is also a factor in temperature controls. In an area where prevailing winds are from large water bodies, temperature changes are rather small. Most islands enjoy fairly constant temperatures. On the other hand, temperature changes are more pronounced where prevailing wind is from dry, barren regions.

Air transfers heat slowly from the surface upward. Thus, temperature changes aloft are more gradual than at the surface. Let's look at temperature changes with altitude.

VARIATION WITH ALTITUDE

In chapter 1, we learned that temperature normally decreases with increasing altitude throughout the troposphere. This *decrease of temperature with altitude* is defined as *lapse rate*. The average decrease of temperature—average lapse rate—in the troposphere is 2° C per 1,000 feet. But since this is an average, the exact value seldom exists. In fact, temperature sometimes increases with height through a layer. *An increase in temperature with altitude is* defined as *an inversion*, i.e., lapse rate is inverted.

An inversion often develops near the ground on clear, cool nights when wind is light. The ground radiates and cools much faster than the overlying air. Air in contact with the ground becomes cold while the temperature a few hundred feet above changes very little. Thus, temperature increases

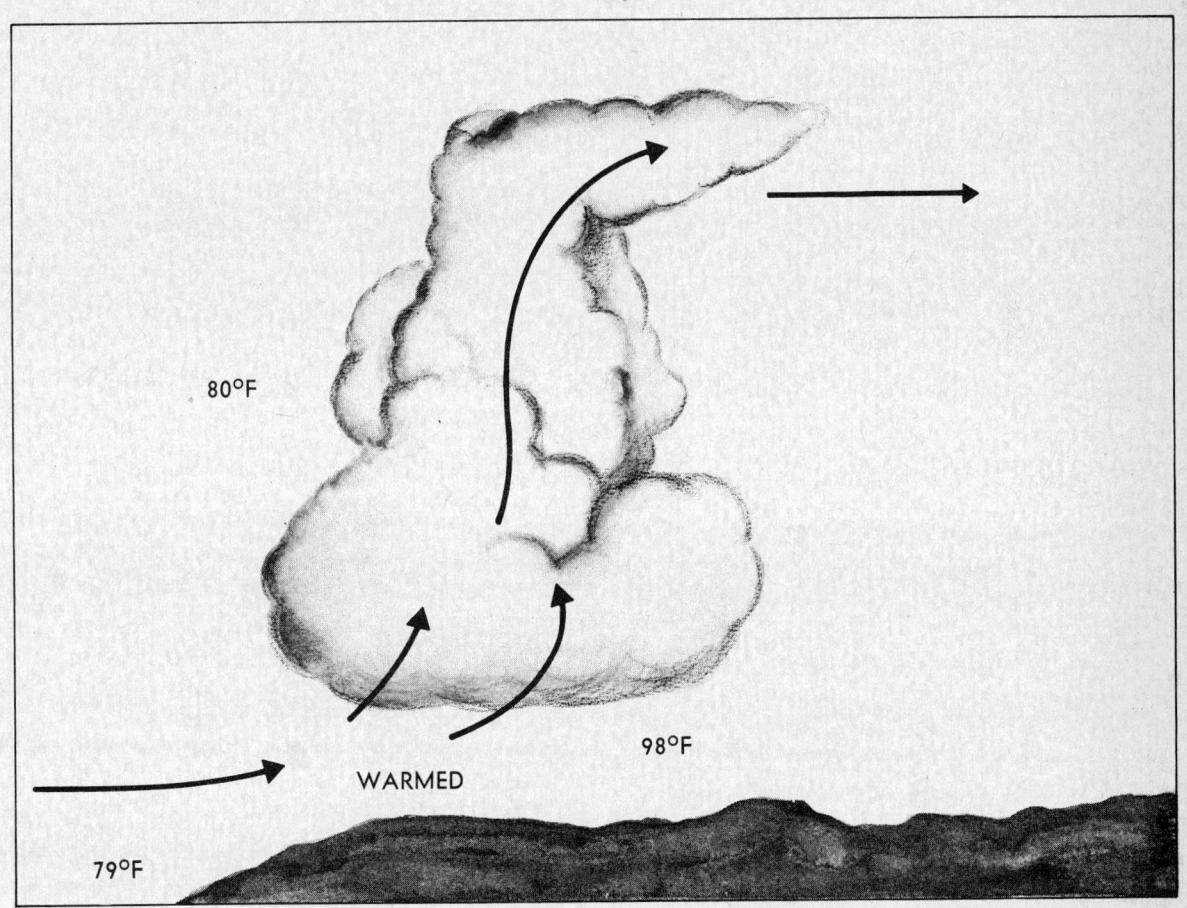

FIGURE 6. Temperature differences create air movement and, at times, cloudiness.

with height. Inversions may also occur at any altitude when conditions are favorable. For example, a current of warm air aloft overrunning cold air near the surface produces an inversion aloft. Figure 7 diagrams temperature inversions both surface and aloft. Inversions are common in the stratosphere.

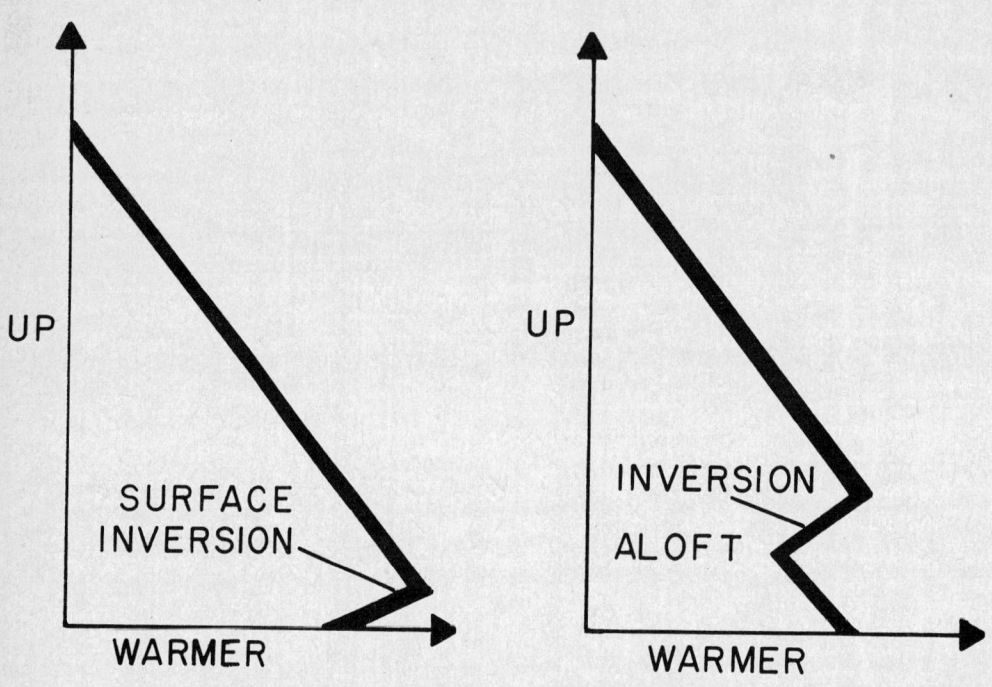

FIGURE 7. Inverted lapse rates or "inversions." A lapse rate is a decrease of temperature with height. An inversion is an increase of temperature with height, i.e., the lapse rate is inverted. Shown here are a surface inversion and an inversion aloft.

IN CLOSING

Temperature affects aircraft performance and is critical to some operations. Following are some operational pointers to remember, and most of them are developed in later chapters:

1. The aircraft thermometer is subject to inaccuracies no matter how good the instrument and its installation. Position of the aircraft relative to the sun can cause errors due to radiation, particularly on a parked aircraft. At high speeds, aerodynamical effects and friction are basically the causes of inaccuracies.
2. High temperature reduces air density and reduces aircraft performance (chapter 3).
3. Diurnal and topographical temperature variations create local winds (chapter 4).
4. Diurnal cooling is conducive to fog (chapter 5).
5. Lapse rate contributes to stability (chapter 6), cloud formation (chapter 7), turbulence (chapter 9), and thunderstorms (chapter 11).
6. An inversion aloft permits warm rain to fall through cold air below. Temperature in the cold air can be critical to icing (chapter 10).
7. A ground based inversion favors poor visibility by trapping fog, smoke, and other restrictions into low levels of the atmosphere (chapter 12).

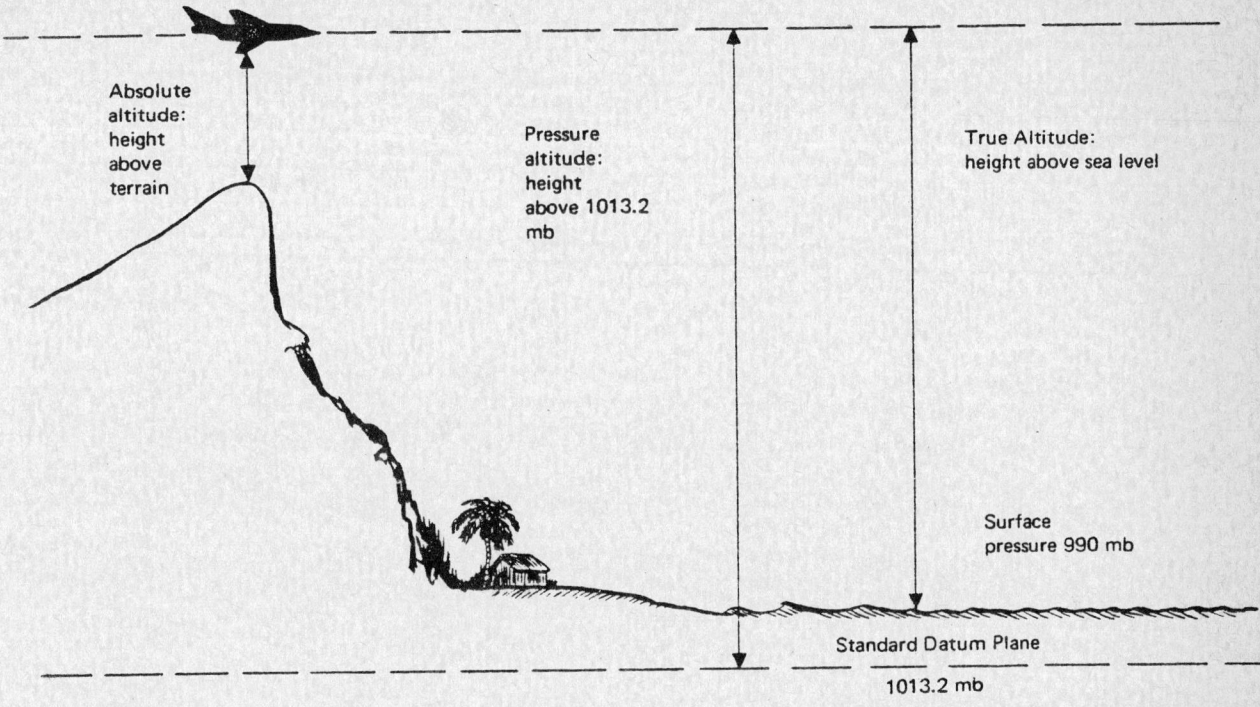

Chapter 3
ATMOSPHERIC PRESSURE AND ALTIMETRY

When you understand pressure, its measurement, and effects of temperature and altitude on pressure, you can more readily grasp the significance of pressure and its application to altimetry.

ATMOSPHERIC PRESSURE

Atmospheric pressure is the force per unit area exerted by the weight of the atmosphere. Since air is not solid, we cannot weigh it with conventional scales. Yet, Toricelli proved three centuries ago that he could weigh the atmosphere by balancing it against a column of mercury. He actually measured pressure converting it directly to weight.

MEASURING PRESSURE

The instrument Toricelli designed for measuring pressure is the barometer. Weather services and the aviation community use two types of barometers in measuring pressure—the mercurial and aneroid.

Atmospheric Pressure and Altimetry

The Mercurial Barometer

The mercurial barometer, diagrammed in figure 8, consists of an open dish of mercury into which we place the open end of an evacuated glass tube. Atmospheric pressure forces mercury to rise in the tube. At stations near sea level, the column of mercury rises on the average to a height of 29.92 inches or 760 millimeters. In other words, a column of mercury of that height weighs the same as a column of air having the same cross section as the column of mercury and extending from sea level to the top of the atmosphere.

Why do we use mercury in the barometer? Mercury is the heaviest substance available which remains liquid at ordinary temperatures. It permits the instrument to be of manageable size. We could use water, but at sea level the water column would be about 34 feet high.

The Aneroid Barometer

Essential features of an aneroid barometer illustrated in figure 9 are a flexible metal cell and the registering mechanism. The cell is partially evacuated and contracts or expands as pressure changes. One end of the cell is fixed, while the other end moves the registering mechanism. The coupling mechanism magnifies movement of the cell driving an indicator hand along a scale graduated in pressure units.

Pressure Units

Pressure is expressed in many ways throughout the world. The term used depends somewhat on its application and the system of measurement. Two popular units are "inches of mercury" or "millimeters of mercury." Since pressure is force per unit area, a more explicit expression of pressure is "pounds per square inch" or "grams per square centimeter." The term "millibar" precisely expresses pressure as a force per unit area, one millibar being a force of 1,000 dynes per square centimeter. The millibar is rapidly becoming a universal pressure unit.

Station Pressure

Obviously, we can measure pressure only at the point of measurement. The pressure measured at a station or airport is "station pressure" or the actual pressure at field elevation. We know that pressure at high altitude is less than at sea level or low altitude. For instance, station pressure at Denver is less than at New Orleans. Let's look more closely at some factors influencing pressure.

PRESSURE VARIATION

Pressure varies with altitude and temperature of the air as well as with other minor influences which we neglect here.

Altitude

As we move upward through the atmosphere, weight of the air above becomes less and less. If we carry a barometer with us, we can measure a decrease in pressure as weight of the air above decreases. Within the lower few thousand feet of the troposphere, pressure decreases roughly one inch for each 1,000 feet increase in altitude. The higher we go, the slower is the rate of decrease with height.

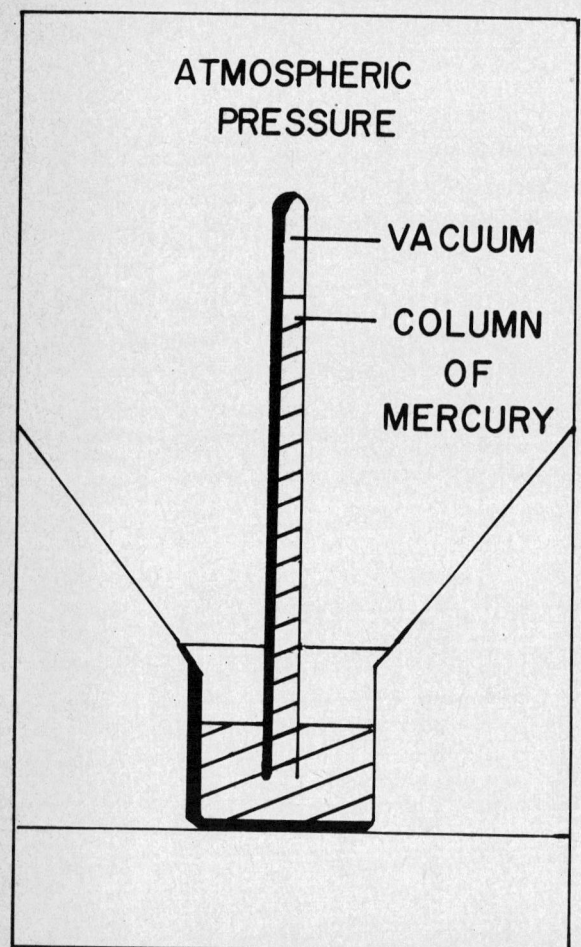

FIGURE 8. The mercurial barometer. Atmospheric pressure forces mercury from the open dish upward into the evacuated glass tube. The height of the mercury column is a measure of atmospheric pressure.

Atmospheric Pressure and Altimetry

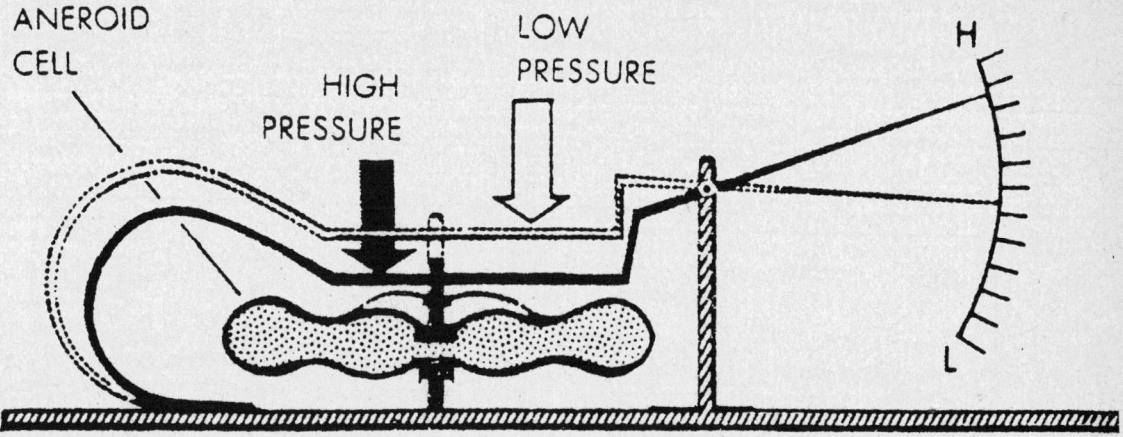

FIGURE 9. The aneroid barometer. The aneroid consists of a partially evacuated metal cell, a coupling mechanism, and an indicator scale. The cell contracts and expands with changing pressure. The coupling mechanism drives the indicator along a scale graduated in pressure units.

Figure 10 shows the pressure decrease with height in the standard atmosphere. These standard altitudes are based on standard temperatures. In the real atmosphere, temperatures are seldom standard, so let's explore temperature effects.

Temperature

Like most substances, air expands as it becomes warmer and shrinks as it cools. Figure 11 shows three columns of air—one colder than standard, one at standard temperature, and one warmer than standard. Pressure is equal at the bottom of each column and equal at the top of each column. Therefore, pressure decrease upward through each column is the same. Vertical expansion of the warm column has made it higher than the column at standard temperature. Shrinkage of the cold column has made it shorter. Since pressure decrease is the same in each column, the *rate of decrease* of pressure with height in warm air is less than standard; the rate of decrease of pressure with height in cold air is greater than standard. You will soon see the importance of temperature in altimetry and weather analysis and on aircraft performance.

Sea Level Pressure

Since pressure varies with altitude, we cannot readily compare station pressures between stations at different altitudes. To make them comparable, we must adjust them to some common level. Mean sea level seems the most feasible common reference. In figure 12, pressure measured at a 5,000-foot station is 25 inches; pressure increases about 1 inch for each 1,000 feet or a total of 5 inches. Sea level pressure is approximately 25 + 5 or 30 inches. The weather observer takes temperature and other effects into account, but this simplified example explains the basic principle of sea level pressure reduction.

We usually express sea level pressure in millibars. Standard sea level pressure is 1013.2 millibars, 29.92 inches of mercury, 760 millimeters of mercury, or about 14.7 pounds per square inch. Figures 23 and 24 in chapter 4 show world-wide averages of sea level pressure for the months of July and January. Pressure changes continually, however, and departs widely from these averages. We use a sequence of weather maps to follow these changing pressures.

15

Atmospheric Pressure and Altimetry

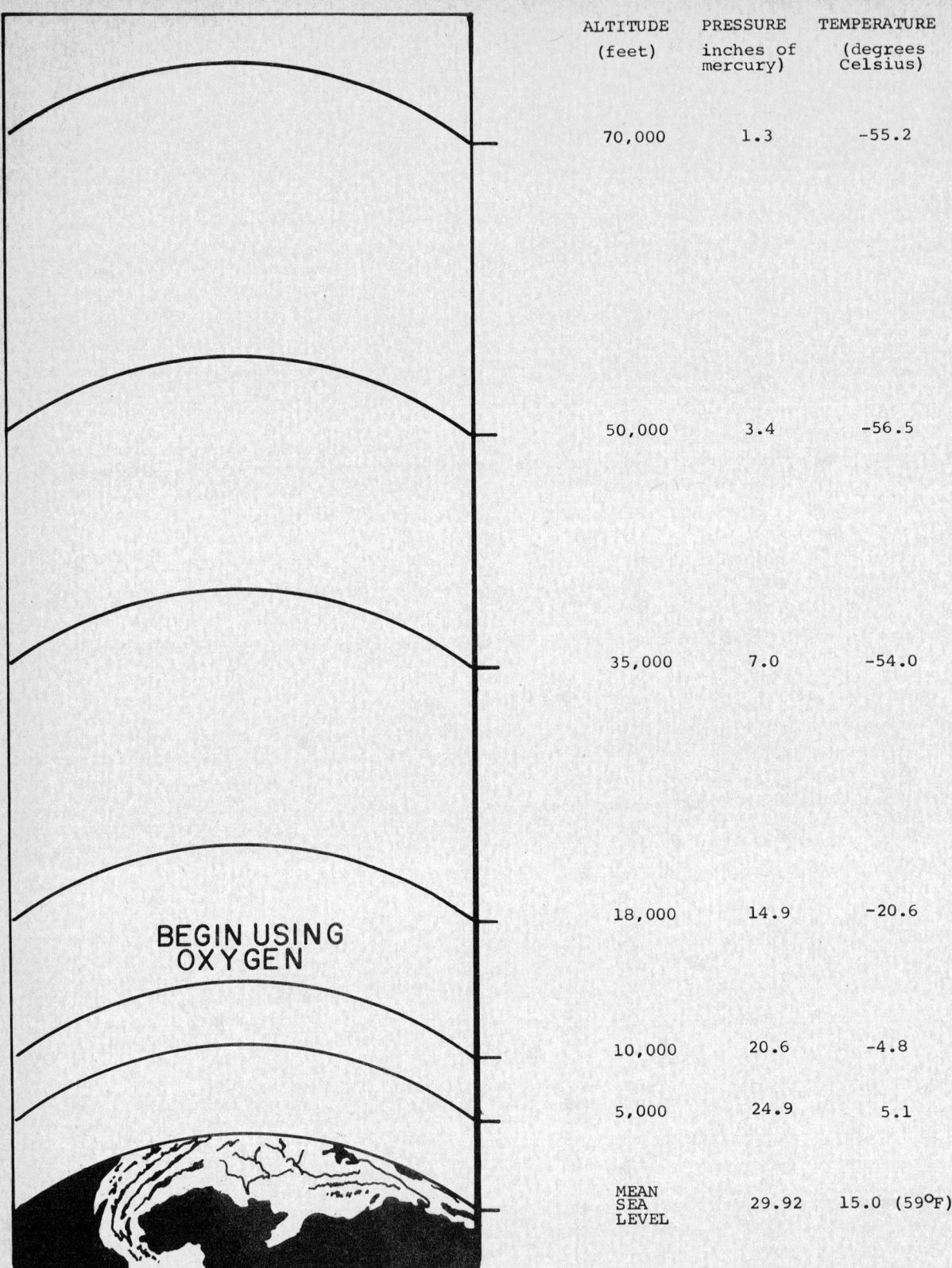

FIGURE 10. The standard atmosphere. Note how pressure decreases with increasing height; the rate of decrease with height is greatest in lower levels.

Atmospheric Pressure and Altimetry

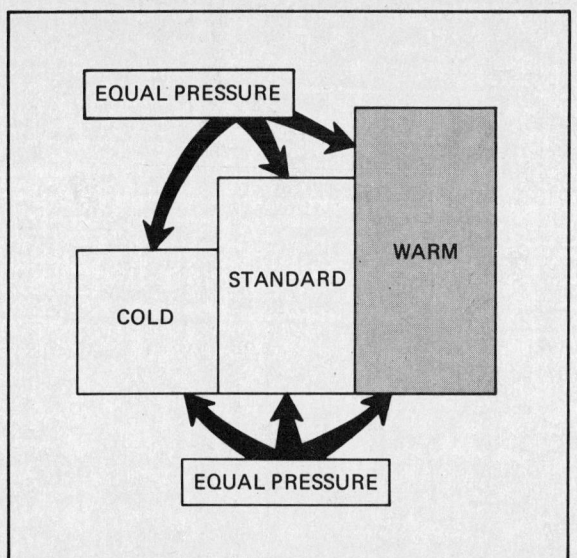

FIGURE 11. Three columns of air showing how decrease of pressure with height varies with temperature. Left column is colder than average and right column, warmer than average. Pressure is equal at the bottom of each column and equal at the top of each column. Pressure decreases most rapidly with height in the cold air and least rapidly in the warm air.

Pressure Analyses

We plot sea level pressures on a map and draw lines connecting points of equal pressure. These lines of equal pressure are *isobars*. Hence, the surface map is an *isobaric analysis* showing identifiable, organized pressure patterns. Five pressure systems are shown in figure 13 and are defined as follow:

1. LOW—a center of pressure surrounded on all sides by higher pressure; also called a cyclone. Cyclonic curvature is the curvature of isobars to the left when you stand with lower pressure to your left.
2. HIGH—a center of pressure surrounded on all sides by lower pressure, also called an anticyclone. Anticyclonic curvature is the curvature of isobars to the right when you stand with lower pressure to your left.
3. TROUGH—an elongated area of low pressure with the lowest pressure along a line marking maximum cyclonic curvature.
4. RIDGE—an elongated area of high pressure with the highest pressure along a line marking maximum anticyclonic curvature.
5. COL—the neutral area between two highs and two lows. It also is the intersection of a trough and a ridge. The col on a pressure surface is analogous to a mountain pass on a topographic surface.

Upper air weather maps reveal these same types of pressure patterns aloft for several levels. They also show temperature, moisture, and wind at each level. In fact, a chart is available for a level within a few thousand feet of your planned cruising altitude. AVIATION WEATHER SERVICES lists the approximate heights of upper air maps and shows details of the surface map and each upper air chart.

FIGURE 12. Reduction of station pressure to sea level. Pressure increases about 1 inch per 1,000 feet from the station elevation to sea level.

17

Atmospheric Pressure and Altimetry

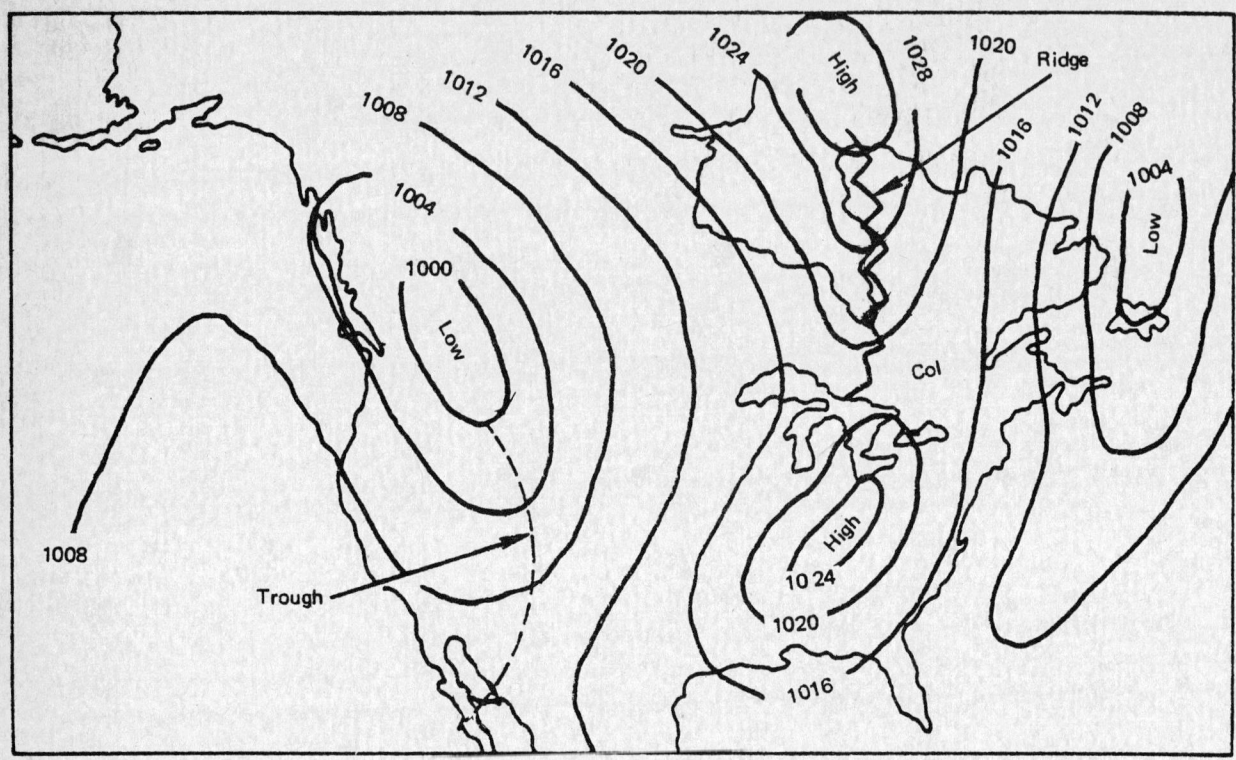

FIGURE 13. Pressure systems.

Chapter 4 of this book ties together the surface chart and upper air charts into a three-dimensional picture.

An upper air map is a *constant pressure analysis*. But, what do we mean by "constant pressure"? Constant pressure simply refers to a specific pressure. Let's arbitrarily choose 700 millibars. Everywhere above the earth's surface, pressure decreases with height; and at some height, it decreases to this constant pressure of 700 millibars. Therefore, there is a "surface" throughout the atmosphere at which pressure is 700 millibars. We call this the 700-millibar constant pressure surface. However, the *height* of this surface is *not* constant. Rising pressure pushes the surface upward into highs and ridges. Falling pressure lowers the height of the surface into lows and troughs. These systems migrate continuously as "waves" on the pressure surface. Remember that we chose this constant pressure surface arbitrarily as a reference. It in no way defines any discrete boundary.

The National Weather Service and military weather services take routine scheduled upper air observations—sometimes called soundings. A balloon carries aloft a radiosonde instrument which consists of miniature radio gear and sensing elements. While in flight, the radiosonde transmits data from which a specialist determines wind, temperature, moisture, and height at selected pressure surfaces.

We routinely collect these observations, plot the heights of a constant pressure surface on a map, and draw lines connecting points of equal height. These lines are *height contours*. But, what is a height contour?

First, consider a topographic map with contours showing variations in elevation. These are height contours of the terrain surface. The Earth surface is a fixed reference and we contour variations in its height.

The same concept applies to height contours on a constant pressure chart, except our reference is a *constant pressure surface*. We simply contour the heights of the pressure surface. For example, a 700-millibar constant pressure analysis is a contour map of the heights of the 700-millibar pressure surface. While the contour map is based on variations in height, these variations are small when compared to flight levels, and for all practical purposes, you may regard the 700-millibar chart as a weather map at approximately 10,000 feet or 3,048 meters.

A contour analysis shows highs, ridges, lows, and troughs aloft just as the isobaric analysis shows such systems at the surface. What we say concerning pressure patterns and systems applies equally to an isobaric or a contour analysis.

Low pressure systems quite often are regions of poor flying weather, and high pressure areas predominantly are regions of favorable flying weather. A word of caution, however—use care in applying the low pressure-bad weather, high pressure-good weather rule of thumb; it all too frequently fails. When planning a flight, gather *all* information possible on expected weather. Pressure patterns also bear a direct relationship to wind which is the subject of the next chapter. But first, let's look at pressure and altimeters.

ALTIMETRY

The altimeter is essentially an aneroid barometer. The difference is the scale. The altimeter is graduated to read increments of height rather than units of pressure. The standard for graduating the altimeter is the standard atmosphere.

ALTITUDE

Altitude seems like a simple term; it means height. But in aviation, it can have many meanings.

True Altitude

Since existing conditions in a real atmosphere are seldom standard, altitude indications on the altimeter are seldom actual or true altitudes. *True altitude is the actual or exact altitude above mean sea level.* If your altimeter does not indicate true altitude, what does it indicate?

Indicated Altitude

Look again at figure 11 showing the effect of mean temperature on the thickness of the three columns of air. Pressures are equal at the bottoms and equal at the tops of the three layers. Since the altimeter is essentially a barometer, altitude indicated by the altimeter at the top of each column would be the same. To see this effect more clearly, study figure 14. Note that in the warm air, you fly at an altitude higher than indicated. In the cold air, you are at an altitude lower than indicated.

Height indicated on the altimeter also changes with changes in surface pressure. A movable scale on the altimeter permits you to adjust for surface pressure, but you have no means of adjusting the instrument for mean temperature of the column of air below you. *Indicated altitude is the altitude above mean sea level indicated on the altimeter when set at the local altimeter setting.* But what is altimeter setting?

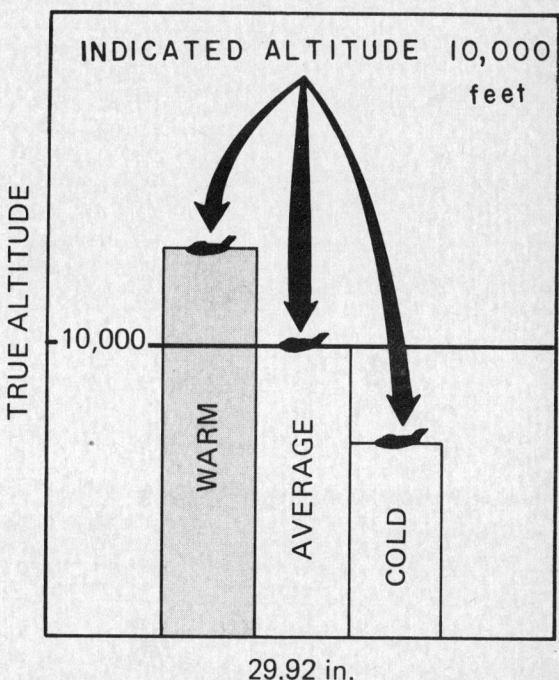

FIGURE 14. Indicated altitude depends on air temperature below the aircraft. Since pressure is equal at the bases and equal at the tops of each column, indicated altitude is the same at the top of each column. When air is colder than average (right), the altimeter reads higher than true altitude. When air is warmer than standard (left), the altimeter reads lower than true altitude.

19

Altimeter Setting

Since the altitude scale is adjustable, you can set the altimeter to read true altitude at some specified height. Takeoff and landing are the most critical phases of flight; therefore, airport elevation is the most desirable altitude for a true reading of the altimeter. *Altimeter setting is the value to which the scale of the pressure altimeter is set so the altimeter indicates true altitude at field elevation.*

In order to ensure that your altimeter reading is compatible with altimeter readings of other aircraft in your vicinity, keep your altimeter setting current. Adjust it frequently in flight to the altimeter setting reported by the nearest tower or weather reporting station. Figure 15 shows the trouble you can encounter if you are lax in adjusting your altimeter in flight. Note that as you fly from high pressure to low pressure, you are lower than your altimeter indicates.

Figure 16 shows that as you fly from warm to cold air, your altimeter reads too high—you are lower than your altimeter indicates. Over flat terrain this lower than true reading is no great problem; other aircraft in the vicinity also are flying indicated rather than true altitude, and your altimeter readings are compatible. If flying in cold weather over mountainous areas, however, you must take this difference between indicated and true altitude into account. You must know that your true altitude assures clearance of terrain, so you compute a correction to indicated altitude.

Corrected (Approximately True) Altitude

If it were possible for a pilot always to determine mean temperature of the column of air between the aircraft and the surface, flight computers would be designed to use this mean temperature in computing true altitude. However, the only guide a pilot has to temperature below him is free air temperature at his altitude. Therefore, the flight computer uses outside air temperature to correct indicated altitude to approximate true altitude. *Corrected altitude is indicated altitude corrected for the temperature of the air column below the aircraft, the correction being based on the estimated departure of the existing temperature from standard atmospheric temperature.* It is a close approximation to true altitude and is labeled *true altitude* on flight computers. It is close enough to

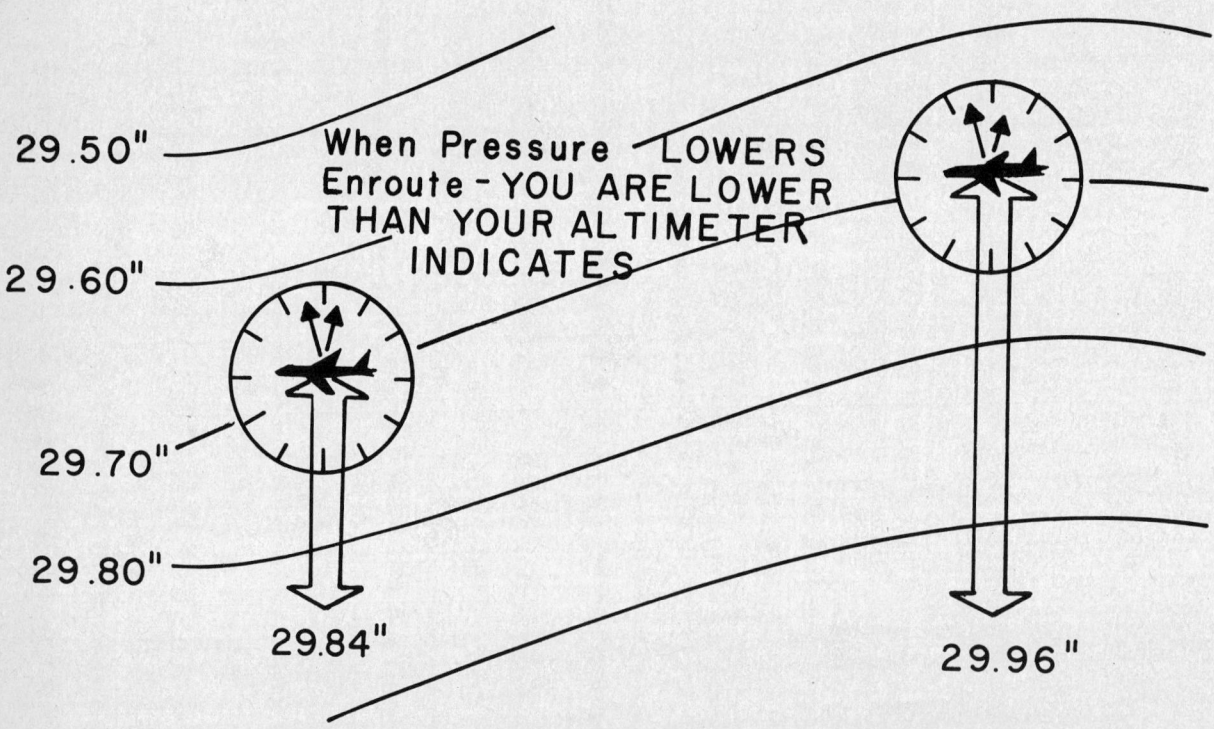

FIGURE 15. When flying from high pressure to lower pressure without adjusting your altimeter, you are losing true altitude.

Atmospheric Pressure and Altimetry

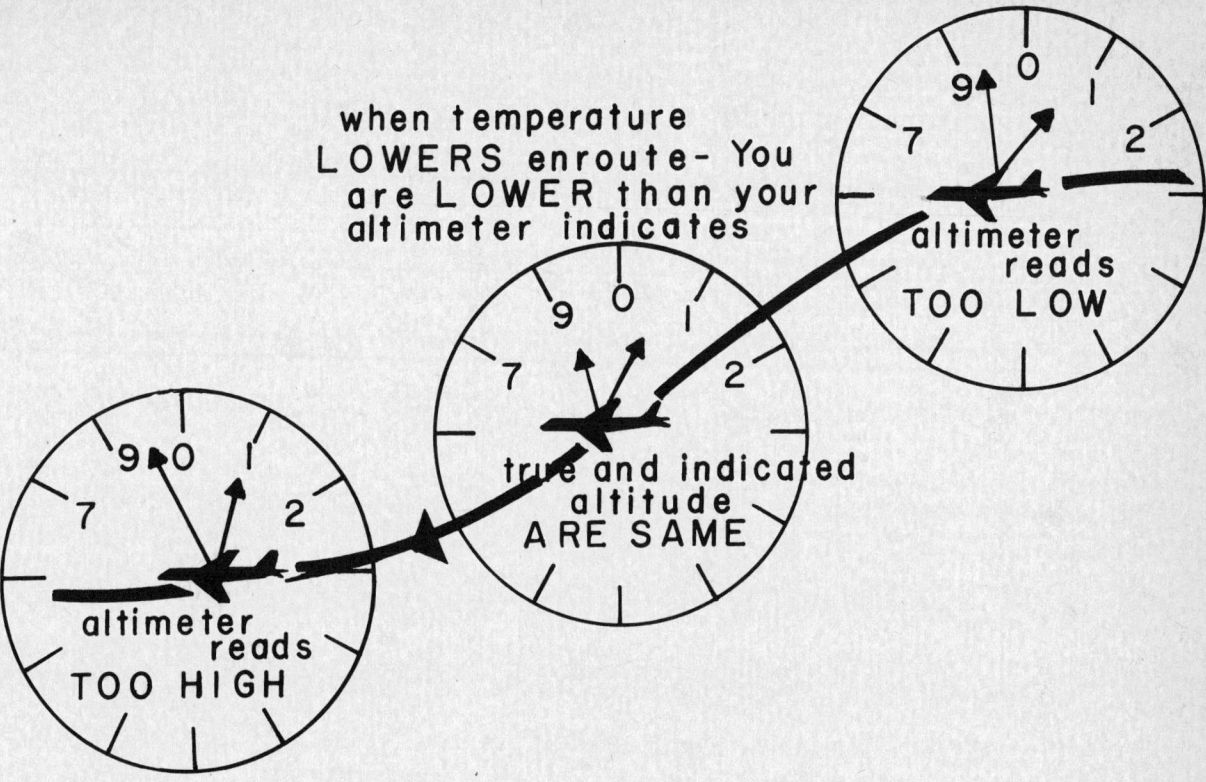

FIGURE 16. Effect of temperature on altitude. When air is warmer than average, you are higher than your altimeter indicates. When temperature is colder than average, you are lower than indicated. When flying from warm to cold air at a constant indicated altitude, you are losing true altitude.

true altitude to be used for terrain clearance provided you have your altimeter set to the value reported from a nearby reporting station.

Pilots have met with disaster because they failed to allow for the difference between indicated and true altitude. In cold weather when you must clear high terrain, take time to compute true altitude.

FAA regulations require you to fly indicated altitude at low levels and pressure altitude at high levels (at or above 18,000 feet at the time this book was printed). What is pressure altitude?

Pressure Altitude

In the standard atmosphere, sea level pressure is 29.92 inches of mercury or 1013.2 millibars. Pressure falls at a fixed rate upward through this hypothetical atmosphere. Therefore, in the standard atmosphere, a given pressure exists at any specified altitude. *Pressure altitude is the altitude in the standard atmosphere where pressure is the same as where you are.* Since at a specific pressure altitude, pressure is everywhere the same, a constant pressure surface defines a constant pressure altitude. When you fly a constant pressure altitude, you are flying a constant pressure surface.

You can always determine pressure altitude from your altimeter whether in flight or on the ground. Simply set your altimeter at the standard altimeter setting of 29.92 inches, and your altimeter indicates pressure altitude.

A conflict sometimes occurs near the altitude separating flights using indicated altitude from those using pressure altitude. Pressure altitude on one aircraft and indicated altitude on another may indicate altitude separation when, actually, the two are at the same true altitude. All flights using pressure altitude at high altitudes are IFR controlled flights. When this conflict occurs, air traffic controllers prohibit IFR flight at the conflicting altitudes.

DENSITY ALTITUDE

What is density altitude? *Density altitude simply is the altitude in the standard atmosphere where air density is the same as where you are.* Pressure, temperature, and humidity determine air density. On a hot day, the air becomes "thinner" or lighter, and its density where you are is equivalent to a higher altitude in the standard atmosphere—thus the term "high density altitude." On a cold day, the air becomes heavy; its density is the same as that at an altitude in the standard atmosphere lower than your altitude—"low density altitude."

Density altitude is not a height reference; rather, it is an index to aircraft performance. Low density altitude increases performance. *High density altitude* is a real hazard since it *reduces aircraft performance*. It affects performance in three ways. (1) It reduces power because the engine takes in less air to support combustion. (2) It reduces thrust because the propeller gets less grip on the light air or a jet has less mass of gases to spit out the exhaust. (3) It reduces lift because the light air exerts less force on the airfoils.

You cannot detect the effect of high density altitude on your airspeed indicator. Your aircraft lifts off, climbs, cruises, glides, and lands at the prescribed indicated airspeeds. But at a specified indicated airspeed, your true airspeed and your groundspeed increase proportionally as density altitude becomes higher.

The net results are that high density altitude lengthens your takeoff and landing rolls and reduces your rate of climb. Before lift-off, you must attain a faster groundspeed, and therefore, you need more runway; your reduced power and thrust add a need for still more runway. You land at a

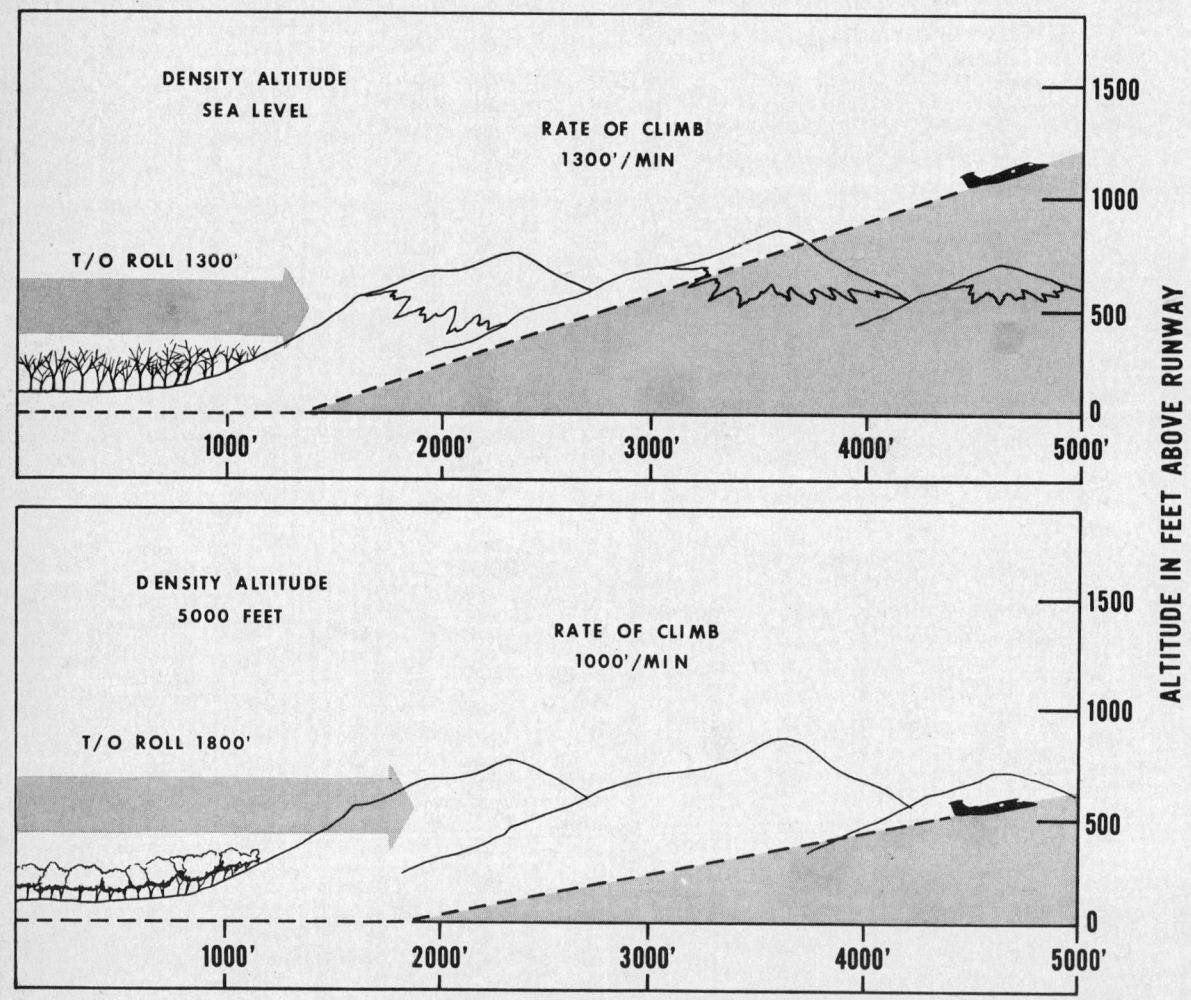

FIGURE 17. Effect of density altitude on takeoff and climb. High density altitude lengthens takeoff roll and reduces rate of climb.

faster groundspeed and, therefore, need more room to stop. At a prescribed indicated airspeed, you are flying at a faster true airspeed, and therefore, you cover more distance in a given time which means climbing at a more shallow angle. Add to this the problems of reduced power and rate of climb, and you are in double jeopardy in your climb. Figure 17 shows the effect of density altitude on takeoff distance and rate of climb.

High density altitude also can be a problem at cruising altitudes. When air is abnormally warm, the high density altitude lowers your service ceiling. For example, if temperature at 10,000 feet pressure altitude is 20° C, density altitude is 12,700 feet. (Check this on your flight computer.) Your aircraft will perform as though it were at 12,700 indicated with a normal temperature of −8° C.

To compute density altitude, set your altimeter at 29.92 inches or 1013.2 millibars and read pressure altitude from your altimeter. Read outside air temperature and then use your flight computer to get density altitude. On an airport served by a weather observing station, you usually can get density altitude for the airport from the observer. Section 16 of AVIATION WEATHER SERVICES has a graph for computing density altitude if you have no flight computer handy.

IN CLOSING

Pressure patterns can be a clue to weather causes and movement of weather systems, but they give only a part of the total weather picture. Pressure decreases with increasing altitude. The altimeter is an aneroid barometer graduated in increments of altitude in the standard atmosphere instead of units of pressure. Temperature greatly affects the rate of pressure decrease with height; therefore, it influences altimeter readings. Temperature also determines the density of air at a given pressure (density altitude). Density altitude is an index to aircraft performance. Always be alert for departures of pressure and temperature from normals and compensate for these abnormalities.

Following are a few operational reminders:
1. Beware of the low pressure-bad weather, high pressure-good weather rule of thumb. It frequently fails. Always get the **complete** weather picture.
2. When flying from high pressure to low pressure at constant indicated altitude and without adjusting the altimeter, you are losing true altitude.
3. When temperature is colder than standard, you are at an altitude *lower* than your altimeter indicates. When temperature is warmer than standard, you are *higher* than your altimeter indicates.
4. When flying cross country, keep your altimeter setting current. This procedure assures more positive altitude separation from other aircraft.
5. When flying over high terrain in cold weather, compute your true altitude to ensure terrain clearance.
6. When your aircraft is heavily loaded, the temperature is abnormally warm, and/or the pressure is abnormally low, compute density altitude. Then check your aircraft manual to ensure that you can become airborne from the available runway. Check further to determine that your rate of climb permits clearance of obstacles beyond the end of the runway. This procedure is **advisable** for any airport regardless of altitude.
7. When planning takeoff or landing at a high altitude airport regardless of load, determine density altitude. The procedure is especially critical when temperature is abnormally warm or pressure abnormally low. Make certain you have sufficient runway for takeoff or landing roll. Make sure you can clear obstacles beyond the end of the runway after takeoff or in event of a go-around.
8. Sometimes the altimeter setting is taken from an instrument of questionable reliability. However, if the instrument can cause an error in altitude reading of more than 20 feet, it is removed from service. When altimeter setting is estimated, be prepared for a possible 10- to 20-foot difference between field elevation and your altimeter reading at touchdown.

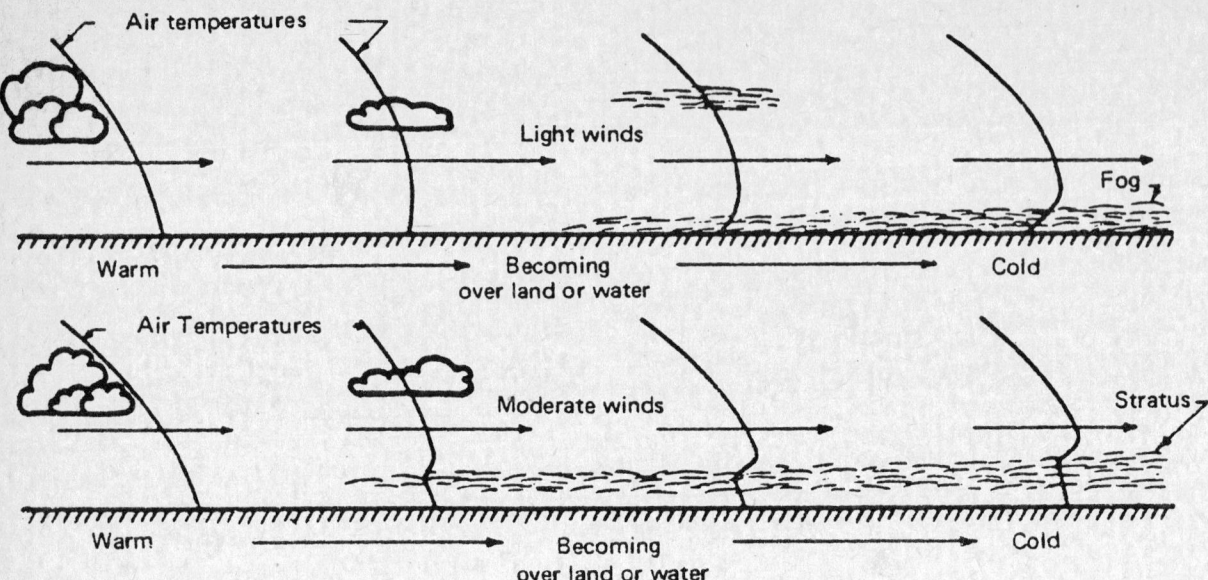

PASSAGE OF WARM AIR OVER COLDER SURFACE

Chapter 4
WIND

Differences in temperature create differences in pressure. These pressure differences drive a complex system of winds in a never ending attempt to reach equilibrium. Wind also transports water vapor and spreads fog, clouds, and precipitation. To help you relate wind to pressure patterns and the movement of weather systems, this chapter explains convection and the pressure gradient force, describes the effects of the Coriolis and frictional forces, relates convection and these forces to the general circulation, discusses local and small-scale wind systems, introduces you to wind shear, and associates wind with weather.

CONVECTION

When two surfaces are heated unequally, they heat the overlying air unevenly. The warmer* air expands and becomes lighter or less dense than the cool* air. The more dense, cool air is drawn to the ground by its greater gravitational force lifting or forcing the warm air upward much as oil is forced to the top of water when the two are mixed. Figure 18 shows the convective process. The rising air spreads and cools, eventually descending to com-

---

*Frequently throughout this book, we refer to air as *warm*, *cool*, or *cold*. These terms refer to relative temperatures and not to any fixed temperature reference or to temperatures as they may affect our comfort. For example, compare air at −10° F to air at 0° F; relative to each other, the −10° F air is *cool* and the 0° F, *warm*. 90° F would be *cool* or *cold* relative to 100° F.

plete the convective circulation. As long as the uneven heating persists, convection maintains a continuous "convective current."

The horizontal air flow in a convective current is "wind." Convection of both large and small scales accounts for systems ranging from hemispheric circulations down to local eddies. This horizontal flow, wind, is sometimes called "advection." However, the term "advection" more commonly applies to the transport of atmospheric properties by the wind, i.e., warm advection; cold advection; advection of water vapor, etc.

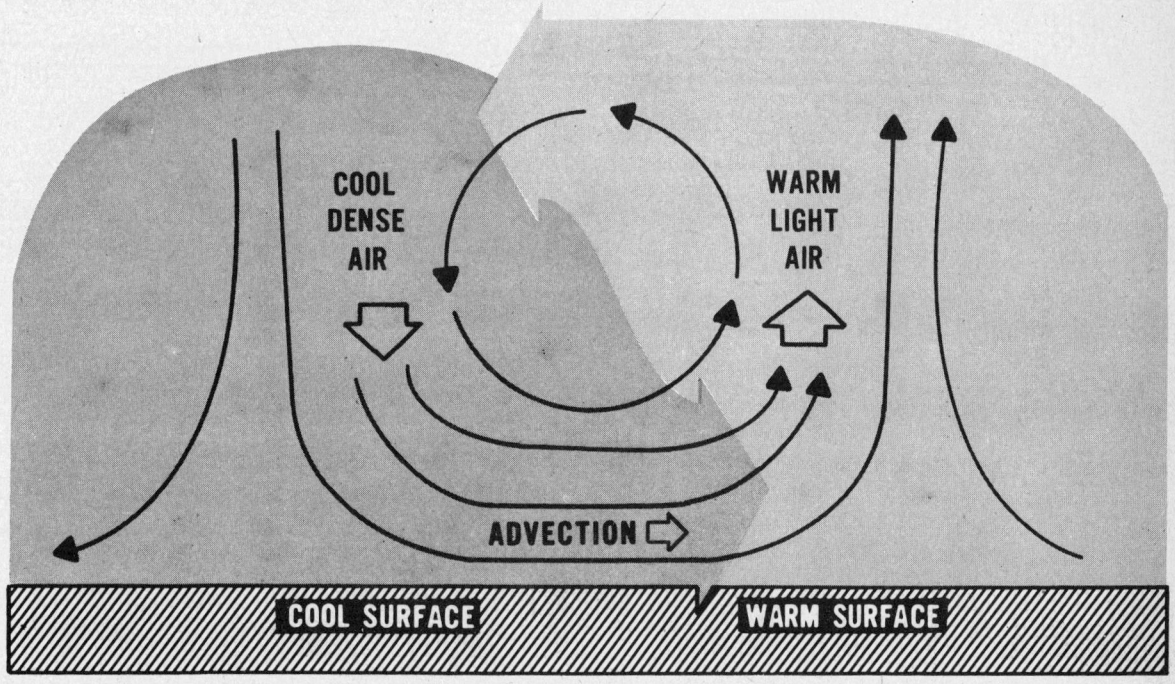

FIGURE 18. Convective current resulting from uneven heating of air by contrasting surface temperatures. The cool, heavier air forces the warmer air aloft establishing a convective cell. Convection continues as long as the uneven heating persists.

PRESSURE GRADIENT FORCE

Pressure differences must create a force in order to drive the wind. This force is the *pressure gradient force*. The force is from higher pressure to lower pressure and is perpendicular to isobars or contours. Whenever a pressure difference develops over an area, the pressure gradient force begins moving the air directly across the isobars. The closer the spacing of isobars, the stronger is the pressure gradient force. The stronger the pressure gradient force, the stronger is the wind. Thus, closely spaced isobars mean strong winds; widely spaced isobars mean lighter wind. From a pressure analysis, you can get a general idea of wind speed from contour or isobar spacing.

Because of uneven heating of the Earth, surface pressure is low in warm equatorial regions and high in cold polar regions. A pressure gradient develops from the poles to the Equator. If the Earth did not rotate, this pressure gradient force would be the only force acting on the wind. Circulation would be two giant hemispheric convective currents as shown in figure 19. Cold air would sink at the poles; wind would blow straight from the poles to the Equator; warm air at the Equator would be forced upward; and high level winds would blow directly toward the poles. However, the Earth does rotate; and because of its rotation, this simple circulation is greatly distorted.

Wind

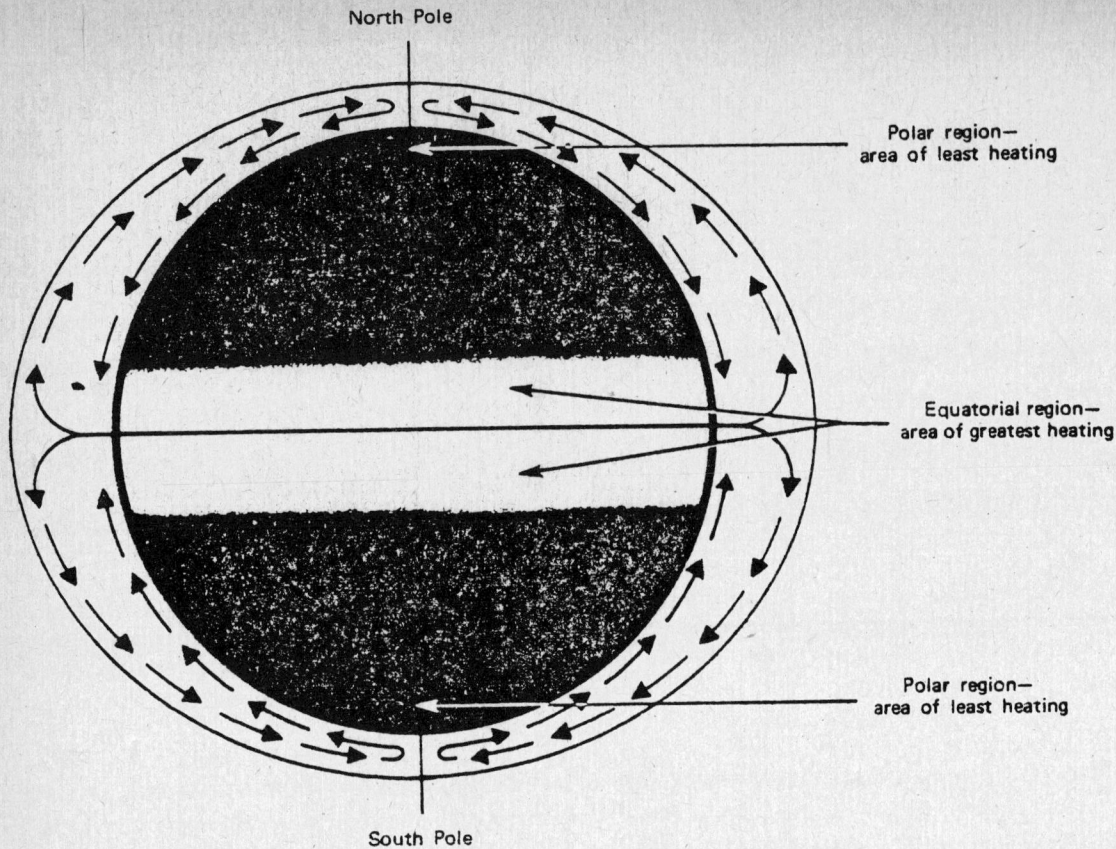

FIGURE 19. Circulation as it would be on a nonrotating globe. Intense heating at the Equator lowers the density. More dense air flows from the poles toward the Equator forcing the less dense air aloft where it flows toward the poles. The circulation would be two giant hemispherical convective currents.

CORIOLIS FORCE

A moving mass travels in a straight line until acted on by some outside force. However, if one views the moving mass from a rotating platform, the path of the moving mass relative to his platform appears to be deflected or curved. To illustrate, start rotating the turntable of a record player. Then using a piece of chalk and a ruler, draw a "straight" line from the center to the outer edge of the turntable. To you, the chalk traveled in a straight line. Now stop the turntable; on it, the line spirals outward from the center as shown in figure 20. To a viewer on the turntable, some "apparent" force deflected the chalk to the right.

A similar apparent force deflects moving particles on the earth. Because the Earth is spherical, the deflective force is much more complex than the simple turntable example. Although the force is termed "apparent," to us on Earth, it is very real. The principle was first explained by a Frenchman, Coriolis, and carries his name—the Coriolis force.

The Coriolis force affects the paths of aircraft; missiles; flying birds; ocean currents; and, most important to the study of weather, air currents. The force deflects air to the right in the Northern Hemisphere and to the left in the Southern Hemisphere. This book concentrates mostly on deflection to the right in the Northern Hemisphere.

Coriolis force is at a right angle to wind direction and directly proportional to wind speed. That is, as wind speed increases, Coriolis force increases. At a given latitude, double the wind speed and you double the Coriolis force. Why at a given latitude?

Coriolis force varies with latitude from zero at the Equator to a maximum at the poles. It influences wind direction everywhere except immediately at the Equator; but the effects are more pronounced in middle and high latitudes.

Remember that the pressure gradient force drives the wind and is perpendicular to isobars. When a pressure gradient force is first established, wind be-

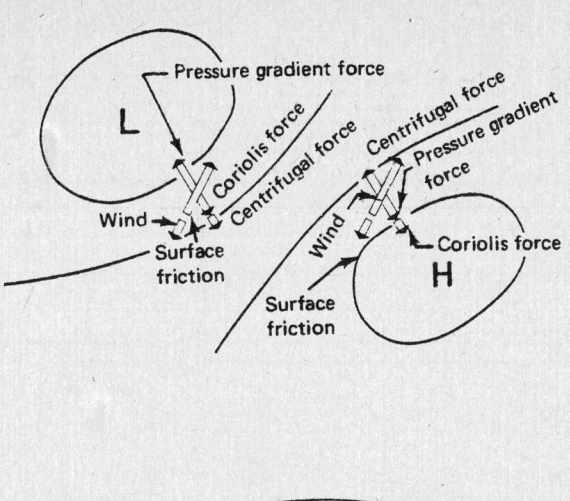

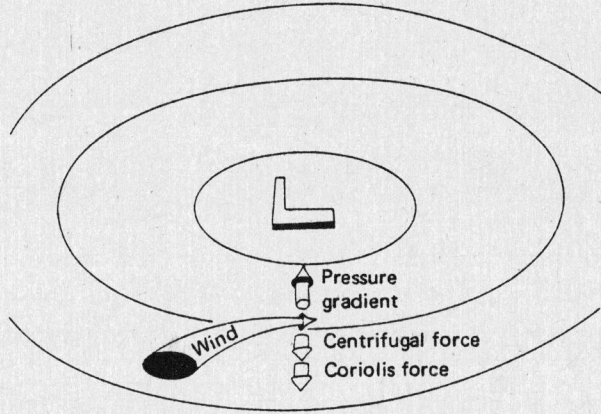

FIGURE 20. Apparent deflective force due to rotation of a horizontal platform. The "space path" is the path taken by a piece of chalk. The "path on the record" is the line traced on the rotating record. Relative to the record, the chalk appeared to curve; in space, it traveled in a straight line.

gins to blow from higher to lower pressure directly across the isobars. However, the instant air begins moving, Coriolis force deflects it to the right. Soon the wind is deflected a full 90° and is parallel to the isobars or contours. At this time, Coriolis force exactly balances pressure gradient force as shown in figure 21. With the forces in balance, wind will remain parallel to isobars or contours. Surface friction disrupts this balance as we discuss later; but first let's see how Coriolis force distorts the fictitious global circulation shown in figure 19.

THE GENERAL CIRCULATION

As air is forced aloft at the Equator and begins its high-level trek northward, the Coriolis force turns it to the right or to the east as shown in figure 22. Wind becomes westerly at about 30° latitude temporarily blocking further northward movement. Similarly, as air over the poles begins its low-level journey southward toward the Equator, it likewise is deflected to the right and becomes an east wind, halting for a while its southerly progress—also shown in figure 22. As a result, air literally "piles up" at about 30° and 60° latitude in both hemispheres. The added weight of the air in-

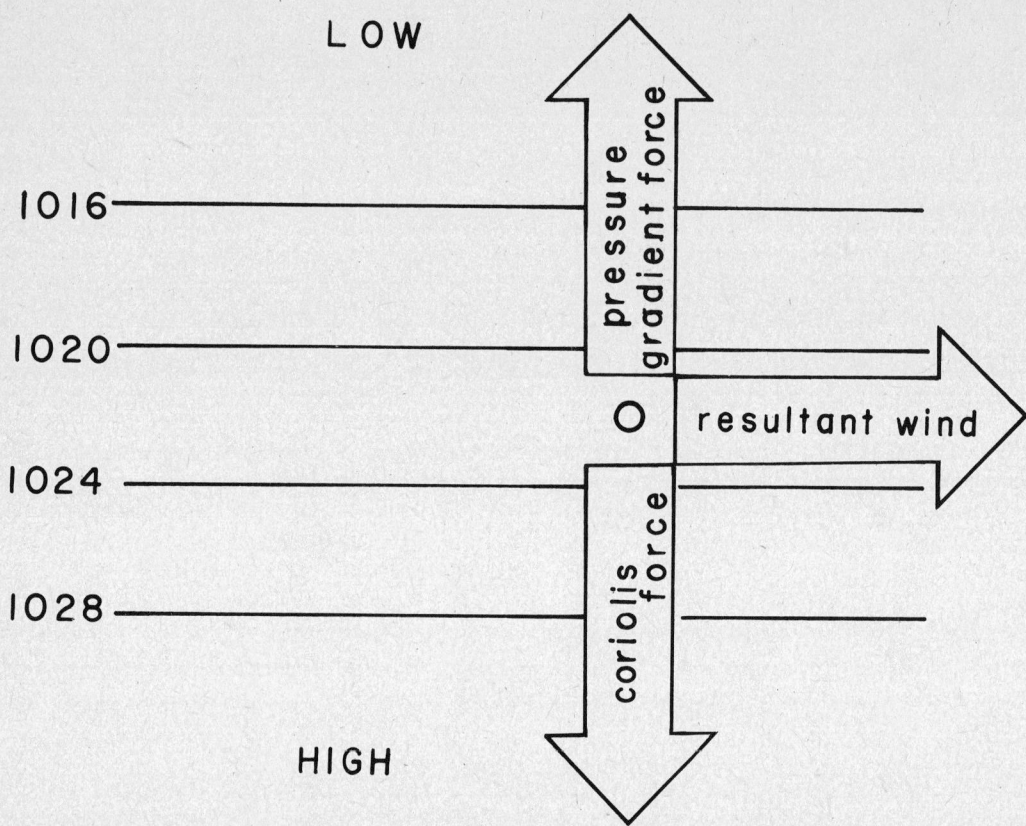

FIGURE 21. Effect of Coriolis force on wind relative to isobars. When Coriolis force deflects the wind until it is parallel to the isobars, pressure gradient balances Coriolis force.

creases the pressure into semipermanent high pressure belts. Figures 23 and 24 are maps of mean surface pressure for the months of July and January. The maps show clearly the subtropical high pressure belts near 30° latitude in both the Northern and Southern Hemispheres.

The building of these high pressure belts creates a temporary impasse disrupting the simple convective transfer between the Equator and the poles. The restless atmosphere cannot live with this impasse in its effort to reach equilibrium. Something has to give. Huge masses of air begin overturning in middle latitudes to complete the exchange.

Large masses of cold air break through the northern barrier plunging southward toward the Tropics. Large midlatitude storms develop between cold outbreaks and carry warm air northward. The result is a midlatitude band of migratory storms with ever changing weather. Figure 25 is an attempt to standardize this chaotic circulation into an average general circulation.

Since pressure differences cause wind, seasonal pressure variations determine to a great extent the areas of these cold air outbreaks and midlatitude storms. But, seasonal pressure variations are largely due to seasonal temperature changes. We have learned that, at the surface, warm temperatures to a great extent determine low pressure and cold temperatures, high pressure. We have also learned that seasonal temperature changes over continents are much greater than over oceans.

During summer, warm continents tend to be

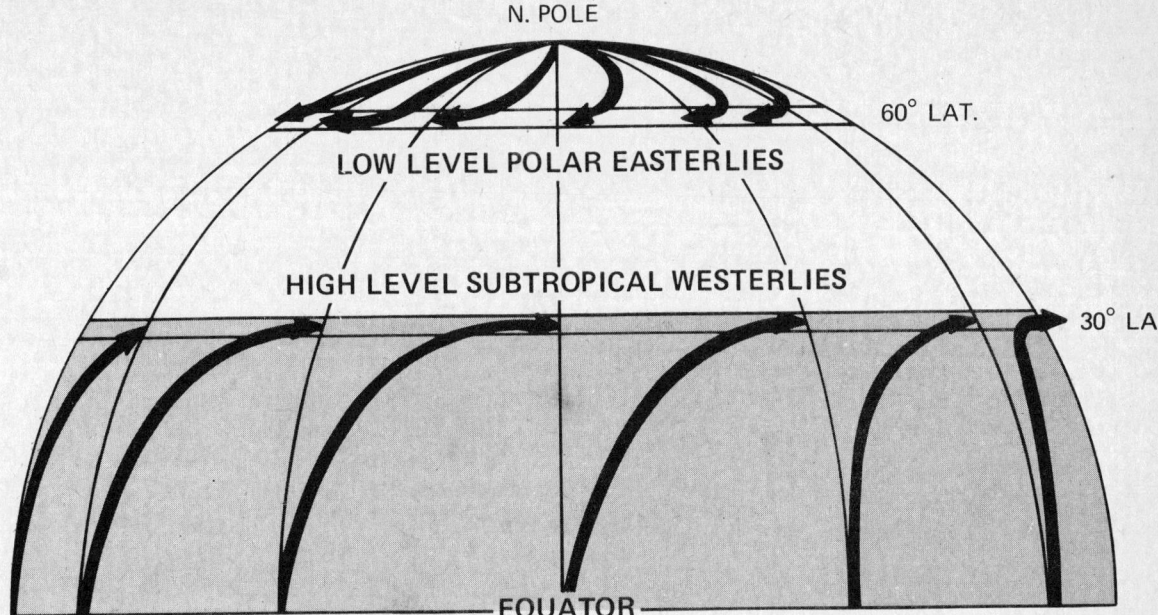

FIGURE 22. In the Northern Hemisphere, Coriolis force turns high level southerly winds to westerlies at about 30° latitude, temporarily halting further northerly progress. Low-level northerly winds from the pole are turned to easterlies, temporarily stopping further southward movement at about 60° latitude. Air tends to "pile up" at these two latitudes creating a void in middle latitudes. The restless atmosphere cannot live with this void; something has to give.

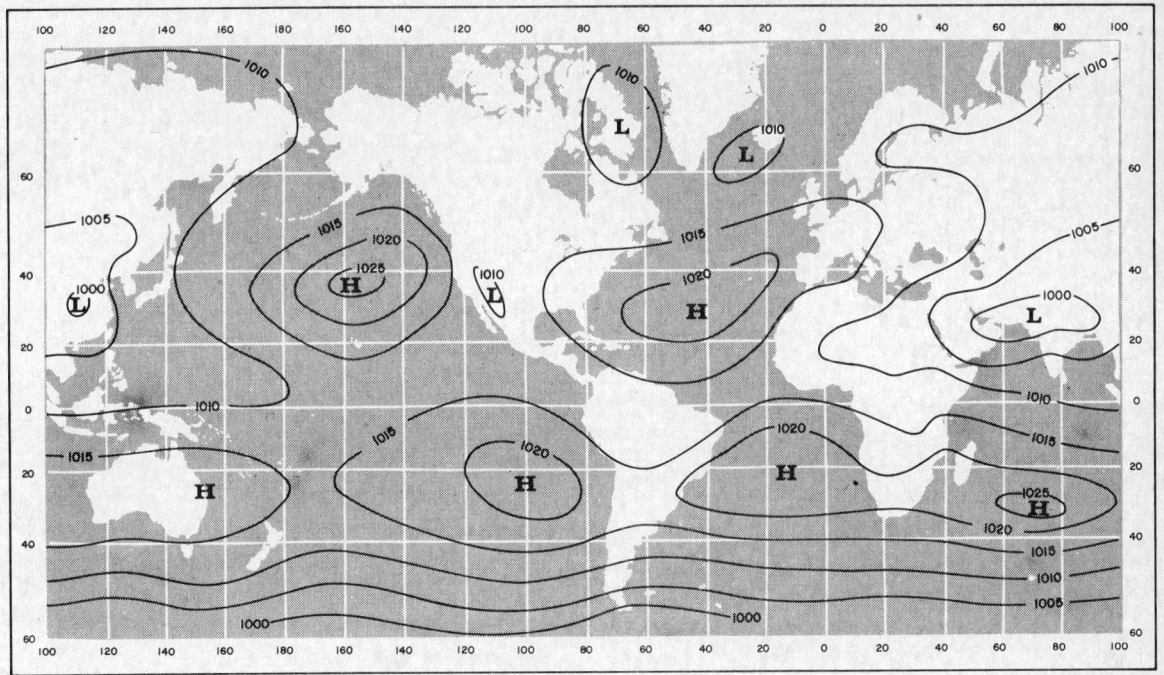

FIGURE 23. Mean world-wide surface pressure distribution in July. In the warm Northern Hemisphere, warm land areas tend to have low pressure, and cool oceanic areas tend to have high pressure. In the cool Southern Hemisphere, the pattern is reversed; cool land areas tend to have high pressure; and water surfaces, low pressure. However, the relationship is not so evident in the Southern Hemisphere because of relatively small amounts of land. The subtropical high pressure belts are clearly evident at about 30° latitude in both hemispheres.

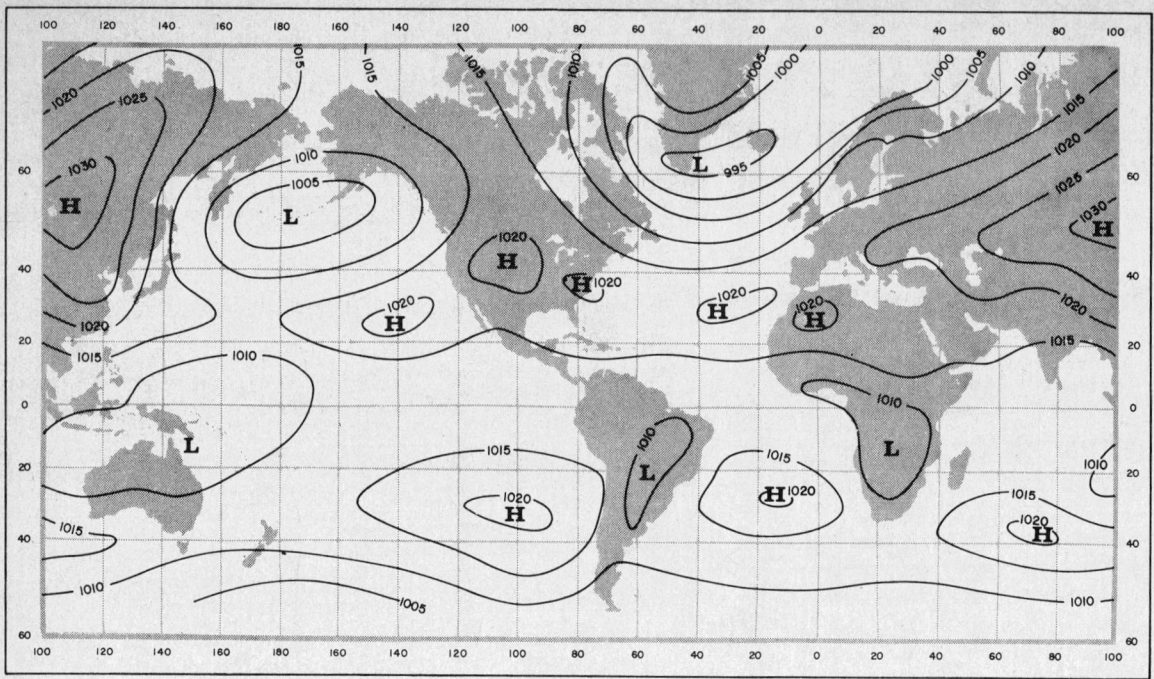

FIGURE 24. Mean world-wide surface pressure distribution in January. In this season, the pattern in figure 23 is reversed. In the cool Northern Hemisphere, cold continental areas are predominantly areas of high pressure while warm oceans tend to be low pressure areas. In the warm Southern Hemisphere, land areas tend to have low pressure; and oceans, high pressure. The subtropical high pressure belts are evident in both hemispheres. Note that the pressure belts shift southward in January and northward in July with the shift in the zone of maximum heating.

areas of low pressure and the relatively cool oceans, high pressure. In winter, the reverse is true—high pressure over the cold continents and low pressure over the relatively warm oceans. Figures 23 and 24 show this seasonal pressure reversal. The same pressure variations occur in the warm and cold seasons of the Southern Hemisphere, although the effect is not as pronounced because of the much larger water areas of the Southern Hemisphere.

Cold outbreaks are strongest in the cold season and are predominantly from cold continental areas. Summer outbreaks are weaker and more likely to originate from cool water surfaces. Since these outbreaks are masses of cool, dense air, they characteristically are high pressure areas.

As the air tries to blow outward from the high pressure, it is deflected to the right by the Coriolis force. Thus, the wind around a high blows clockwise. The high pressure with its associated wind system is an *anticyclone*.

The storms that develop between high pressure systems are characterized by low pressure. As winds try to blow inward toward the center of low pressure, they also are deflected to the right. Thus, the wind around a low is counterclockwise. The low pressure and its wind system is a *cyclone*. Figure 26 shows winds blowing parallel to isobars (contours on upper level charts). The winds are clockwise around highs and counterclockwise around lows.

The high pressure belt at about 30° north latitude forces air outward at the surface to the north and to the south. The northbound air becomes entrained into the midlatitude storms. The southward moving air is again deflected by the Coriolis force becoming the well-known subtropical northeast trade winds. In midlatitudes, high level winds are predominantly from the west and are known as the prevailing westerlies. Polar easterlies dominate low-level circulation north of about 60° latitude.

These three major wind belts are shown in figure 25. Northeasterly trade winds carry tropical storms from east to west. The prevailing westerlies drive midlatitude storms generally from west to east. Few major storm systems develop in the comparatively small Arctic region; the chief influence of the polar easterlies is their contribution to the development of midlatitude storms.

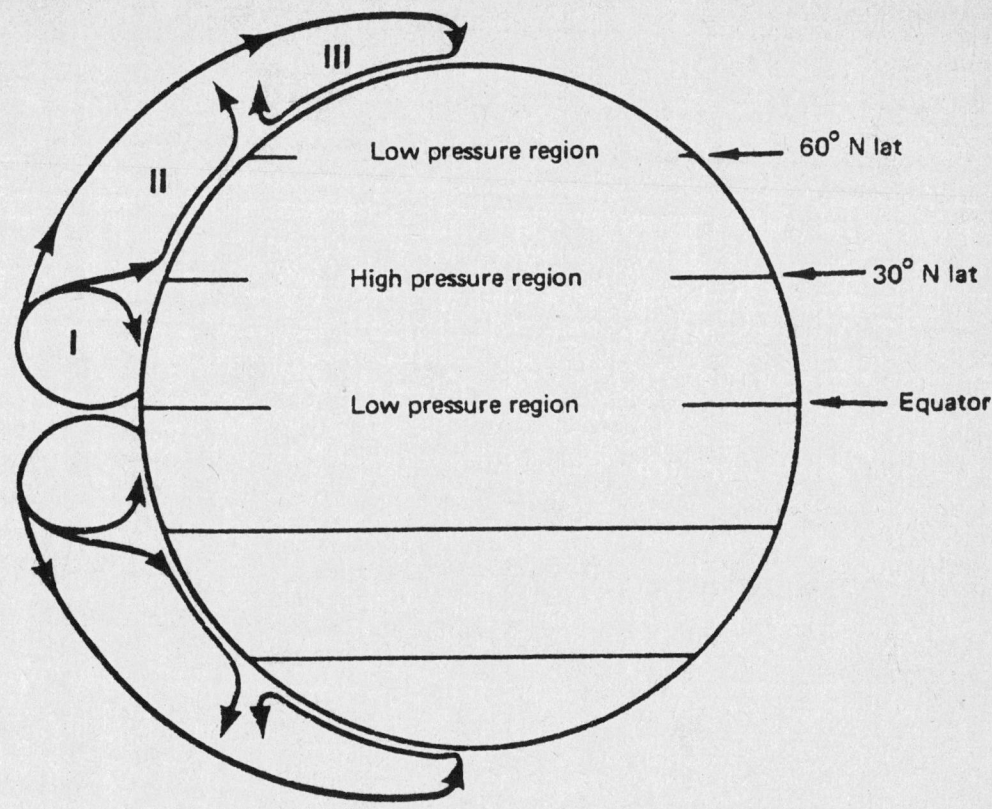

FIGURE 25. General average circulation in the Northern Hemisphere. Note the three belts of prevailing winds, the polar easterlies, the prevailing westerlies in middle latitudes, and the northeasterly "trade" winds. The belt of prevailing westerlies is a mixing zone between the North Pole and the Equator characterized by migrating storms.

Our discussion so far has said nothing about friction. Wind flow patterns aloft follow isobars or contours where friction has little effect. We cannot, however, neglect friction near the surface.

FRICTION

Friction between the wind and the terrain surface slows the wind. The rougher the terrain, the greater is the frictional effect. Also, the stronger the wind speed, the greater is the friction. One may not think of friction as a force, but it is a very real and effective force always acting opposite to wind direction.

As frictional force slows the windspeed, Coriolis force decreases. However, friction does not affect pressure gradient force. Pressure gradient and Coriolis forces are no longer in balance. The stronger pressure gradient force turns the wind at an angle across the isobars toward lower pressure until the three forces balance as shown in figure 27. Frictional and Coriolis forces combine to just balance pressure gradient force. Figure 28 shows how surface wind spirals outward from high pressure into low pressure crossing isobars at an angle.

The angle of surface wind to isobars is about 10° over water increasing with roughness of terrain. In mountainous regions, one often has difficulty relating surface wind to pressure gradient because of immense friction and also because of local terrain effects on pressure.

Wind

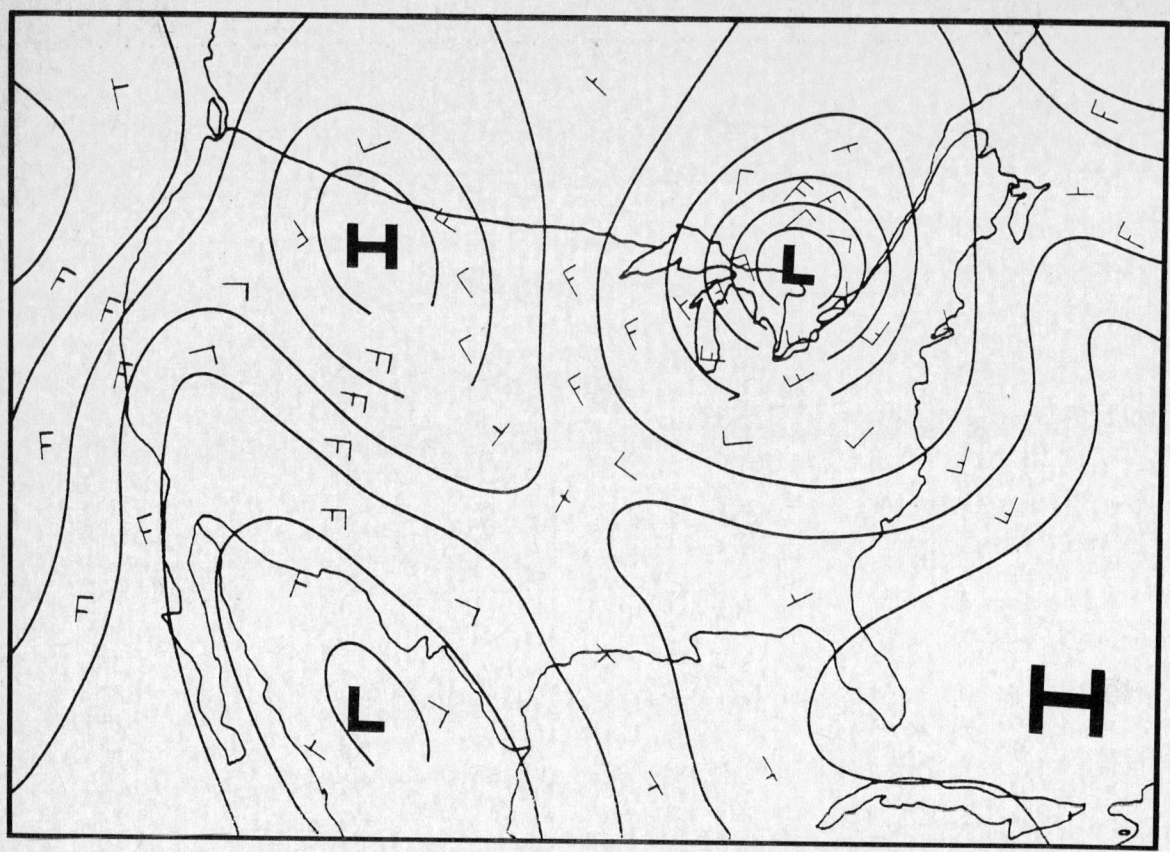

SURFACE WEATHER MAP SHOWING PRESSURE DISTRIBUTION AND WINDS

THE JET STREAM

A discussion of the general circulation is incomplete when it does not mention the "jet stream." Winds on the average increase with height throughout the troposphere culminating in a maximum near the level of the tropopause. These maximum winds tend to be further concentrated in narrow bands. A jet stream, then, is a narrow band of strong winds meandering through the atmosphere at a level near the tropopause. Since it is of interest primarily to high level flight, further discussion of the jet stream is reserved for chapter 13, "High Altitude Weather."

LOCAL AND SMALL SCALE WINDS

Until now, we have dealt only with the general circulation and major wind systems. Local terrain features such as mountains and shore lines influence local winds and weather.

MOUNTAIN AND VALLEY WINDS

In the daytime, air next to a mountain slope is heated by contact with the ground as it receives radiation from the sun. This air usually becomes warmer than air at the same altitude but farther from the slope.

Colder, denser air in the surroundings settles downward and forces the warmer air near the ground up the mountain slope. This wind is a "valley wind" so called because the air is flowing up out of the valley.

At night, the air in contact with the mountain slope is cooled by terrestrial radiation and becomes heavier than the surrounding air. It sinks along the

Wind

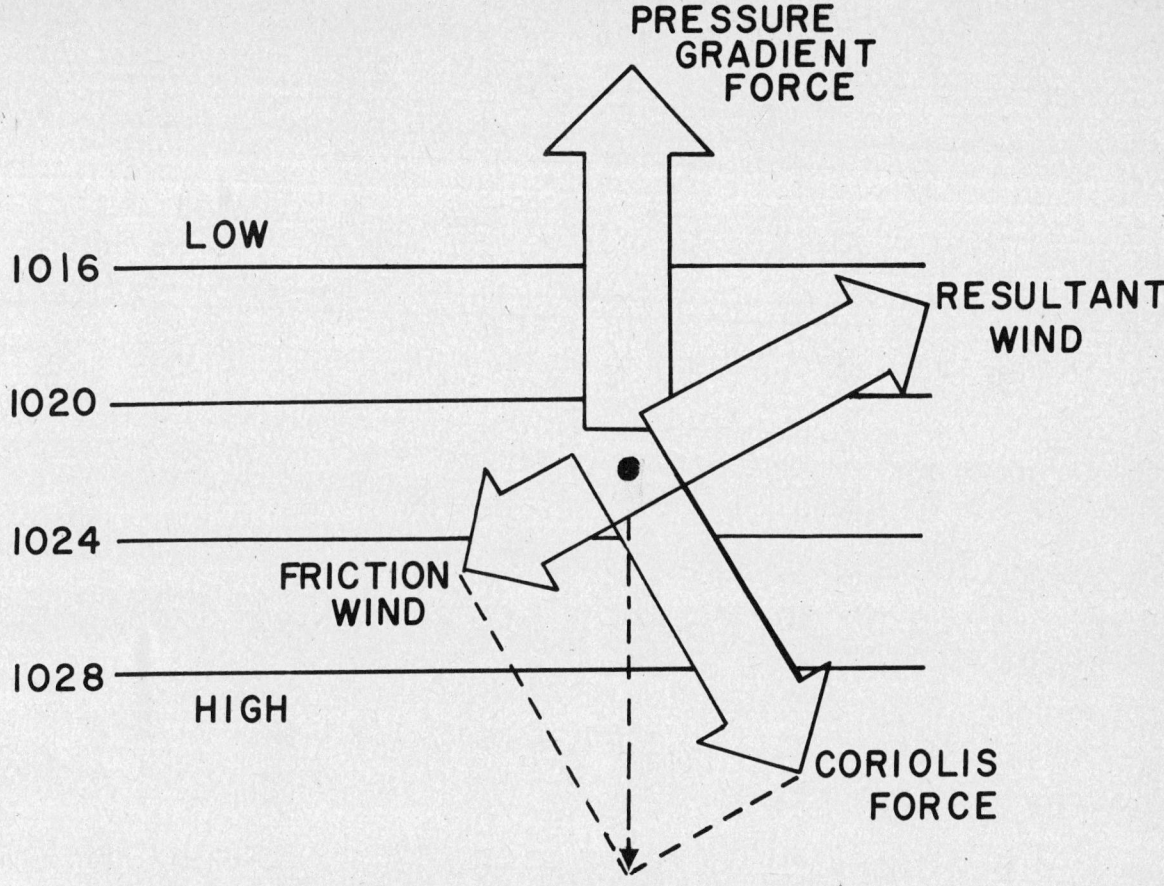

FIGURE 27. Surface friction slows the wind and reduces Coriolis force but does not affect pressure gradient force; winds near the surface are deflected across the isobars toward lower pressure.

slope, producing the "mountain wind" which flows like water down the mountain slope. Mountain winds are usually stronger than valley winds, especially in winter. The mountain wind often continues down the more gentle slopes of canyons and valleys, and in such cases takes the name "drainage wind." It can become quite strong over some terrain conditions and in extreme cases can become hazardous when flowing through canyon restrictions as discussed in chapter 9.

KATABATIC WIND

A katabatic wind is any wind blowing down an incline when the incline is influential in causing the wind. Thus, the mountain wind is a katabatic wind. Any katabatic wind originates because cold, heavy air spills down sloping terrain displacing warmer, less dense air ahead of it. Air is heated and dried as it flows down slope as we will study in later chapters. Sometimes the descending air becomes warmer than the air it replaces.

Many katabatic winds recurring in local areas have been given colorful names to highlight their dramatic, local effect. Some of these are the Bora, a cold northerly wind blowing from the Alps to the Mediterranean coast; the Chinook, figure 29, a warm wind down the east slope of the Rocky Mountains often reaching hundreds of miles into the high plains; the Taku, a cold wind in Alaska blowing off the Taku glacier; and the Santa Ana, a warm wind descending from the Sierras into the Santa Ana Valley of California.

LAND AND SEA BREEZES

As frequently stated earlier, land surfaces warm and cool more rapidly than do water surfaces; therefore, land is warmer than the sea during the

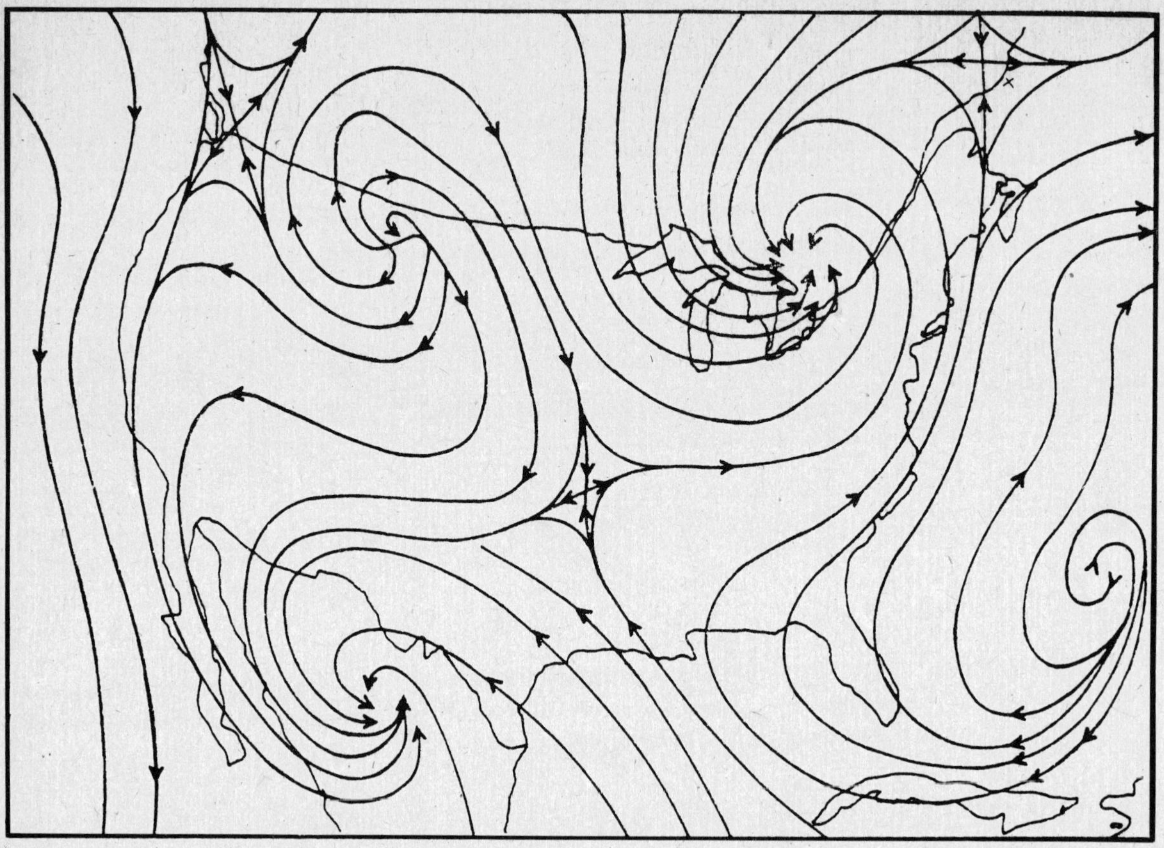

STREAMLINE ANALYSIS

FIGURE 29. The "Chinook" is a katabatic (downslope) wind. Air cools as it moves upslope and warms as it blows downslope. The Chinook occasionally produces dramatic warming over the plains just east of the Rocky Mountains.

Wind

day; wind blows from the cool water to warm land—the "sea breeze" so called because it blows from the sea. At night, the wind reverses, blows from cool land to warmer water, and creates a "land breeze." Figure 30 diagrams land and sea breezes.

Land and sea breezes develop only when the overall pressure gradient is weak. Wind with a stronger pressure gradient mixes the air so rapidly that local temperature and pressure gradients do not develop along the shore line.

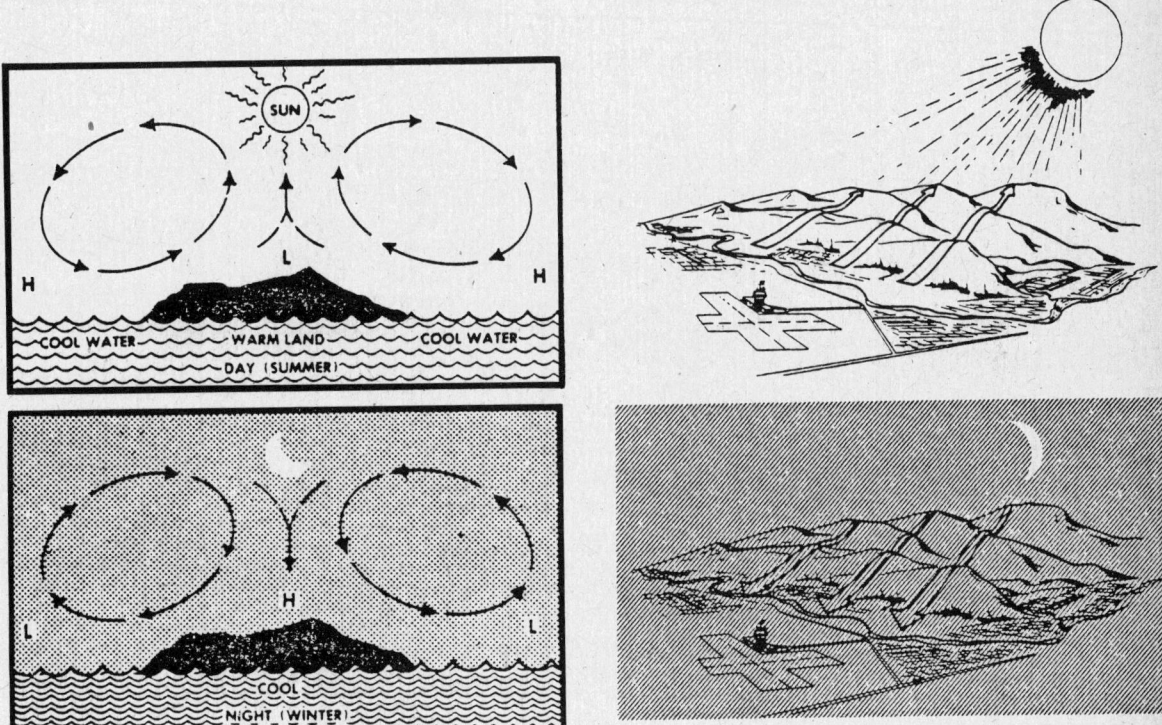

FIGURE 30. Land and sea breezes. At night, cool air from the land flows toward warmer water—the land breeze. During the day, wind blows from the water to the warmer land—the sea breeze.

WIND SHEAR

Rubbing two objects against each other creates friction. If the objects are solid, no exchange of mass occurs between the two. However, if the objects are fluid currents, friction creates eddies along a common shallow mixing zone, and a mass transfer takes place in the shallow mixing layer. This zone of induced eddies and mixing is a shear zone. Figure 31 shows two adjacent currents of air and their accompanying shear zone. Chapter 9 relates wind shear to turbulence.

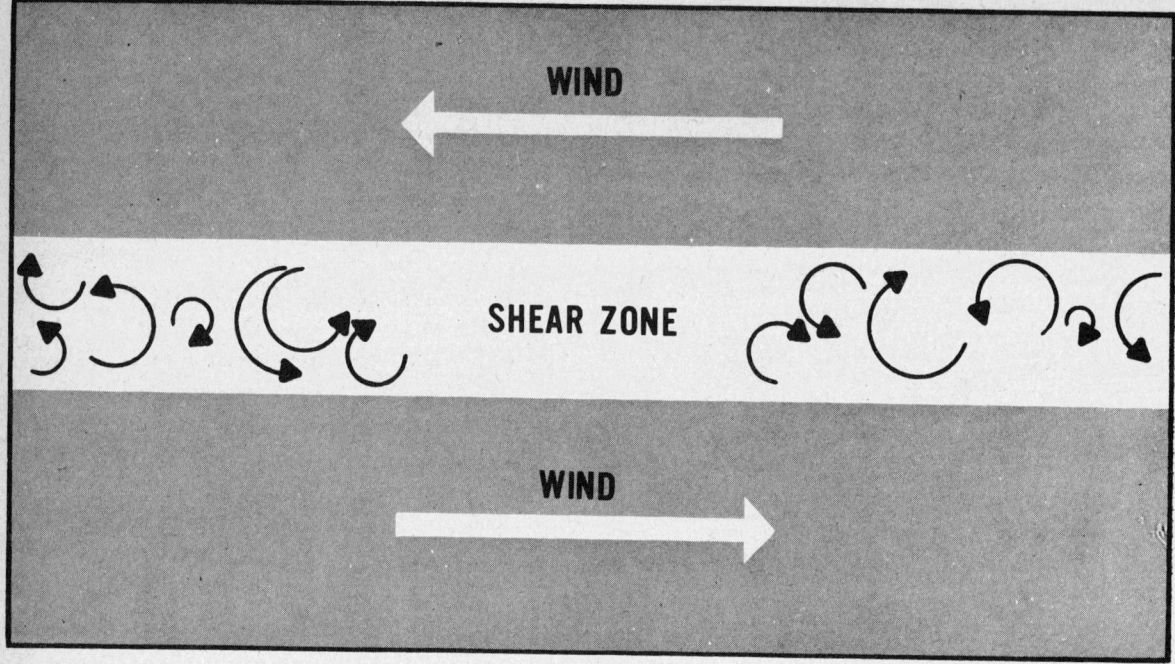

FIGURE 31. Wind shear. Air currents of differing velocities create friction or "shear" between them. Mixing in the shear zone results in a snarl of eddies and whirls.

WIND, PRESSURE SYSTEMS, AND WEATHER

We already have shown that wind speed is proportional to the spacing of isobars or contours on a weather map. However, with the same spacing, wind speed at the surface will be less than aloft because of surface friction.

You also can determine wind direction from a weather map. If you face along an isobar or contour with lower pressure on your left, wind will be blowing in the direction you are facing. On a surface map, wind will cross the isobar at an angle toward lower pressure; on an upper air chart, it will be parallel to the contour.

Wind blows counterclockwise (Northern Hemisphere) around a low and clockwise around a high. At the surface where winds cross the isobars at an angle, you can see a transport of air from high to low pressure. Although winds are virtually parallel to contours on an upper air chart, there still is a slow transport of air from high to low pressure.

At the surface when air converges into a low, it cannot go outward against the pressure gradient, nor can it go downward into the ground; it must go upward.* Therefore, a low or trough is an area of rising air.

Rising air is conducive to cloudiness and precipitation; thus we have the general association of low pressure–bad weather. Reasons for the inclement weather are developed in later chapters.

By similar reasoning, air moving out of a high or ridge depletes the quantity of air. Highs and ridges, therefore, are areas of descending air. Descending air favors dissipation of cloudiness; hence the association, high pressure–good weather.

Many times weather is more closely associated with an upper air pattern than with features shown by the surface map. Although features on the two charts are related, they seldom are identical. A

---

*You may recall that earlier we said air "piles up" in the vicinity of 30° latitude increasing pressure and forming the subtropical high pressure belt. Why, then, does not air flowing into a low or trough increase pressure and fill the system? Dynamic forces maintain the low or trough; and these forces differ from the forces that maintain the subtropical high.

weak surface system often loses its identity in the upper air pattern, while another system may be more evident on the upper air chart than on the surface map.

Widespread cloudiness and precipitation often develop in advance of an upper trough or low. A line of showers and thunderstorms is not uncommon with a trough aloft even though the surface pressure pattern shows little or no cause for the development.

On the other hand, downward motion in a high or ridge places a "cap" on convection, preventing any upward motion. Air may become stagnant in a high, trap moisture and contamination in low levels, and restrict ceiling and visibility. Low stratus, fog, haze, and smoke are not uncommon in high pressure areas. However, a high or ridge aloft with moderate surface winds most often produces good flying weather.

Highs and lows tend to *lean* from the surface into the upper atmosphere. Due to this slope, winds aloft often blow across the associated surface systems. Upper winds tend to steer surface systems in the general direction of the upper wind flow.

An intense, cold, low pressure vortex *leans less* than does a weaker system. The intense low becomes oriented almost vertically and is clearly evident on both surface and upper air charts. Upper winds encircle the surface low and do not blow across it. Thus, the storm moves very slowly and usually causes an extensive and persistent area of clouds, precipitation, strong winds, and generally adverse flying weather. The term *cold low* sometimes used by the weatherman describes such a system.

A contrasting analogy to the cold low is the *thermal low*. A dry, sunny region becomes quite warm from intense surface heating thus generating a surface low pressure area. The warm air is carried to high levels by convection, but cloudiness is scant because of lack of moisture. Since in warm air, pressure decreases slowly with altitude, the warm surface low is not evident at upper levels. Unlike the cold low, the thermal low is relatively shallow with weak pressure gradients and no well defined cyclonic circulation. It generally supports good flying weather. However, during the heat of the day, one must be alert for high density altitude and convective turbulence.

We have cited three exceptions to the low pressure–bad weather, high pressure–good weather rule: (1) cloudiness and precipitation with an upper air trough or low not evident on the surface chart; (2) the contaminated high; and (3) the thermal low. As this book progresses, you can further relate weather systems more specifically to flight operations.

Moisture, Cloud Formation, and Precipitation

FIGURE 8-8. cP AIR MOVING SOUTHWARD

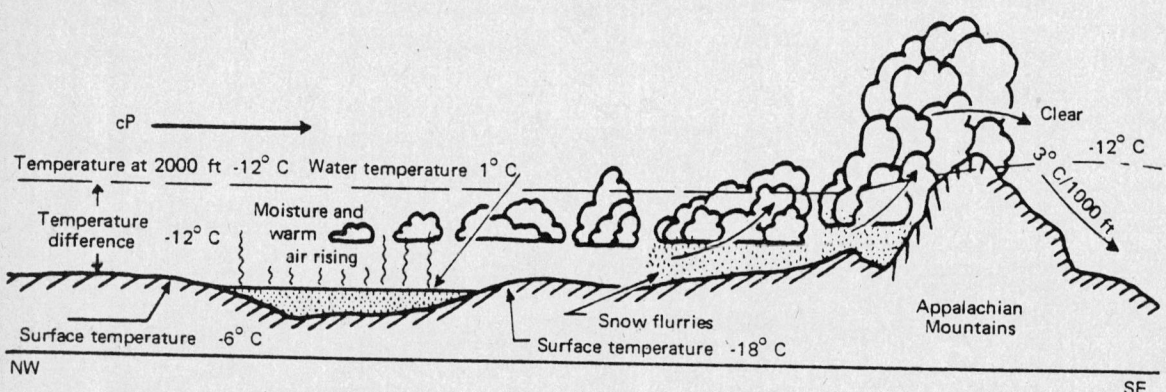

Chapter 5
MOISTURE, CLOUD FORMATION, AND PRECIPITATION

Imagine, if you can, how easy flying would be if skies everywhere were clear! But, flying isn't always that easy; moisture in the atmosphere creates a variety of hazards unmatched by any other weather element. Within Earth's climatic range, water is in the frozen, liquid, and gaseous states.

WATER VAPOR

Water evaporates into the air and becomes an ever-present but variable constituent of the atmosphere. Water vapor is invisible just as oxygen and other gases are invisible. However, we can readily measure water vapor and express it in different ways. Two commonly used terms are (1) relative humidity, and (2) dew point.

Moisture, Cloud Formation, and Precipitation

RELATIVE HUMIDITY

Relative humidity routinely is expressed in percent. As the term suggests, *relative humidity is "relative."* It relates *the actual water vapor present to that which could be present.*

Temperature largely determines the maximum amount of water vapor air can hold. As figure 32 shows, warm air can hold more water vapor than cool air. Figure 33 relates water vapor, temperature, and relative humidity. Actually, relative humidity expresses the degree of saturation. Air with 100% relative humidity is saturated; less than 100% is unsaturated.

If a given volume of air is cooled to some specific temperature, it can hold no more water vapor than is actually present, relative humidity becomes 100%, and saturation occurs. What is that temperature?

DEW POINT

Dew point is the temperature to which air must be cooled to become saturated by the water vapor already present in the air. Aviation weather reports normally include the air temperature and dew point temperature. Dew point when related to air temperature reveals qualitatively how close the air is to saturation.

TEMPERATURE—DEW POINT SPREAD

The difference between air temperature and dew point temperature is popularly called the "spread." As spread becomes less, relative humidity increases, and it is 100% when temperature and dew point are the same. Surface temperature–dew point spread is important in anticipating fog but has little bearing on precipitation. To support precipitation, air must be saturated through thick layers aloft.

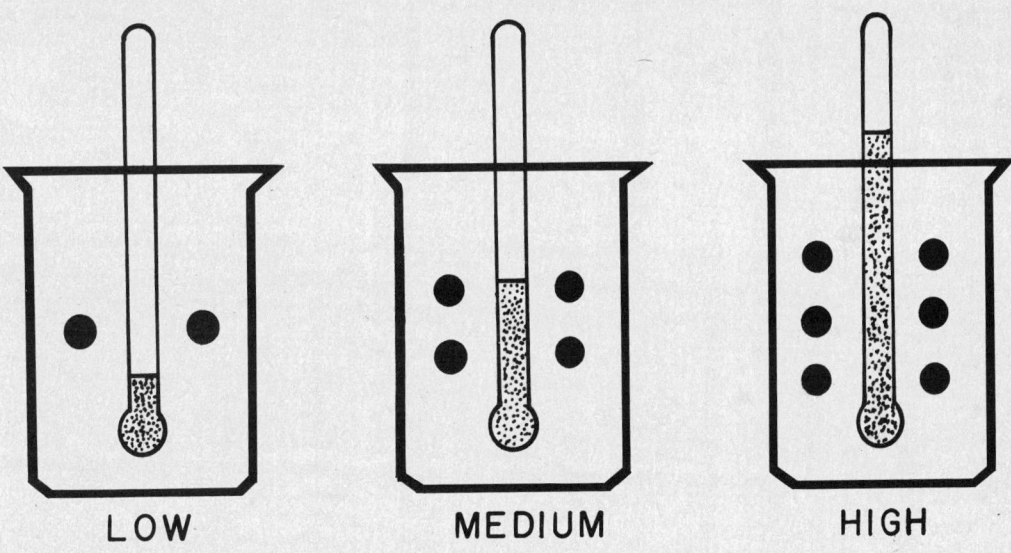

FIGURE 32. Blue dots illustrate the increased water vapor capacity of warmer air. At each temperature, air can hold a specific amount of water vapor—no more.

Moisture, Cloud Formation, and Precipitation

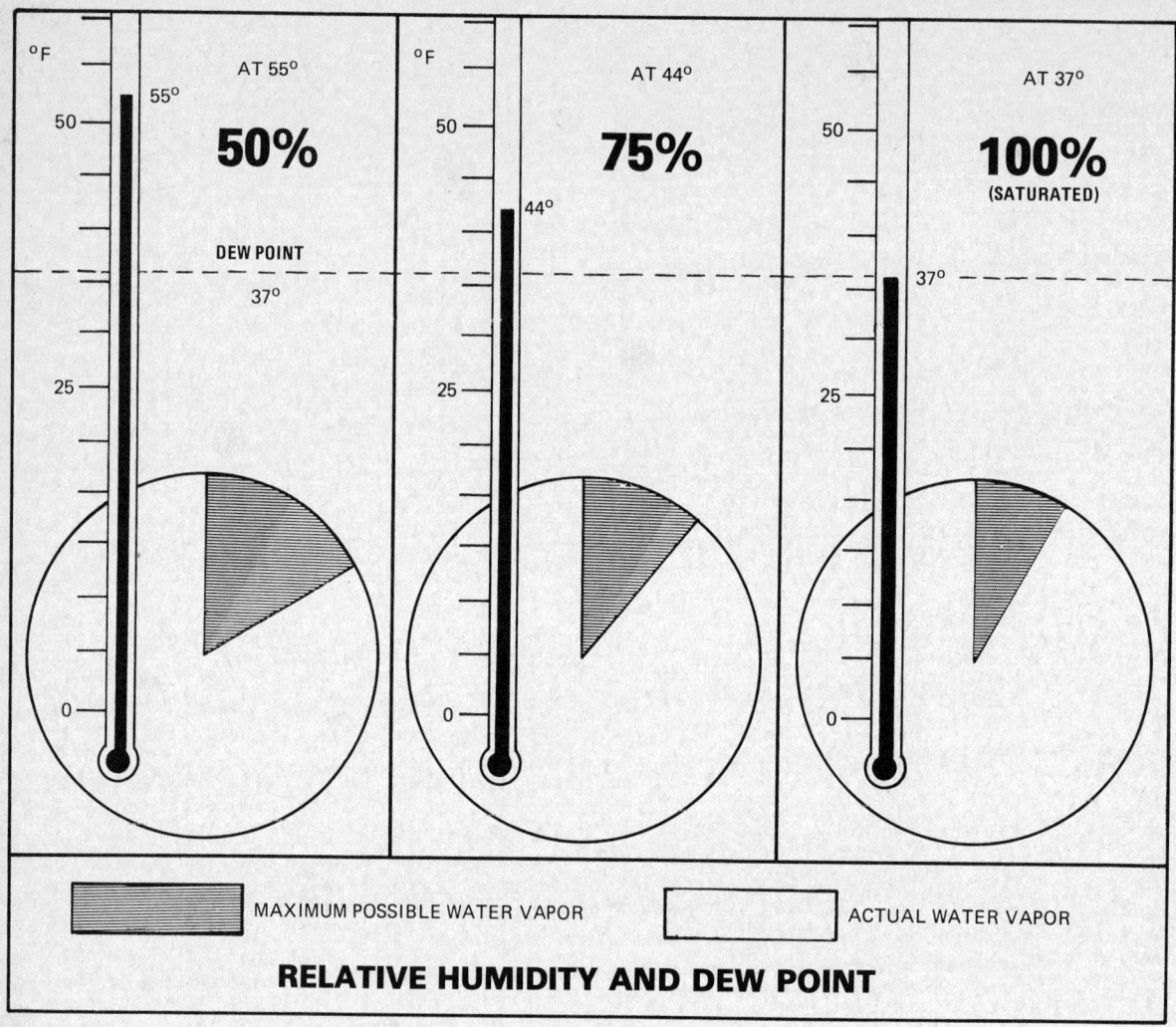

FIGURE 33. Relative humidity depends on both temperature and water vapor. In this figure, water vapor is constant but temperature varies. On the left, relative humidity is 50%; the warmer air could hold twice as much water vapor as is actually present. As the air cools, center and right, relative humidity increases. As the air cools to 37° F, its capacity to hold water vapor is reduced to the amount actually present. Relative humidity is 100% and the air is now "saturated." Note that at 100% humidity, temperature and dew point are the same. The air cooled to saturation, i.e., it cooled to the dew point.

Sometimes the spread at ground level may be quite large, yet at higher altitudes the air is saturated and clouds form. Some rain may reach the ground or it may evaporate as it falls into the drier air. Figure 34 is a photograph of "virga"—streamers of precipitation trailing beneath clouds but evaporating before reaching the ground. Our never ending weather cycle involves a continual reversible change of water from one state to another. Let's take a closer look at change of state.

CHANGE OF STATE

Evaporation, condensation, sublimation, freezing, and melting are changes of state. Evaporation is the changing of liquid water to invisible water vapor. Condensation is the reverse process. Sublimation is the changing of ice directly to water vapor, or water vapor to ice, bypassing the liquid

Moisture, Cloud Formation, and Precipitation

FIGURE 34. Virga. Precipitation from the cloud evaporates in drier air below and does not reach the ground.

state in each process. Snow or ice crystals result from the sublimation of water vapor directly to the solid state. We are all familiar with freezing and melting processes.

LATENT HEAT

Any change of state involves a heat transaction with no change in temperature. Figure 35 diagrams the heat exchanges between the different states. Evaporation requires heat energy that comes from the nearest available heat source. This heat energy is known as the "latent heat of vaporization," and its removal cools the source it comes from. An example is the cooling of your body by evaporation of perspiration.

What becomes of this heat energy used by evaporation? Energy cannot be created or destroyed, so it is hidden or stored in the invisible water vapor. When the water vapor condenses to liquid water or sublimates directly to ice, energy originally used in the evaporation reappears as heat and is released to the atmosphere. This energy is "latent heat" and is quite significant as we learn in later chapters. Melting and freezing involve the exchange of "latent heat of fusion" in a similar manner. The latent heat of fusion is much less than that of condensation and evaporation; however, each in its own way plays an important role in aviation weather.

As air becomes saturated, water vapor begins to condense on the nearest available surface. What surfaces are in the atmosphere on which water vapor may condense?

CONDENSATION NUCLEI

The atmosphere is never completely clean; an abundance of microscopic solid particles suspended in the air are condensation surfaces. These particles, such as salt, dust, and combustion byproducts

Moisture, Cloud Formation, and Precipitation

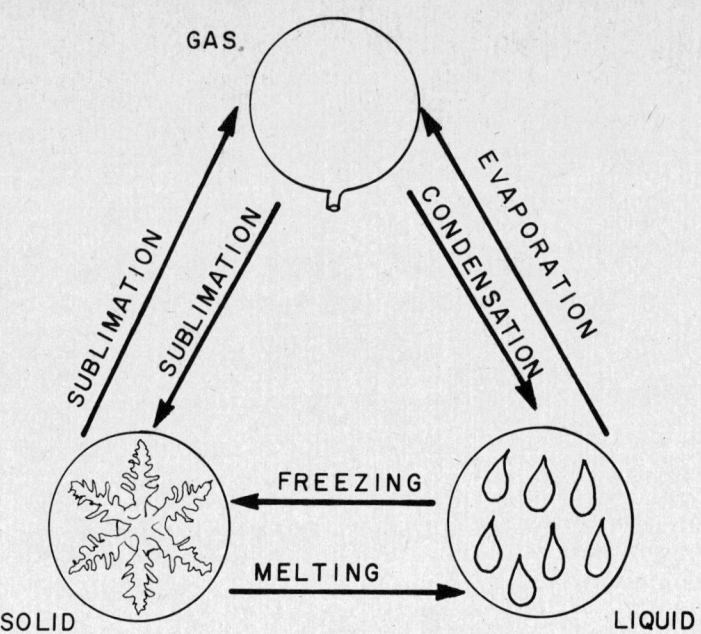

FIGURE 35. Heat transactions when water changes state.

are "condensation nuclei." Some condensation nuclei have an affinity for water and can induce condensation or sublimation even when air is almost but not completely saturated.

As water vapor condenses or sublimates on condensation nuclei, liquid or ice particles begin to grow. Whether the particles are liquid or ice does not depend entirely on temperature. Liquid water may be present at temperatures well below freezing.

SUPERCOOLED WATER

Freezing is complex and liquid water droplets often condense or persist at temperatures colder than 0° C. Water droplets colder than 0° C are supercooled. When they strike an exposed object, the impact induces freezing. Impact freezing of supercooled water can result in aircraft icing.

Supercooled water drops very often are in abundance in clouds at temperatures between 0° C and $-15°$ C with decreasing amounts at colder temperatures. Usually, at temperatures colder than $-15°$ C, sublimation is prevalent; and clouds and fog may be mostly ice crystals with a lesser amount of supercooled water. However, strong vertical currents may carry supercooled water to great heights where temperatures are much colder than $-15°$ C. Supercooled water has been observed at temperatures colder than $-40°$ C.

DEW AND FROST

During clear nights with little or no wind, vegetation often cools by radiation to a temperature at or below the dew point of the adjacent air. Moisture then collects on the leaves just as it does on a pitcher of ice water in a warm room. Heavy dew often collects on grass and plants when none collects on pavements or large solid objects. These more massive objects absorb abundant heat during the day, lose it slowly during the night, and cool below the dew point only in rather extreme cases.

Frost forms in much the same way as dew. The difference is that the dew point of surrounding air must be colder than freezing. Water vapor then sublimates directly as ice crystals or frost rather than condensing as dew. Sometimes dew forms and later freezes; however, frozen dew is easily distinguished from frost. Frozen dew is hard and transparent while frost is white and opaque.

To now, we have said little about clouds. What brings about the condensation or sublimation that results in cloud formation?

CLOUD FORMATION

Normally, air must become saturated for condensation or sublimation to occur. Saturation may result from cooling temperature, increasing dew point, or both. Cooling is far more predominant.

COOLING PROCESSES

Three basic processes may cool air to saturation. They are (1) air moving over a colder surface, (2) stagnant air overlying a cooling surface, and (3) expansional cooling in upward moving air. Expansional cooling is the major cause of cloud formation. Chapter 6, "Stable and Unstable Air," discusses expansional cooling in detail.

CLOUDS AND FOG

A cloud is a visible aggregate of minute water or ice particles suspended in air. If the cloud is on the ground, it is fog. When entire layers of air cool to saturation, fog or sheet-like clouds result. Saturation of a localized updraft produces a towering cloud. A cloud may be composed entirely of liquid water, of ice crystals, or a mixture of the two.

PRECIPITATION

Precipitation is an all inclusive term denoting drizzle, rain, snow, ice pellets, hail, and ice crystals. Precipitation occurs when these particles grow in size and weight until the atmosphere no longer can suspend them and they fall. These particles grow primarily in two ways.

PARTICLE GROWTH

Once a water droplet or ice crystal forms, it continues to grow by added condensation or sublimation directly onto the particle. This is the slower of the two methods and usually results in drizzle or very light rain or snow.

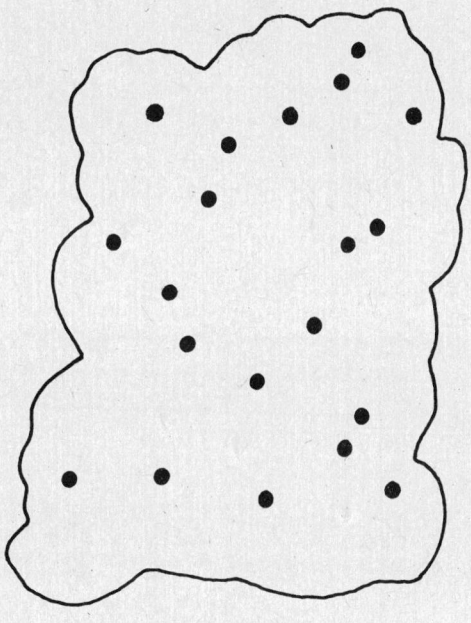
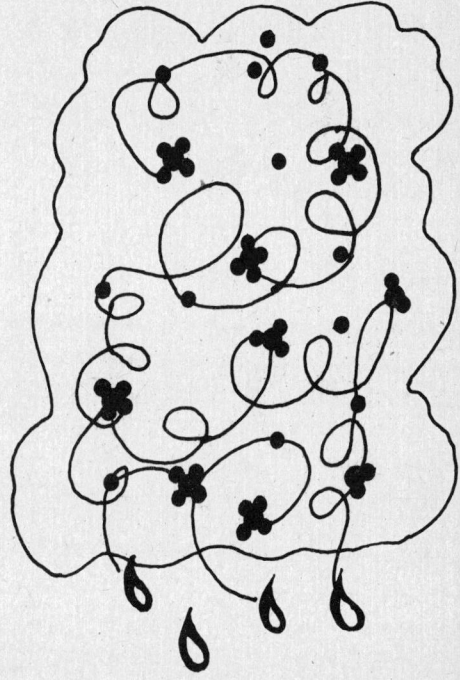

FIGURE 36. Growth of raindrops by collision of cloud droplets.

Moisture, Cloud Formation, and Precipitation

Cloud particles collide and merge into a larger drop in the more rapid growth process. This process produces larger precipitation particles and does so more rapidly than the simple condensation growth process. Upward currents enhance the growth rate and also support larger drops as shown in figure 36. Precipitation formed by merging drops with mild upward currents can produce light to moderate rain and snow. Strong upward currents support the largest drops and build clouds to great heights. They can produce heavy rain, heavy snow, and hail.

LIQUID, FREEZING, AND FROZEN

Precipitation forming and remaining liquid falls as rain or drizzle. Sublimation forms snowflakes, and they reach the ground as snow if temperatures remain below freezing.

Precipitation can change its state as the temperature of its environment changes. Falling snow may melt in warmer layers of air at lower altitudes to form rain. Rain falling through colder air may become supercooled, freezing on impact as freezing rain; or it may freeze during its descent, falling as ice pellets. Ice pellets always indicate freezing rain at higher altitude.

Sometimes strong upward currents sustain large supercooled water drops until some freeze; subsequently, other drops freeze to them forming hailstones.

PRECIPITATION VERSUS CLOUD THICKNESS

To produce significant precipitation, clouds usually are 4,000 feet thick or more. The heavier the precipitation, the thicker the clouds are likely to be. When arriving at or departing from a terminal reporting precipitation of light or greater intensity, expect clouds to be more than 4,000 feet thick.

LAND AND WATER EFFECTS

Land and water surfaces underlying the atmosphere greatly affect cloud and precipitation development. Large bodies of water such as oceans and large lakes add water vapor to the air. Expect the greatest frequency of low ceilings, fog, and precipitation in areas where prevailing winds have an over-water trajectory. Be especially alert for these hazards when moist winds are blowing upslope.

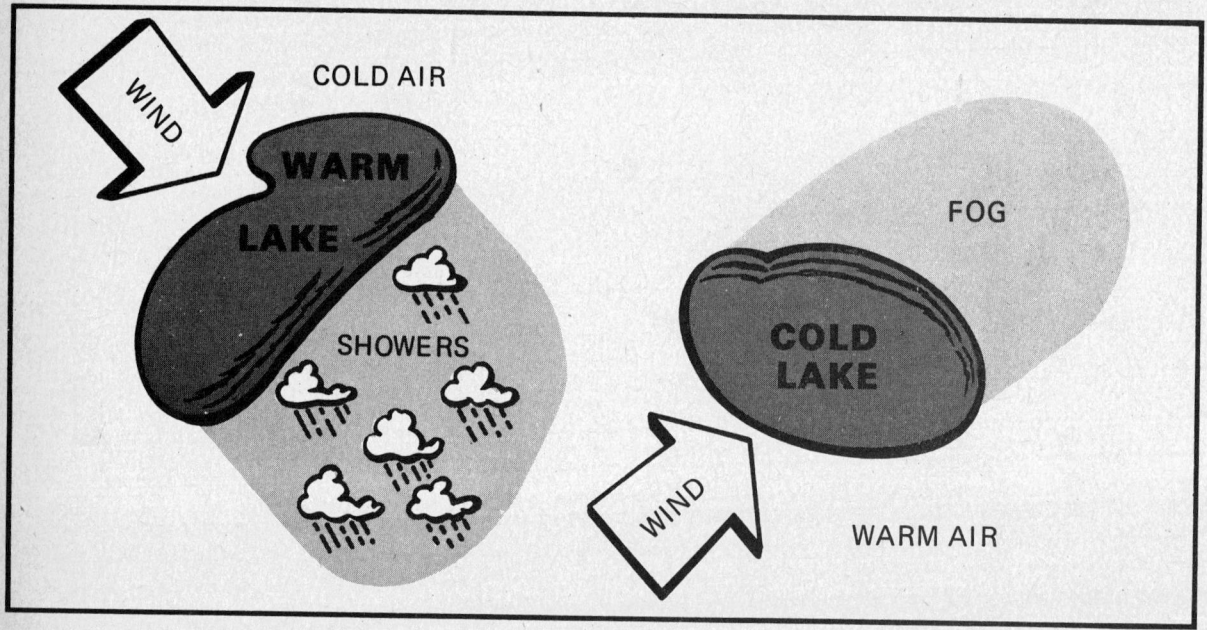

FIGURE 37. Lake effects. Air moving across a sizeable lake absorbs water vapor. Showers may appear on the leeward side if the air is colder than the water. When the air is warmer than the water, fog often develops on the lee side.

Moisture, Cloud Formation, and Precipitation

In winter, cold air frequently moves over relatively warm lakes. The warm water adds heat and water vapor to the air causing showers to the lee of the lakes. In other seasons, the air may be warmer than the lakes. When this occurs, the air may become saturated by evaporation from the water while also becoming cooler in the low levels by contact with the cool water. Fog often becomes extensive and dense to the lee of a lake. Figure 37 illustrates movement of air over both warm and cold lakes. Strong cold winds across the Great Lakes often carry precipitation to the Appalachians as shown in figure 38.

A lake only a few miles across can influence convection and cause a diurnal fluctuation in cloudiness. During the day, cool air over the lake blows toward the land, and convective clouds form over the land as shown in figure 39, a photograph of Lake Okeechobee in Florida. At night, the pattern reverses; clouds tend to form over the lake as cool air from the land flows over the lake creating convective clouds over the water.

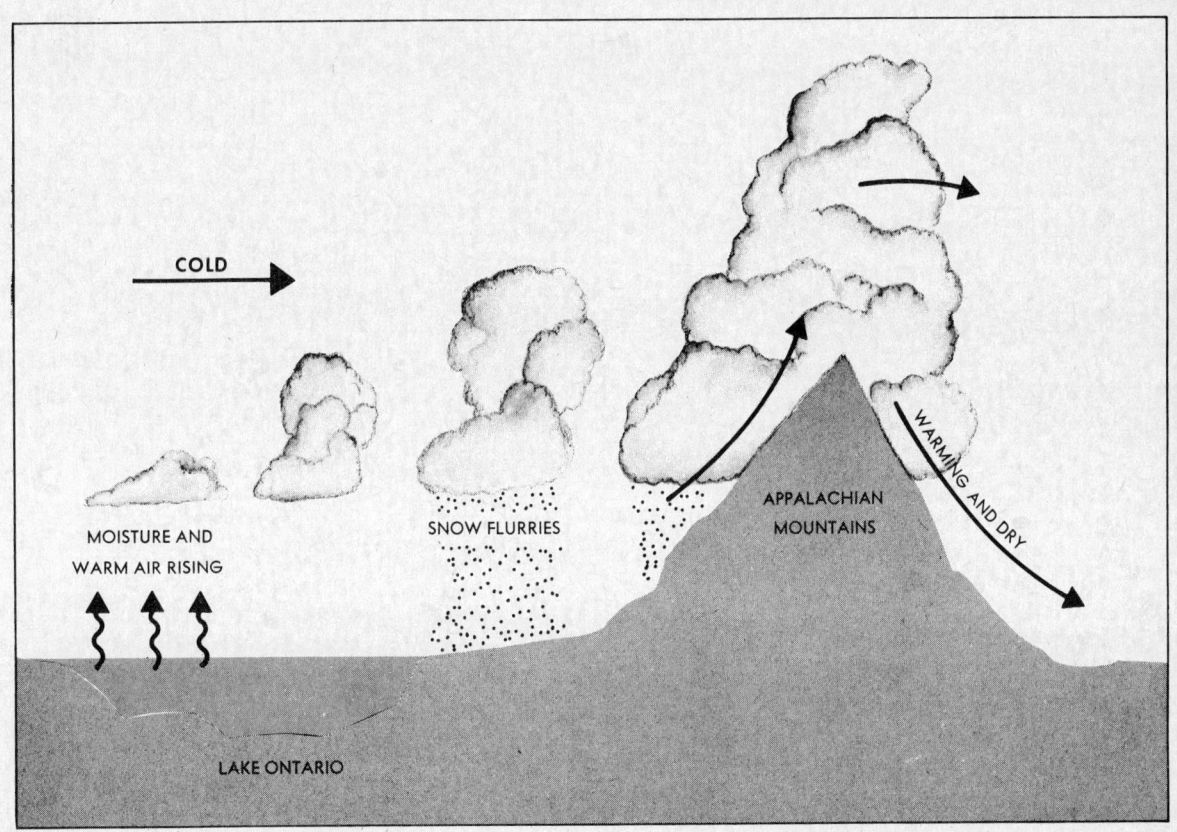

FIGURE 38. Strong cold winds across the Great Lakes absorb water vapor and may carry showers as far eastward as the Appalachians.

FIGURE 39. A view of clouds from 27,000 feet over Lake Okeechobee in southern Florida. Note the lake effect. During daytime, cool air from the lake flows toward the warmer land forming convective clouds over the land.

IN CLOSING

Water exists in three states—solid, liquid, and gaseous. Water vapor is an invisible gas. Condensation or sublimation of water vapor creates many common aviation weather hazards. You may anticipate:

1. Fog when temperature-dew point spread is 5° F or less and decreasing.
2. Lifting or clearing of low clouds and fog when temperature-dew point spread is increasing.
3. Frost on a clear night when temperature-dew point spread is 5° F or less, is decreasing, and dew point is colder than 32° F.
4. More cloudiness, fog, and precipitation when wind blows from water than when it blows from land.
5. Cloudiness, fog, and precipitation over higher terrain when moist winds are blowing uphill.
6. Showers to the lee of a lake when air is cold and the lake is warm. Expect fog to the lee of the lake when the air is warm and the lake is cold.
7. Clouds to be at least 4,000 feet thick when significant precipitation is reported. The heavier the precipitation, the thicker the clouds are likely to be.
8. Icing on your aircraft when flying through liquid clouds or precipitation with temperature freezing or colder.

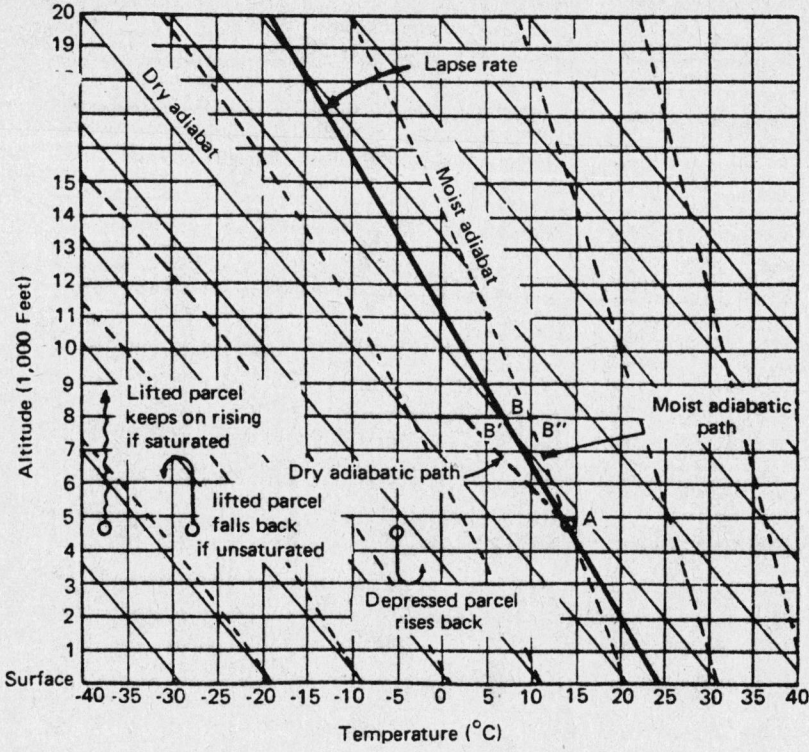

Chapter 6
STABLE AND UNSTABLE AIR

To a pilot, the stability of his aircraft is a vital concern. A stable aircraft, when disturbed from straight and level flight, returns by itself to a steady balanced flight. An unstable aircraft, when disturbed, continues to move away from a normal flight attitude.

So it is with the atmosphere. A *stable* atmosphere resists any upward or downward displacement. An *unstable* atmosphere allows an upward or downward disturbance to grow into a vertical or convective current.

This chapter first examines fundamental changes in upward and downward moving air and then relates stable and unstable air to clouds, weather, and flying.

CHANGES WITHIN UPWARD AND DOWNWARD MOVING AIR

Anytime air moves upward, it expands because of decreasing atmospheric pressure as shown in figure 40. Conversely, downward moving air is compressed by increasing pressure. But as pressure and volume change, temperature also changes.

When air expands, it cools; and when compressed, it warms. These changes are *adiabatic*, meaning that no heat is removed from or added to the air. We frequently use the terms *expansional* or *adiabatic cooling* and *compressional* or *adiabatic*

Stable and Unstable Air

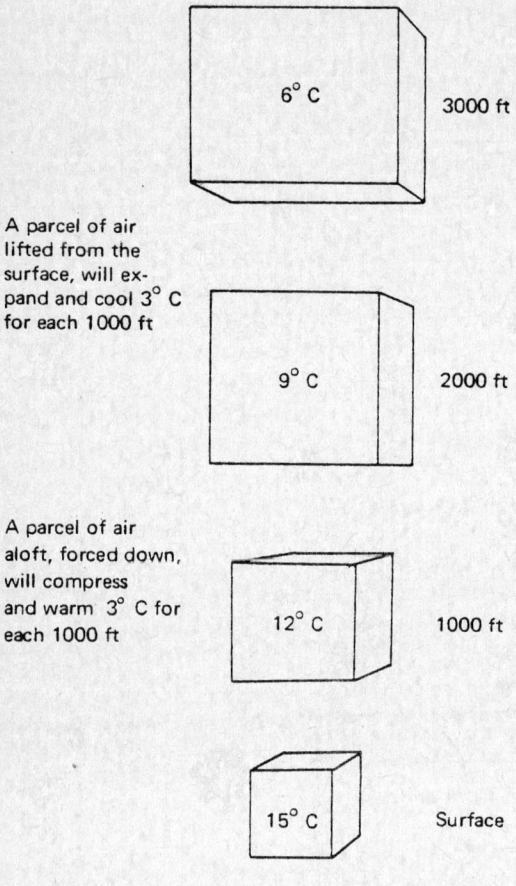

A parcel of air lifted from the surface, will expand and cool 3° C for each 1000 ft

A parcel of air aloft, forced down, will compress and warm 3° C for each 1000 ft

FIGURE 40.

heating. The adiabatic rate of change of temperature is virtually fixed in unsaturated air but varies in saturated air.

UNSATURATED AIR

Unsaturated air moving upward and downward cools and warms at about 3.0° C (5.4° F) per 1,000 feet. This rate is *the "dry adiabatic rate of temperature change"* and *is independent of the temperature of the mass of air through which the vertical movements occur*. Figure 41 illustrates a

"Chinook Wind"—an excellent example of dry adiabatic warming.

SATURATED AIR

Condensation occurs when *saturated* air moves upward. Latent heat released through condensation (chapter 5) partially offsets the expansional cooling. Therefore, *the saturated adiabatic rate of cooling is slower than the dry adiabatic rate*. The saturated rate depends on saturation temperature or dew point of the air. Condensation of copious moisture in saturated warm air releases more latent heat to offset expansional cooling than does the scant moisture in saturated cold air. Therefore, *the saturated adiabatic rate of cooling is less in warm air than in cold air*.

When saturated air moves downward, it heats at the same rate as it cools on ascent *provided* liquid water evaporates rapidly enough to maintain saturation. Minute water droplets evaporate at virtually this rate. Larger drops evaporate more slowly and complicate the moist adiabatic process in downward moving air.

ADIABATIC COOLING AND VERTICAL AIR MOVEMENT

If we force a sample of air upward into the atmosphere, we must consider two possibilities:

(1) The air may become colder than the surrounding air, or
(2) Even though it cools, the air may remain warmer than the surrounding air.

If the upward moving air becomes colder than surrounding air, it sinks; but if it remains warmer, it is accelerated upward as a convective current. Whether it sinks or rises depends on the ambient or existing temperature lapse rate (chapter 2).

Do not confuse existing lapse rate with adiabatic rates of cooling in vertically moving air.* The difference between the existing lapse rate of a given mass of air and the adiabatic rates of cooling in upward moving air determines if the air is stable or unstable.

---

*Sometimes you will hear the dry and moist adiabatic rates of cooling called the dry adiabatic lapse rate and the moist adiabatic lapse rate. In this book, *lapse rate* refers exclusively to the existing, or actual, decrease of temperature with height in a real atmosphere. The dry or moist adiabatic lapse rate signifies a prescribed rate of expansional cooling or compressional heating. An adiabatic lapse rate becomes real *only* when it becomes a condition brought about by vertically moving air.

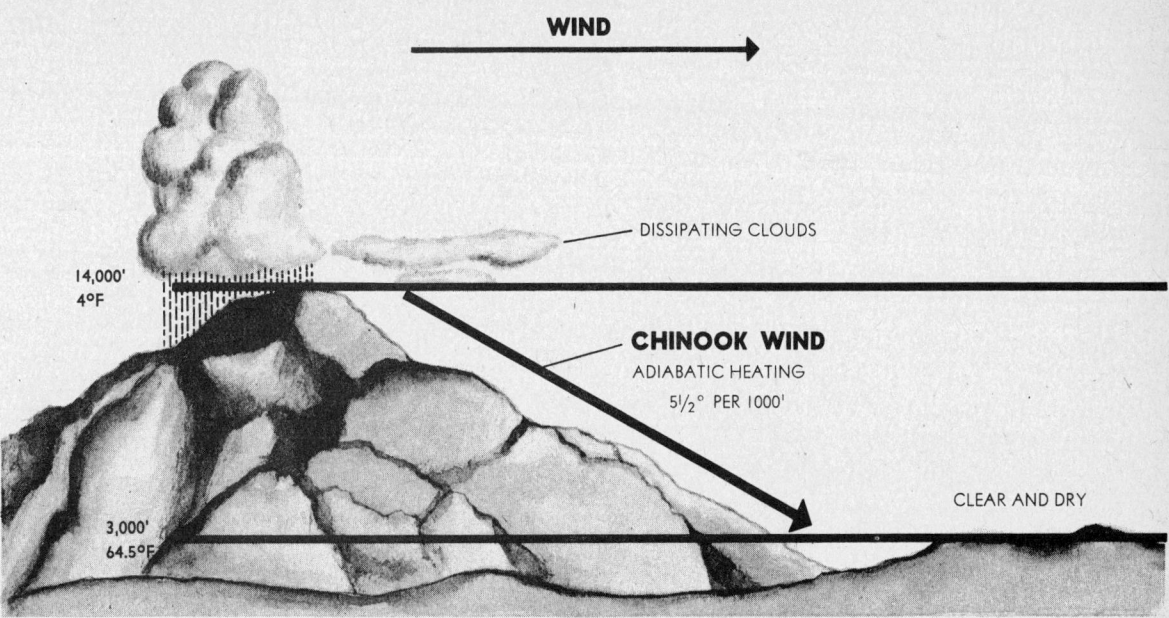

FIGURE 41. Adiabatic warming of downward moving air produces the warm Chinook wind.

STABILITY AND INSTABILITY

Let's use a balloon to demonstrate stability and instability. In figure 42 we have, for three situations, filled a balloon at sea level with air at 31° C —the same as the ambient temperature. We have carried the balloon to 5,000 feet. In each situation, the air in the balloon expanded and cooled at the dry adiabatic rate of 3° C for each 1,000 feet to a temperature of 16° C at 5,000 feet.

In the first situation (left), air inside the balloon, even though cooling adiabatically, remains warmer than surrounding air. Vertical motion is favored. The colder, more dense surrounding air forces the balloon on upward. This air is unstable, and a convective current develops.

In situation two (center) the air aloft is warmer. Air inside the balloon, cooling adiabatically, now becomes colder than the surrounding air. The balloon sinks under its own weight returning to its original position when the lifting force is removed. The air is stable, and spontaneous convection is impossible.

In the last situation, temperature of air inside the balloon is the same as that of surrounding air. The balloon will remain at rest. This condition is neutrally stable; that is, the air is neither stable nor unstable.

Note that, *in all three situations, temperature of air in the expanding balloon cooled at a fixed rate.* The differences in the three conditions depend, therefore, on the temperature differences between the surface and 5,000 feet, that is, on the ambient lapse rates.

HOW STABLE OR UNSTABLE?

Stability runs the gamut from absolutely stable to absolutely unstable, and the atmosphere usually is in a delicate balance somewhere in between. A change in ambient temperature lapse rate of an air mass can tip this balance. For example, surface heating or cooling aloft can make the air more unstable; on the other hand, surface cooling or warming aloft often tips the balance toward greater stability.

Air may be stable or unstable in layers. A stable layer may overlie and cap unstable air; or, conversely, air near the surface may be stable with unstable layers above.

Stable and Unstable Air

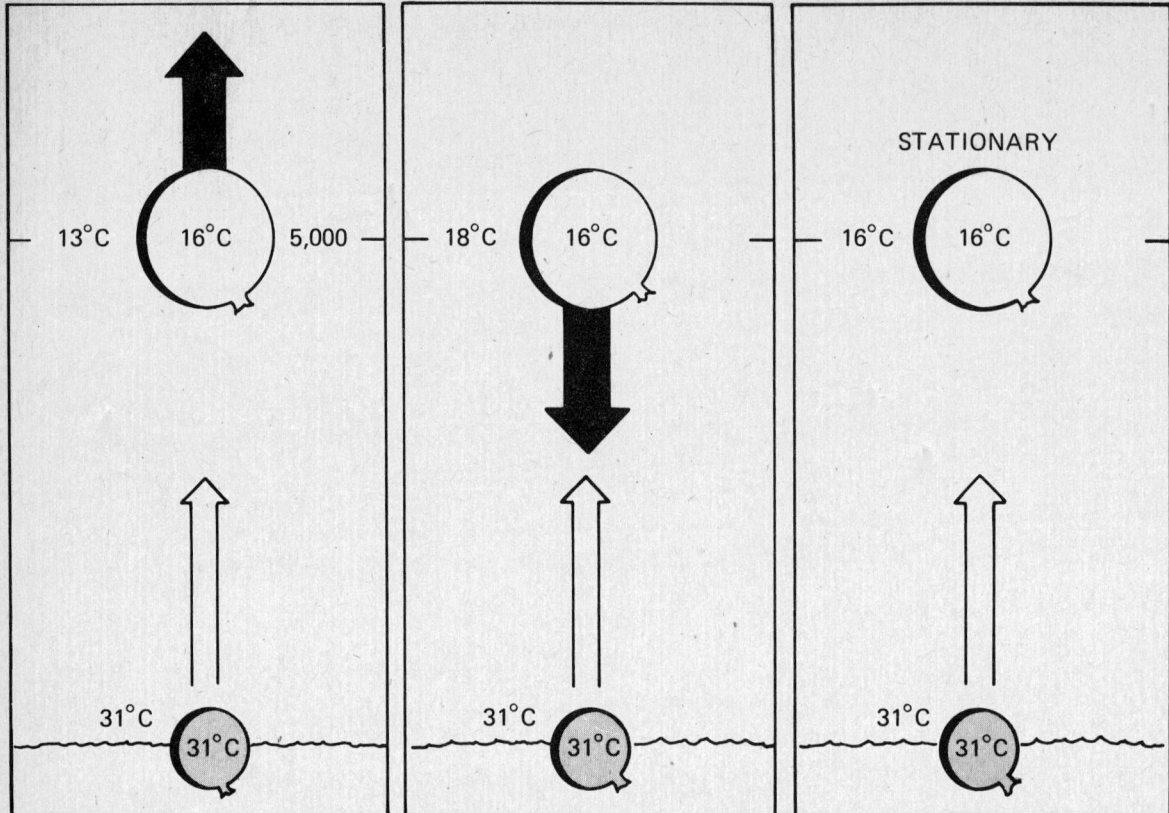

FIGURE 42. Stability related to temperatures aloft and adiabatic cooling. In each situation, the balloon is filled at sea level with air at 31° C, carried manually to 5,000 feet, and released. In each case, air in the balloon expands and cools to 16° C (at the dry adiabatic rate of 3° C per 1,000 feet). But, the temperature of the surrounding air aloft in each situation is different. The balloon on the left will rise. Even though it cooled adiabatically, the balloon remains warmer and lighter than the surrounding cold air; when released, it will continue upward spontaneously. The air is unstable; it favors vertical motion. In the center, the surrounding air is warmer. The cold balloon will sink. It resists our forced lifting and cannot rise spontaneously. The air is stable—it resists upward motion. On the right, surrounding air and the balloon are at the same temperature. The balloon remains at rest since no density difference exists to displace it vertically. The air is neutrally stable, i.e., it neither favors nor resists vertical motion. A mass of air in which the temperature decreases rapidly with height favors instability; but, air tends to be stable if the temperature changes little or not at all with altitude.

CLOUDS—STABLE OR UNSTABLE?

Chapter 5 states that when air is cooling and first becomes saturated, condensation or sublimation begins to form clouds. Chapter 7 explains cloud types and their significance as "signposts in the sky." Whether the air is stable or unstable within a layer largely determines cloud structure.

Stratiform Clouds

Since stable air resists convection, clouds in stable air form in horizontal, sheet-like layers or "strata." Thus, within a *stable* layer, clouds are *stratiform*. Adiabatic cooling may be by upslope flow as illustrated in figure 43; by lifting over cold, more dense air; or by converging winds. Cooling by an underlying cold surface is a stabilizing process and may produce fog. If clouds are to remain stratiform, the layer must remain stable after condensation occurs.

Cumuliform Clouds

Unstable air favors convection. A "cumulus" cloud, meaning "heap," forms in a convective updraft and builds upward, also shown in figure 43. Thus, within an *unstable* layer, clouds are *cumuliform*; and the vertical extent of the cloud depends on the depth of the unstable layer.

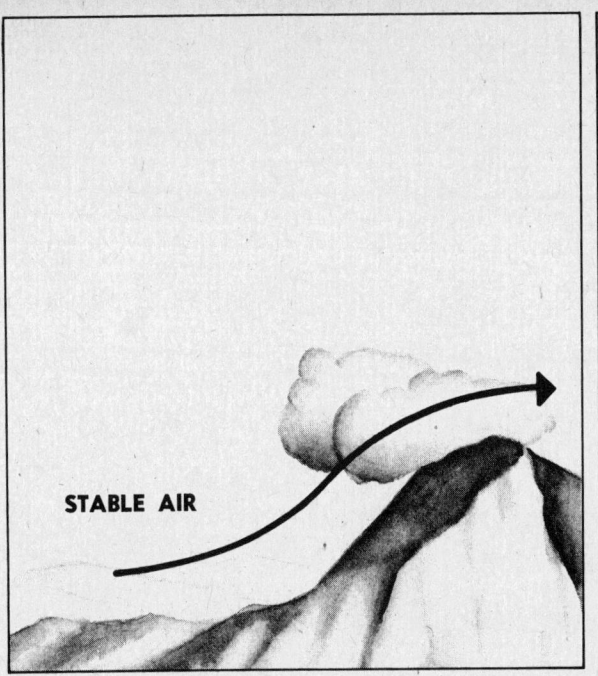 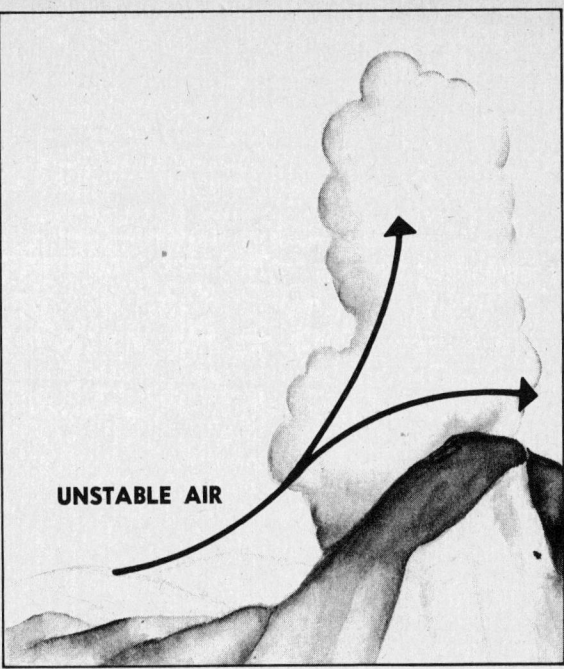

FIGURE 43. When stable air (left) is forced upward, the air tends to retain horizontal flow, and any cloudiness is flat and stratified. When unstable air is forced upward, the disturbance grows, and any resulting cloudiness shows extensive vertical development.

Initial lifting to trigger a cumuliform cloud may be the same as that for lifting stable air. In addition, convection may be set off by surface heating (chapter 4). Air may be unstable or slightly stable before condensation occurs; but for convective cumuliform clouds to develop, it must be unstable after saturation. Cooling in the updraft is now at the slower moist adiabatic rate because of the release of latent heat of condensation. Temperature in the saturated updraft is warmer than ambient temperature, and convection is spontaneous. Updrafts accelerate until temperature within the cloud cools below the ambient temperature. This condition occurs where the unstable layer is capped by a stable layer often marked by a temperature inversion. Vertical heights range from the shallow fair weather cumulus to the giant thunderstorm cumulonimbus—the ultimate in atmospheric instability capped by the tropopause.

You can estimate height of cumuliform cloud bases using surface temperature-dew point spread. Unsaturated air in a convective current cools at about 5.4° F (3.0° C) per 1,000 feet; dew point decreases at about 1° F (5/9° C). Thus, in a convective current, temperature and dew point converge at about 4.4° F (2.5° C) per 1,000 feet as illustrated in figure 44. We can get a quick *estimate* of a convective cloud base in thousands of feet by rounding these values and dividing into the spread or by multiplying the spread by their reciprocals. When using Fahrenheit, divide by 4 or multiply by .25; when using Celsius, divide by 2.2 or multiply by .45. This method of estimating is reliable only with instability clouds and during the warmer part of the day.

When unstable air lies above stable air, convective currents aloft sometimes form middle and high level cumuliform clouds. In relatively shallow layers they occur as altocumulus and ice crystal cirrocumulus clouds. Altocumulus castellanus clouds develop in deeper midlevel unstable layers.

Merging Stratiform and Cumuliform

A layer of stratiform clouds may sometimes form in a mildly stable layer while a few ambitious convective clouds penetrate the layer thus merging stratiform with cumuliform. Convective clouds may be almost or entirely embedded in a massive stratiform layer and pose an unseen threat to instrument flight.

WHAT DOES IT ALL MEAN?

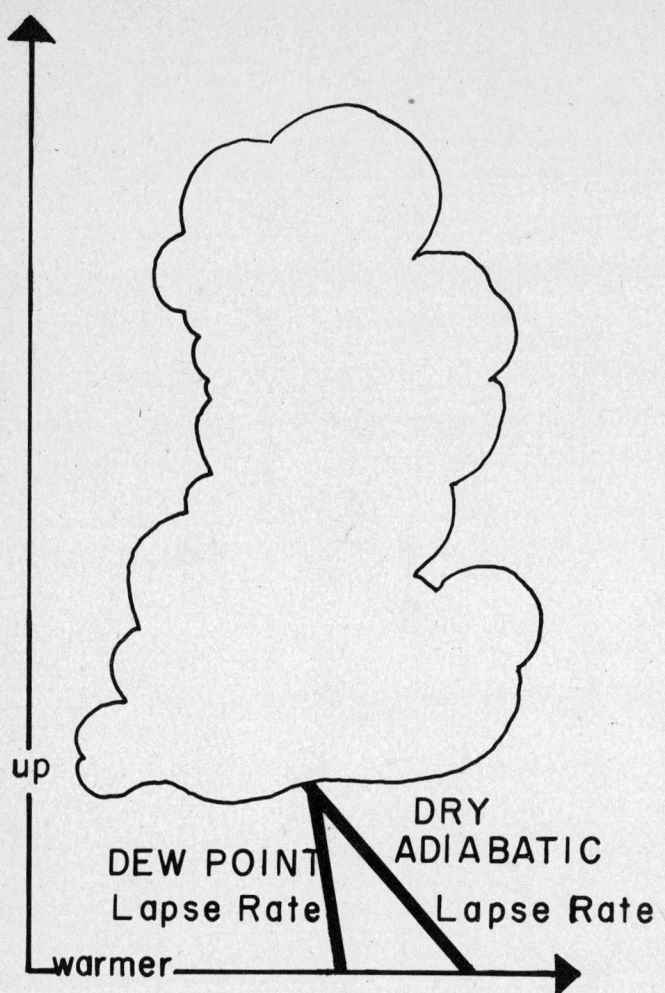

FIGURE 44. Cloud base determination. Temperature and dew point in upward moving air converge at a rate of about 4° F or 2.2° C per 1,000 feet.

1. Thunderstorms are sure signs of violently unstable air. Give these storms a wide berth.
2. Showers and clouds towering upward with great ambition indicate strong updrafts and rough (turbulent) air. Stay clear of these clouds.
3. Fair weather cumulus clouds often indicate bumpy turbulence beneath and in the clouds. The cloud tops indicate the approximate upper limit of convection; flight above is usually smooth.
4. Dust devils are a sign of dry, unstable air, usually to considerable height. Your ride may be fairly rough unless you can get above the instability.
5. Stratiform clouds indicate stable air. Flight generally will be smooth, but low ceiling and visibility might require IFR.
6. Restricted visibility at or near the surface over large areas usually indicates stable air. Expect a smooth ride, but poor visibility may require IFR.
7. Thunderstorms may be embedded in stratiform clouds posing an unseen threat to instrument flight.
8. Even in clear weather, you have some clues to stability, viz.:
   a. When temperature decreases uniformly and rapidly as you climb (approaching 3° C per 1,000 feet), you have an indication of unstable air.
   b. If temperature remains unchanged or decreases only slightly with altitude, the air tends to be stable.
   c. If the temperature increases with altitude through a layer—an inversion—the layer is stable and convection is suppressed. Air may be unstable beneath the inversion.
   d. When air near the surface is warm and moist, suspect instability. Surface heating, cooling aloft, converging or upslope winds, or an invading mass of colder air may lead to instability and cumuliform clouds.

Can we fly in unstable air? Stable air? Certainly we can and ordinarily do since air is seldom neutrally stable. The usual convection in unstable air gives a "bumpy" ride; only at times is it violent enough to be hazardous. In stable air, flying is usually smooth but sometimes can be plagued by low ceiling and visibility. It behooves us in preflight planning to take into account stability or instability and any associated hazards. Certain observations you can make on your own:

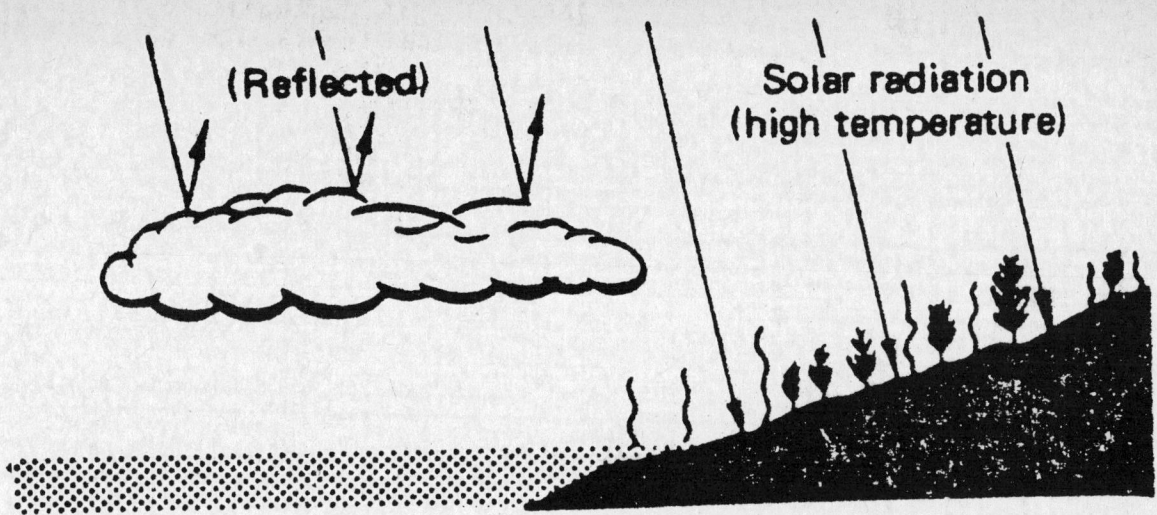

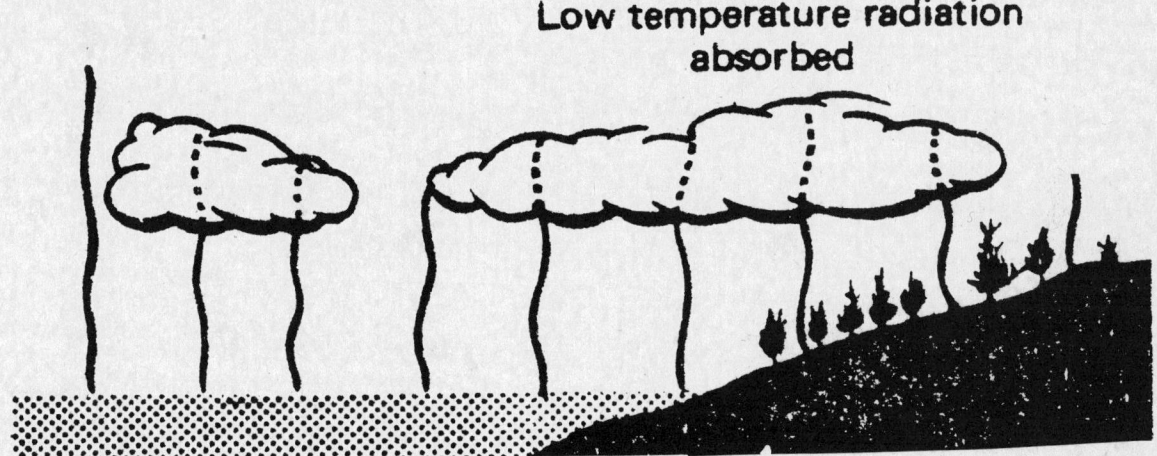

Chapter 7
CLOUDS

Clouds, to almost everyone, have some meaning. But to you as a pilot, clouds are your weather "signposts in the sky." They give you an indication of air motion, stability, and moisture. Clouds help you visualize weather conditions and potential weather hazards you might encounter in flight. Let's examine these "signposts" and how to identify them.

IDENTIFICATION

For identification purposes, you need be concerned only with the more basic cloud types, which are divided into four "families." The families are: high clouds, middle clouds, low clouds, and clouds with extensive vertical development. The first three families are further classified according to the way they are formed. Clouds formed by vertical currents in unstable air are *cumulus* meaning *accumulation* or *heap;* they are characterized by their lumpy, billowy appearance. Clouds formed by the cooling of a stable layer are *stratus* meaning *stratified* or *layered;* they are characterized by their uniform, sheet-like appearance.

In addition to the above, the prefix *nimbo* or the suffix *nimbus* means raincloud. Thus, stratified

Clouds

FIGURE 45. CIRRUS. Cirrus are thin, feather-like ice crystal clouds in patches or narrow bands. Larger ice crystals often trail downward in well-defined wisps called "mares' tails." Wispy, cirrus-like, these contain no significant icing or turbulence. Dense, banded cirrus, which often are turbulent, are discussed in chapter 13.

clouds from which rain is falling are *nimbostratus*. A heavy, swelling cumulus type cloud which produces precipitation is a *cumulonimbus*. Clouds broken into fragments are often identified by adding the suffix *fractus;* for example, fragmentary cumulus is *cumulus fractus*.

HIGH CLOUDS

The high cloud family is cirriform and includes cirrus, cirrocumulus, and cirrostratus. They are composed almost entirely of ice crystals. The height of the bases of these clouds ranges from about 16,500 to 45,000 feet in middle latitudes. Figures 45 through 47 are photographs of high clouds.

MIDDLE CLOUDS

In the middle cloud family are the altostratus, altocumulus, and nimbostratus clouds. These clouds are primarily water, much of which may be supercooled. The height of the bases of these clouds ranges from about 6,500 to 23,000 feet in middle latitudes. Figures 48 through 52 are photographs of middle clouds.

LOW CLOUDS

In the low cloud family are the stratus, stratocumulus, and fair weather cumulus clouds. Low clouds are almost entirely water, but at times the water may be supercooled. Low clouds at subfreezing temperatures can also contain snow and ice particles. The bases of these clouds range from near the surface to about 6,500 feet in middle latitudes. Figures 53 through 55 are photographs of low clouds.

CLOUDS WITH EXTENSIVE VERTICAL DEVELOPMENT

The vertically developed family of clouds includes towering cumulus and cumulonimbus. These clouds usually contain supercooled water above the freezing level. But when a cumulus grows to great heights, water in the upper part of the cloud freezes into ice crystals forming a cumulonimbus. The heights of cumuliform cloud bases range from 1,000 feet or less to above 10,000 feet. Figures 56 and 57 are photographs of clouds with extensive vertical development.

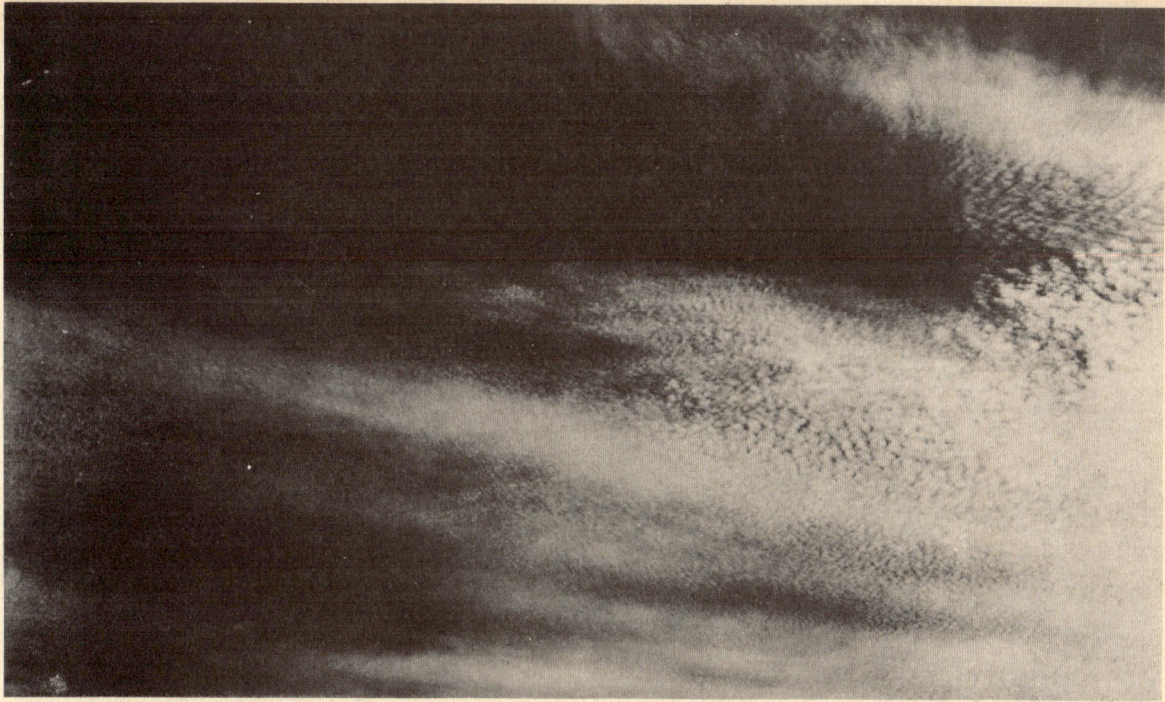

FIGURE 46. CIRROCUMULUS. Cirrocumulus are thin clouds, the individual elements appearing as small white flakes or patches of cotton. May contain highly supercooled water droplets. Some turbulence and icing.

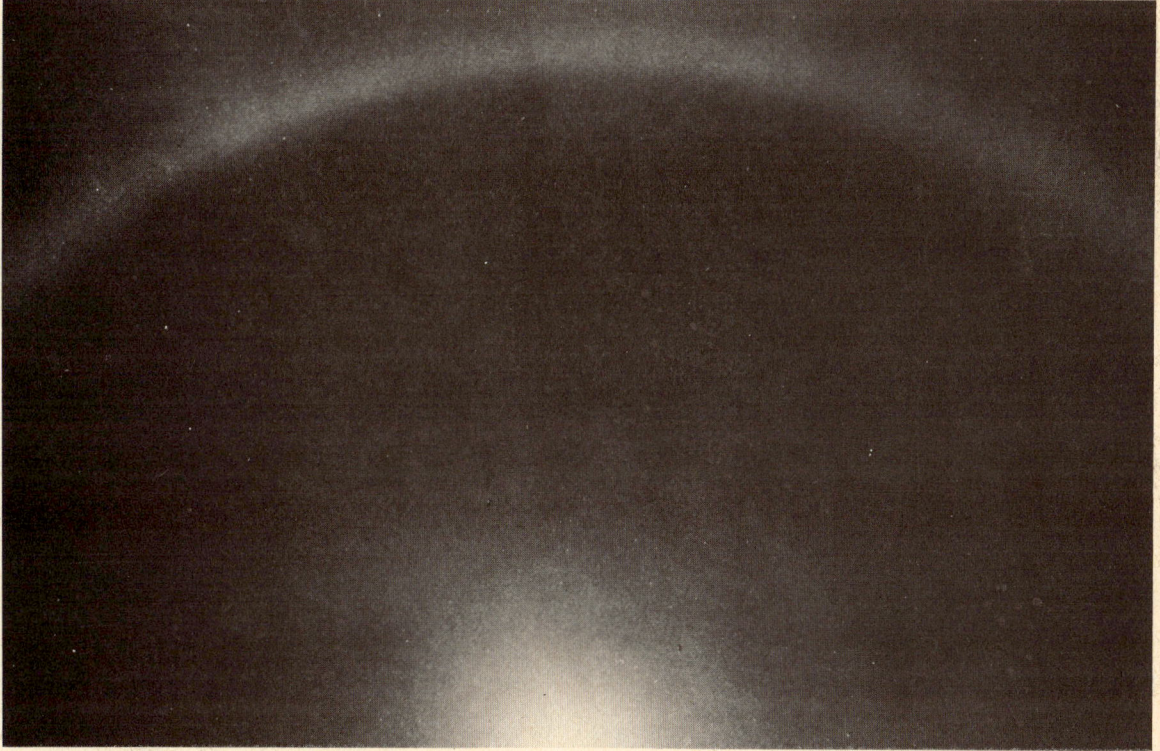

FIGURE 47. CIRROSTRATUS. Cirrostratus is a thin whitish cloud layer appearing like a sheet or veil. Cloud elements are diffuse, sometimes partially striated or fibrous. Due to their ice crystal makeup, these clouds are associated with halos—large luminous circles surrounding the sun or moon. No turbulence and little if any icing. The greatest problem flying in cirriform clouds is restriction to visibility. They can make the strict use of instruments mandatory.

Clouds

FIGURE 48. ALTOCUMULUS. Altocumulus are composed of white or gray colored layers or patches of solid cloud. The cloud elements may have a waved or roll-like appearance. Some turbulence and small amounts of icing.

FIGURE 49. ALTOSTRATUS. Altostratus is a bluish veil or layer of clouds. It is often associated with altocumulus and sometimes gradually merges into cirrostratus. The sun may be dimly visible through it. Little or no turbulence with moderate amounts of ice.

FIGURE 50. ALTOCUMULUS CASTELLANUS. Altocumulus castellanus are middle level convective clouds. They are characterized by their billowing tops and comparatively high bases. They are a good indication of mid-level instability. Rough turbulence with some icing.

FIGURE 51. STANDING LENTICULAR ALTOCUMULUS CLOUDS. Standing lenticular altocumulus clouds are formed on the crests of waves created by barriers in the wind flow. The clouds show little movement, hence the name *standing*. Wind, however, can be quite strong blowing through such clouds. They are characterized by their smooth, polished edges. The presence of these clouds is a good indication of very strong turbulence and should be avoided. Chapter 9, "Turbulence," further explains the significance of this cloud.

Clouds

FIGURE 52. NIMBOSTRATUS. Nimbostratus is a gray or dark massive cloud layer, diffused by more or less continuous rain, snow, or ice pellets. This type is classified as a middle cloud although it may merge into very low stratus or stratocumulus. Very little turbulence, but can pose a serious icing problem if temperatures are near or below freezing.

FIGURE 53. STRATUS. Stratus is a gray, uniform, sheet-like cloud with relatively low bases. When associated with fog or precipitation, the combination can become troublesome for visual flying. Little or no turbulence, but temperatures near or below freezing can create hazardous icing conditions.

Clouds

FIGURE 54. STRATOCUMULUS. Stratocumulus bases are globular masses or rolls unlike the flat, sometimes indefinite, bases of stratus. They usually form at the top of a layer mixed by moderate surface winds. Sometimes, they form from the breaking up of stratus or the spreading out of cumulus. Some turbulence, and possible icing at subfreezing temperatures. Ceiling and visibility usually better than with low stratus.

FIGURE 55. CUMULUS. Fair weather cumulus clouds form in convective currents and are characterized by relatively flat bases and dome-shaped tops. Fair weather cumulus do not show extensive vertical development and do not produce precipitation. More often, fair weather cumulus indicates a shallow layer of instability. Some turbulence and no significant icing.

Clouds

FIGURE 56. TOWERING CUMULUS. Towering cumulus signifies a relatively deep layer of unstable air. It shows considerable vertical development and has billowing *cauliflower* tops. Showers can result from these clouds. Very strong turbulence; some clear icing above the freezing level.

FIGURE 57. CUMULONIMBUS. Cumulonimbus are the ultimate manifestation of instability. They are vertically developed clouds of large dimensions with dense *boiling* tops often crowned with thick veils of dense cirrus (the anvil). Nearly the entire spectrum of flying hazards are contained in these clouds including violent turbulence. They should be avoided at all times! This cloud is the thunderstorm cloud and is discussed in detail in chapter 11, "Thunderstorms."

Clouds

SIGNPOSTS IN THE SKY

The photographs illustrate some of the basic cloud types. The caption with each photograph describes the type and its significance to flight. In closing, we suggest you take a second look at the cloud photographs. Study the descriptions and potential hazards posed by each type and learn to use the clouds as "signposts in the sky."

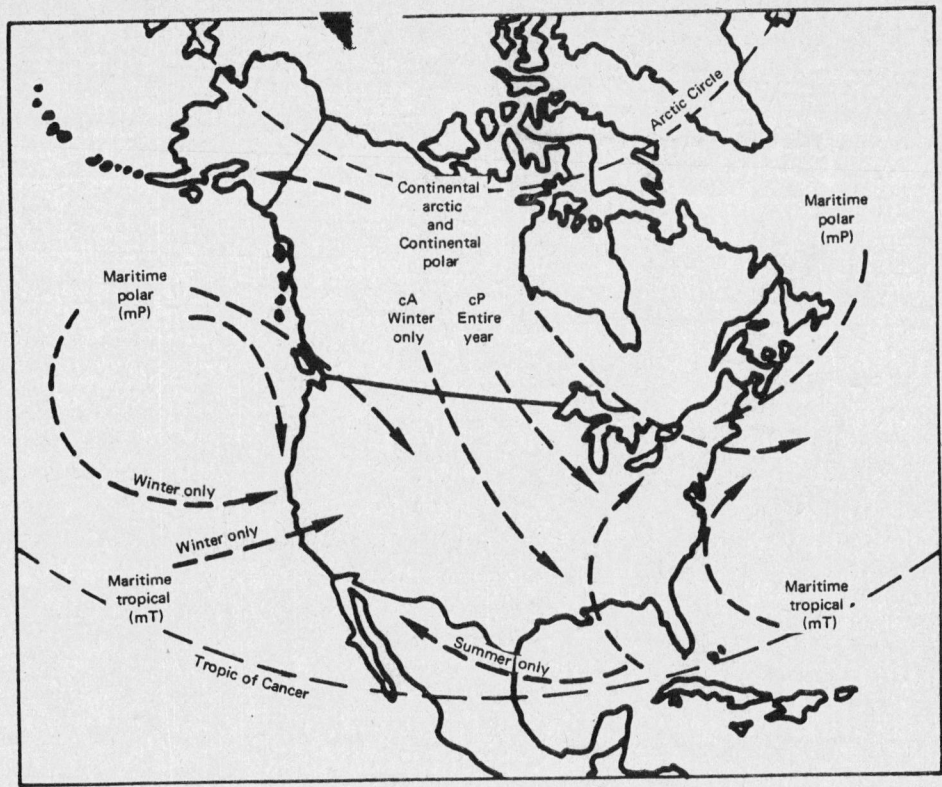

Chapter 8
AIR MASSES AND FRONTS

Why is weather today clear and cold over Oklahoma while it is warm and moist over Alabama? What caused the line of thunderstorms that you circumnavigated over eastern Arkansas? Air masses and fronts provide the answer. You can better plan the safety and economy of flight when you can evaluate the expected effects of air masses and fronts. This chapter explains air masses and fronts and relates them to weather and flight planning.

AIR MASSES

When a body of air comes to rest or moves slowly over an extensive area having fairly uniform properties of temperature and moisture, the air takes on those properties. Thus, the air over the area becomes somewhat of an entity as illustrated in figure 58 and has fairly uniform horizontal distribution of its properties. The area over which the air mass acquires its identifying distribution of moisture and temperature is its "source region."

Source regions are many and varied, but the best source regions for air masses are large snow or ice-covered polar regions, cold northern oceans, tropical oceans, and large desert areas. Midlatitudes are poor source regions because transitional disturbances dominate these latitudes giving little opportunity for air masses to stagnate and take on the properties of the underlying region.

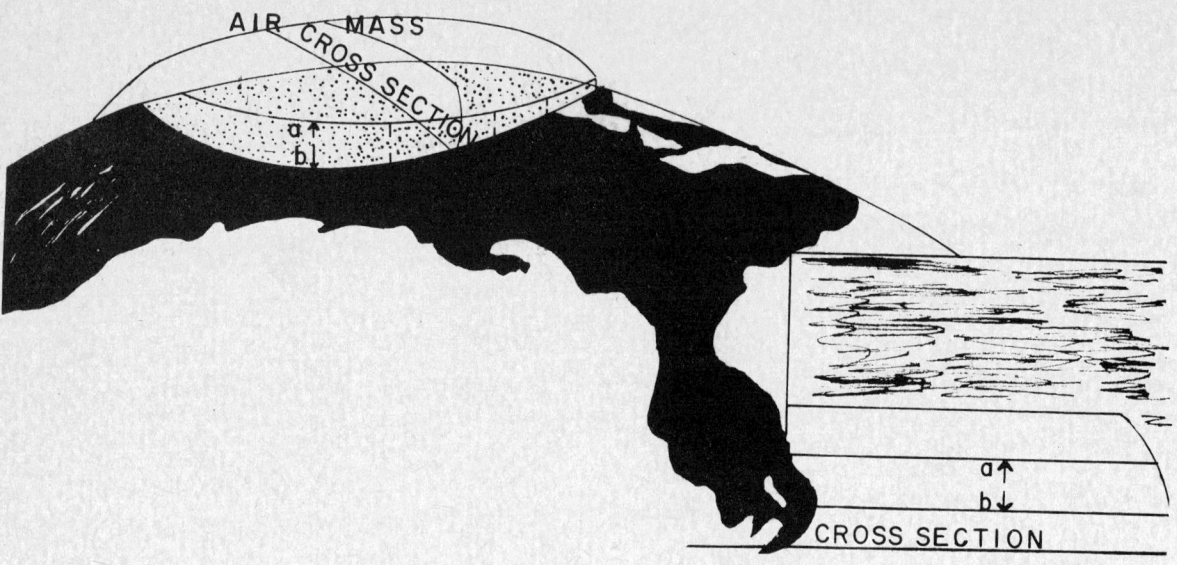

Figure 58. Horizontal uniformity of an air mass

AIR MASS MODIFICATION

Just as an air mass took on the properties of its source region, it tends to take on properties of the underlying surface when it moves away from its source region, thus becoming modified.

The degree of modification depends on the speed with which the air mass moves, the nature of the region over which it moves, and the temperature difference between the new surface and the air mass. Some ways air masses are modified are: (1) warming from below, (2) cooling from below, (3) addition of water vapor, and (4) subtraction of water vapor:

1. Cool air moving over a warm surface is heated from below, generating instability and increasing the possibility of showers.
2. Warm air moving over a cool surface is cooled from below, increasing stability. If air is cooled to its dew point, stratus and/or fog forms.
3. Evaporation from water surfaces and falling precipitation adds water vapor to the air. When the water is warmer than the air, evaporation can raise the dew point sufficiently to saturate the air and form stratus or fog.
4. Water vapor is removed by condensation and precipitation.

STABILITY

Stability of an air mass determines its typical weather characteristics. When one type of air mass overlies another, conditions change with height. Characteristics typical of an unstable and a stable air mass are as follows:

Unstable Air	*Stable Air*
Cumuliform clouds	Stratiform clouds and fog
Showery precipitation	Continuous precipitation
Rough air (turbulence)	Smooth air
Good visibility, except in blowing obstructions	Fair to poor visibility in haze and smoke

FRONTS

As air masses move out of their source regions, they come in contact with other air masses of different properties. The zone between two different air masses is a frontal zone or front. Across this zone, temperature, humidity and wind often change rapidly over short distances.

DISCONTINUITIES

When you pass through a front, the change from the properties of one air mass to those of the other is sometimes quite abrupt. Abrupt changes indicate a narrow frontal zone. At other times, the change of properties is very gradual indicating a broad and diffuse frontal zone.

Temperature

Temperature is one of the most easily recognized discontinuities across a front. At the surface, the passage of a front usually causes noticeable temperature change. When flying through a front, you note a significant change in temperature, especially at low altitudes. Remember that the temperature change, even when gradual, is faster and more pronounced than a change during a flight wholly within one air mass. Thus, for safety, obtain a new altimeter setting after flying through a front. Chapter 3 discussed the effect of a temperature change on the aircraft altimeter.

Dew Point

As you learned in Chapter 5, dew point temperature is a measure of the amount of water vapor in the air. Temperature–dew point spread is a measure of the degree of saturation. Dew point and temperature–dew point spread usually differ across a front. The difference helps identify the front and may give a clue to differences of cloudiness and/or fog.

Wind

Wind always changes across a front. Wind discontinuity may be in direction, in speed, or in both. Be alert for a wind shift when flying in the vicinity of a frontal surface; if the wind shift catches you unaware it can get you off course or even lost in a short time. The relatively sudden change in wind also creates wind shear, and you will study its significance in the next chapter, "Turbulence."

Pressure

A front lies in a pressure trough, and pressure generally is higher in the cold air. Thus, when you cross a front directly into colder air, pressure usually rises abruptly. When you approach a front toward warm air, pressure generally falls until you cross the front and then remains steady or falls slightly in the warm air. However, pressure patterns vary widely across fronts, and your course may not be directly across a front. The important thing to remember is that when crossing a front, you will encounter a difference in the rate of pressure change; be especially alert in keeping your altimeter setting current.

TYPES OF FRONTS

The three principal types of fronts are the cold front, the warm front, and the stationary front.

Cold Front

The leading edge of an advancing cold air mass is a cold front. At the surface, cold air is overtaking and replacing warmer air. Cold fronts move at about the speed of the wind component perpendicular to the front just above the frictional layer. Figure 59 shows the vertical cross section of a cold front and the symbol depicting it on a surface weather chart. A shallow cold air mass or a slow moving cold front may have a frontal slope more like a warm front shown in figure 60.

Warm Front

The edge of an advancing warm air mass is a warm front—warmer air is overtaking and replacing colder air. Since the cold air is denser than the warm air, the cold air hugs the ground. The warm air slides up and over the cold air and lacks direct push on the cold air. Thus, the cold air is slow to retreat in advance of the warm air. This slowness of the cold air to retreat produces a frontal slope that is more gradual than the cold frontal slope as shown in figure 60. Consequently, warm fronts on the surface are seldom as well marked as cold fronts, and they usually move about half as fast when the general wind flow is the same in each case.

Stationary Fronts

When neither air mass is replacing the other, the front is stationary. Figure 61 shows a cross section of a stationary front and its symbol on a surface chart. The opposing forces exerted by adjacent air masses of different densities are such that the frontal surface between them shows little or no movement. In such cases, the surface winds tend to blow parallel to the frontal zone. Slope of a stationary front is normally shallow, although it may be steep depending on wind distribution and density difference.

Air Masses and Fronts

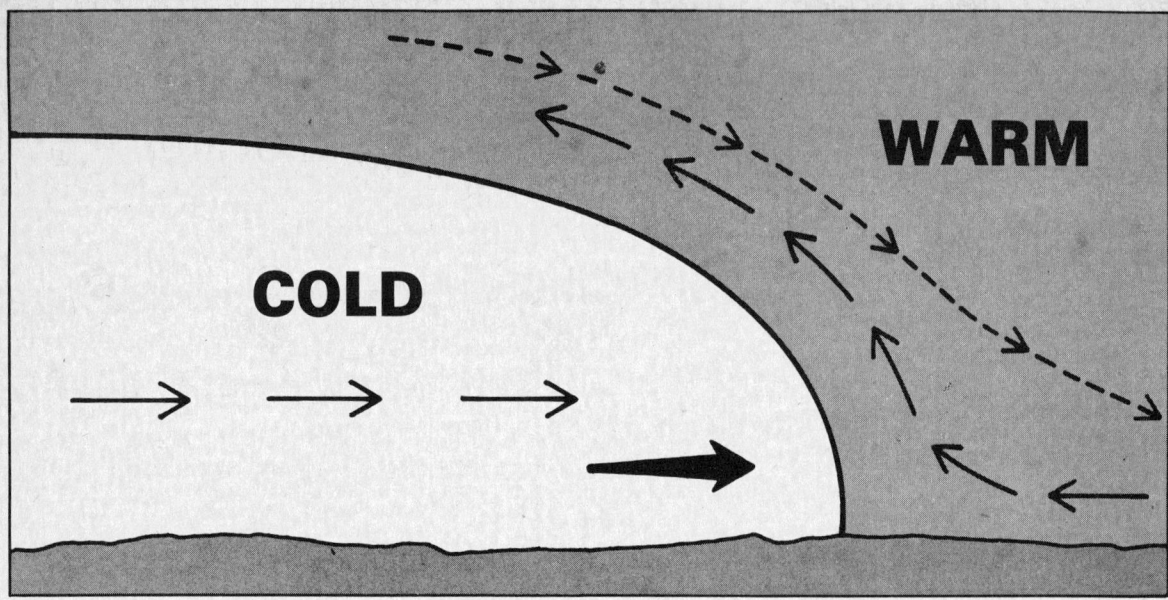

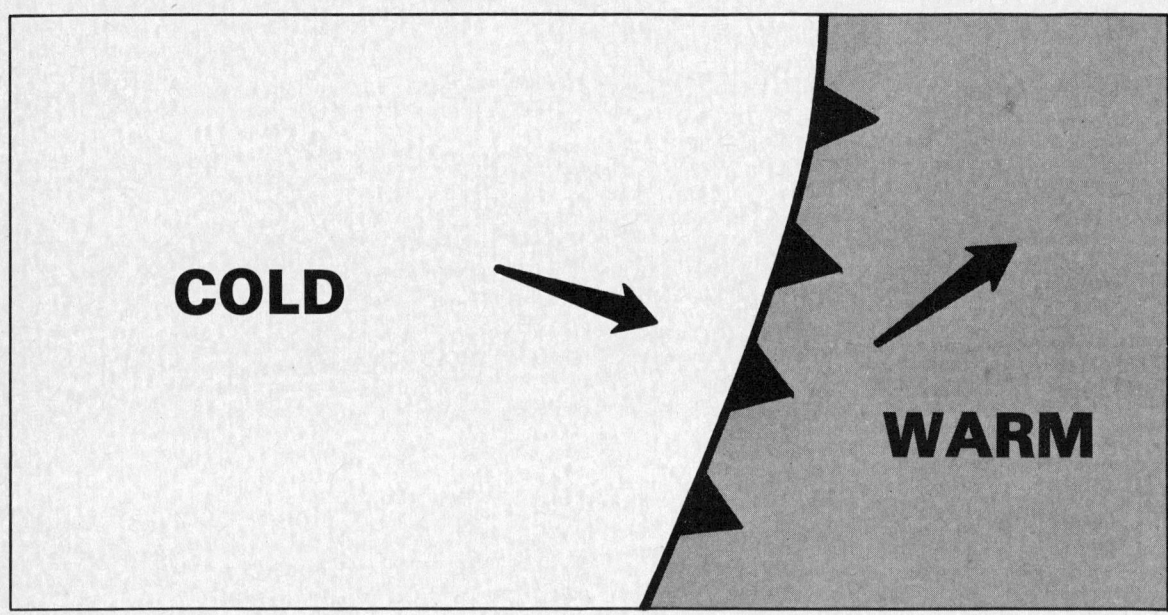

FIGURE 59. Cross section of a cold front (above) with the weather map symbol (below). The symbol is a line with pointed barbs pointing in the direction of movement. If a map is in color, a blue line represents the cold front. The vertical scale is expanded in the top illustration to show the frontal slope. The frontal slope is steep near the leading edge as cold air replaces warm air. The solid heavy arrow shows movement of the front. Warm air may descend over the front as indicated by the dashed arrows; but more commonly, the cold air forces warm air upward over the frontal surface as shown by the solid arrows.

FRONTAL WAVES AND OCCLUSION

Frontal waves and cyclones (areas of low pressure) usually form on slow-moving cold fronts or on stationary fronts. The life cycle and movement of a cyclone is dictated to a great extent by the upper wind flow.

In the initial condition of frontal wave development in figure 62, the winds on both sides of the front are blowing parallel to the front (A). Small disturbances then may start a wavelike bend in the front (B).

If this tendency persists and the wave increases in size, a cyclonic (counterclockwise) circulation

Air Masses and Fronts

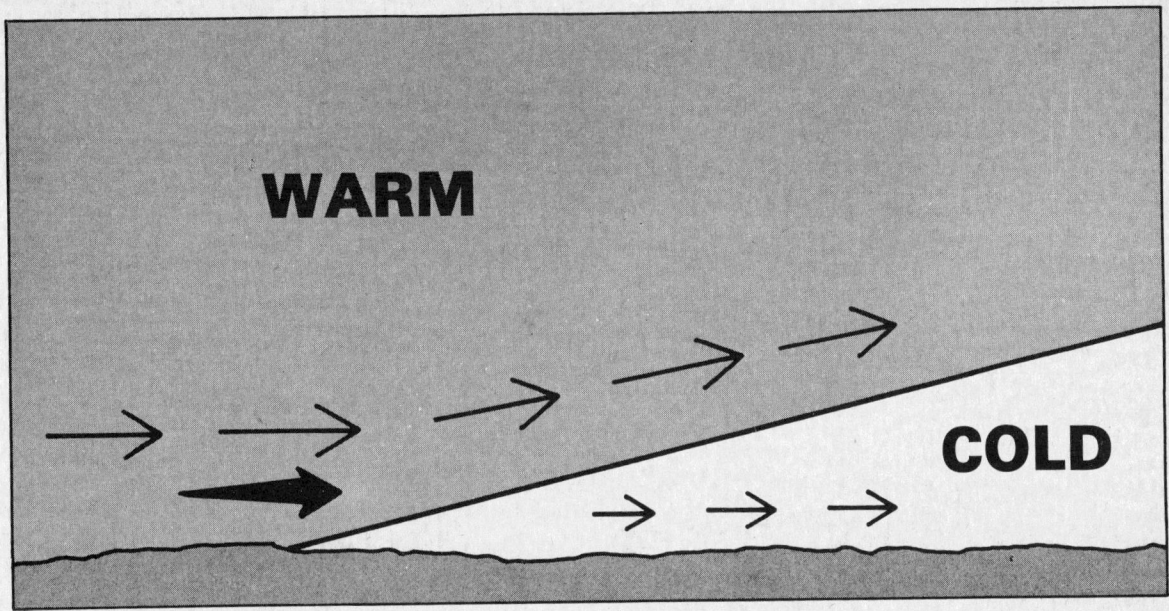

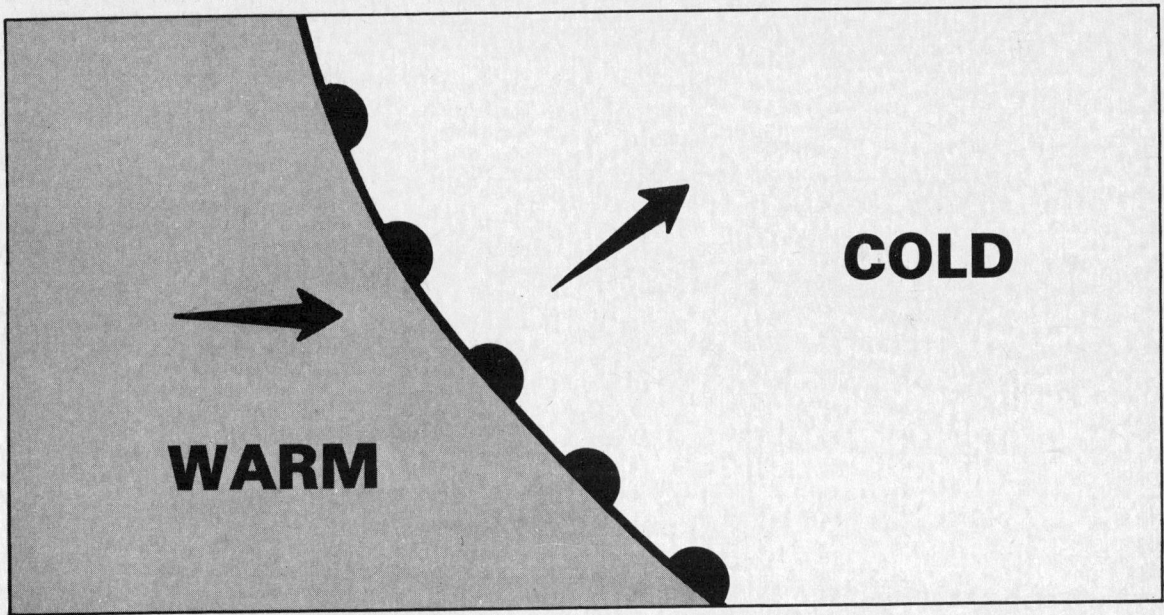

FIGURE 60. Cross section of a warm front (top) with the weather map symbol (bottom). The symbol is a line with rounded barbs pointing in the direction of movement. If a map is in color, a red line represents the warm front. Slope of a warm front generally is more shallow than slope of a cold front. Movement of a warm front shown by the heavy black arrow is slower than the wind in the warm air represented by the light solid arrows. The warm air gradually erodes the cold air.

develops. One section of the front begins to move as a warm front, while the section next to it begins to move as a cold front (C). This deformation is a frontal wave.

The pressure at the peak of the frontal wave falls, and a low-pressure center forms. The cyclonic circulation becomes stronger, and the surface winds are now strong enough to move the fronts; the cold front moves faster than the warm front (D). When the cold front catches up with the warm front, the two of them *occlude* (close together). The result is an *occluded front* or, for brevity, an *occlusion* (E). This is the time of maximum intensity for the wave cyclone. Note that the symbol depicting the occlusion is a combination of the symbols for the warm and cold fronts.

Air Masses and Fronts

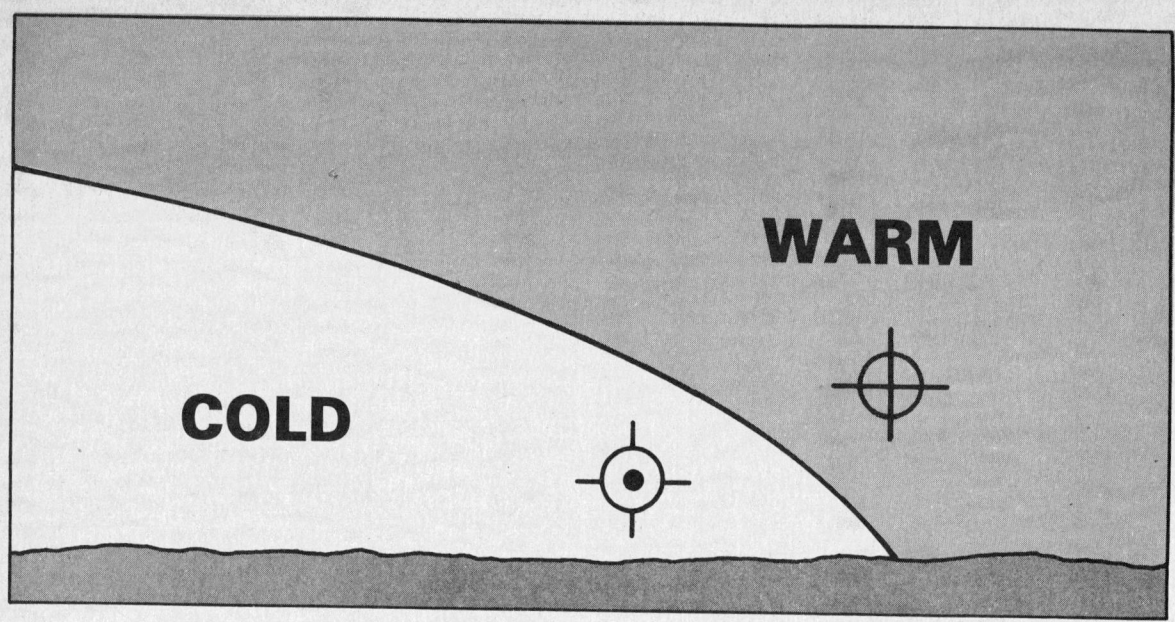

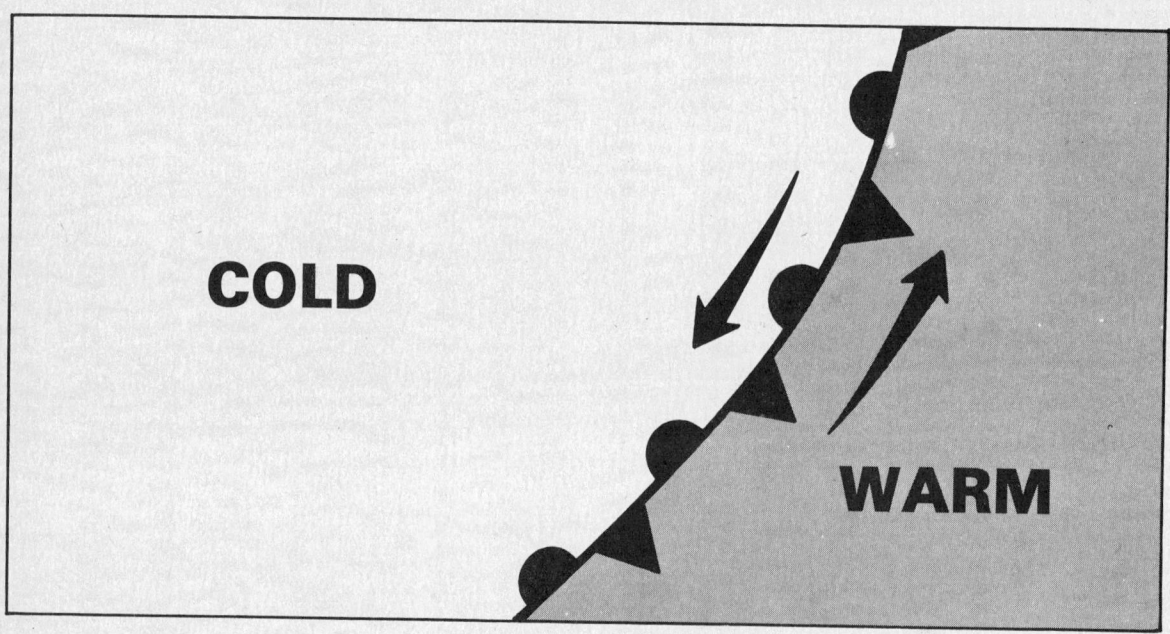

FIGURE 61. Cross section of a stationary front (top) and its weather map symbol (bottom). The symbol is a line with alternating pointed and rounded barbs on opposite sides of the line, the pointed barbs pointing away from the cold air and the rounded barbs away from the warm air. If a map is in color, the symbol is a line of alternating red and blue segments. The front has little or no movement and winds are nearly parallel to the front. The symbol in the warm air is the tail of a wind arrow into the page. The symbol in the cold air is the point of a wind arrow out of the page. Slope of the front may vary considerably depending on wind and density differences across the front.

As the occlusion continues to grow in length, the cyclonic circulation diminishes in intensity and the frontal movement slows down (F). Sometimes a new frontal wave begins to form on the long westward-trailing portion of the cold front (F,G), or a secondary low pressure system forms at the apex where the cold front and warm front come together to form the occlusion. In the final stage, the two fronts may have become a single stationary front again. The low center with its remnant of the occlusion is disappearing (G).

Figure 63 indicates a warm-front occlusion in vertical cross section. This type of occlusion occurs when the air is colder in advance of the warm

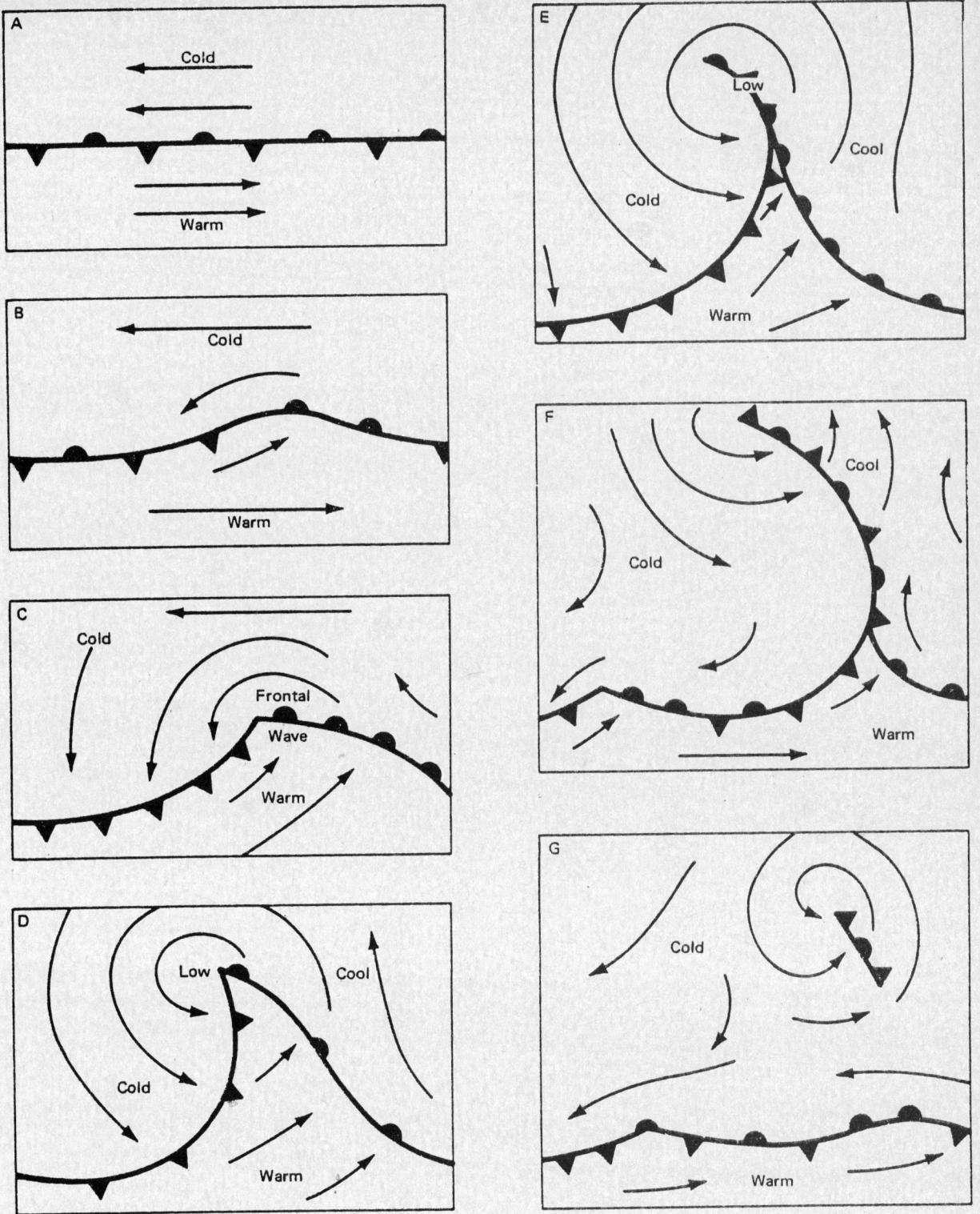

FIGURE 62. The life cycle of a frontal wave.

Air Masses and Fronts

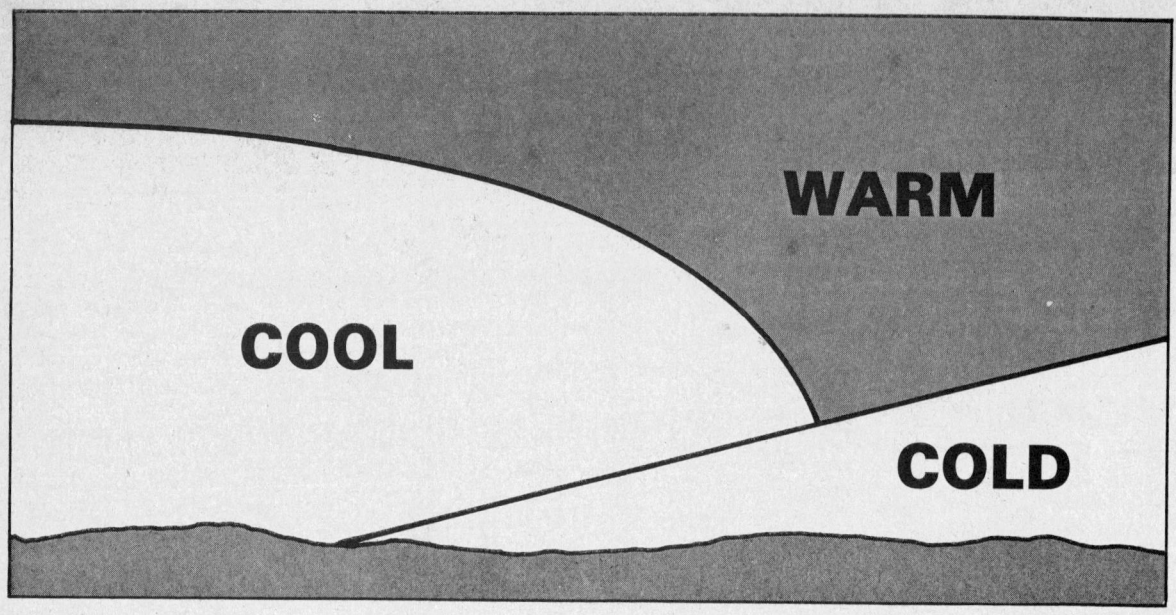

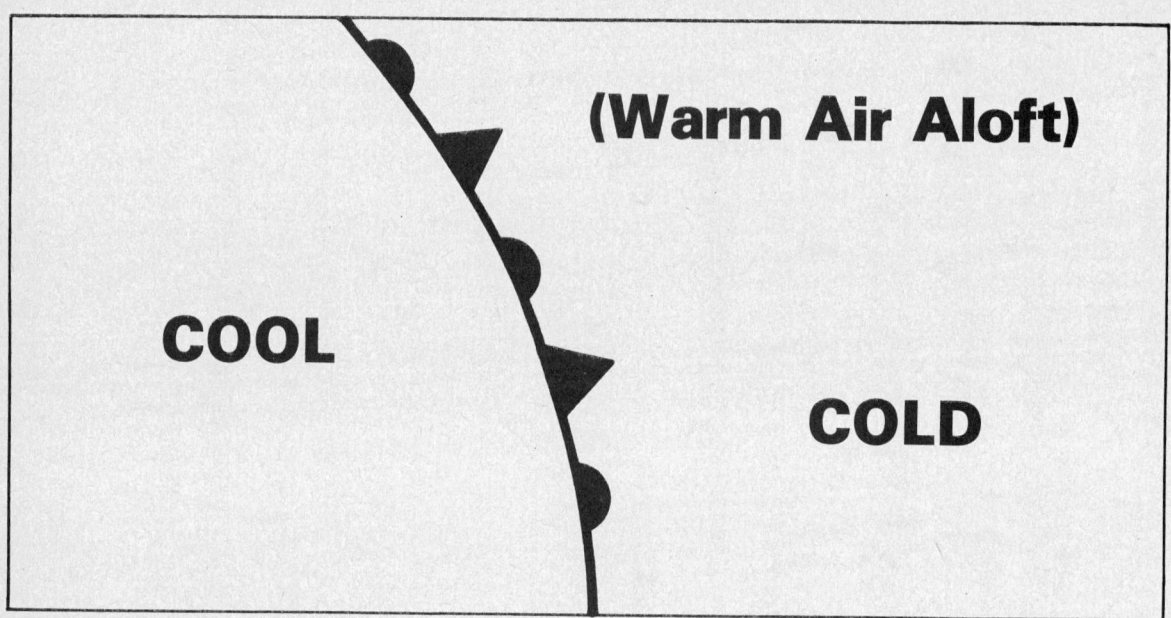

FIGURE 63. Cross section of a warm-front occlusion (top) and its weather map symbol (bottom). The symbol is a line with alternating pointed and rounded barbs on the same side of the line pointing in the direction of movement. Shown in color on a weather map, the line is purple. In the warm front occlusion, air under the cold front is not as cold as air ahead of the warm front; and when the cold front overtakes the warm front, the less cold air rides over the colder air. In a warm front occlusion, cool air replaces cold air at the surface.

front than behind the cold front, lifting the cold front aloft.

Figure 64 indicates a cold-front occlusion in vertical cross section. This type of occlusion occurs when the air behind the cold front is colder than the air in advance of the warm front, lifting the warm front aloft.

NON-FRONTAL LOWS

Since fronts are boundaries between air masses of different properties, fronts are not associated with lows lying solely in a homogeneous air mass. Nonfrontal lows are infrequent east of the Rocky Mountains in midlatitudes but do occur occasion-

Air Masses and Fronts

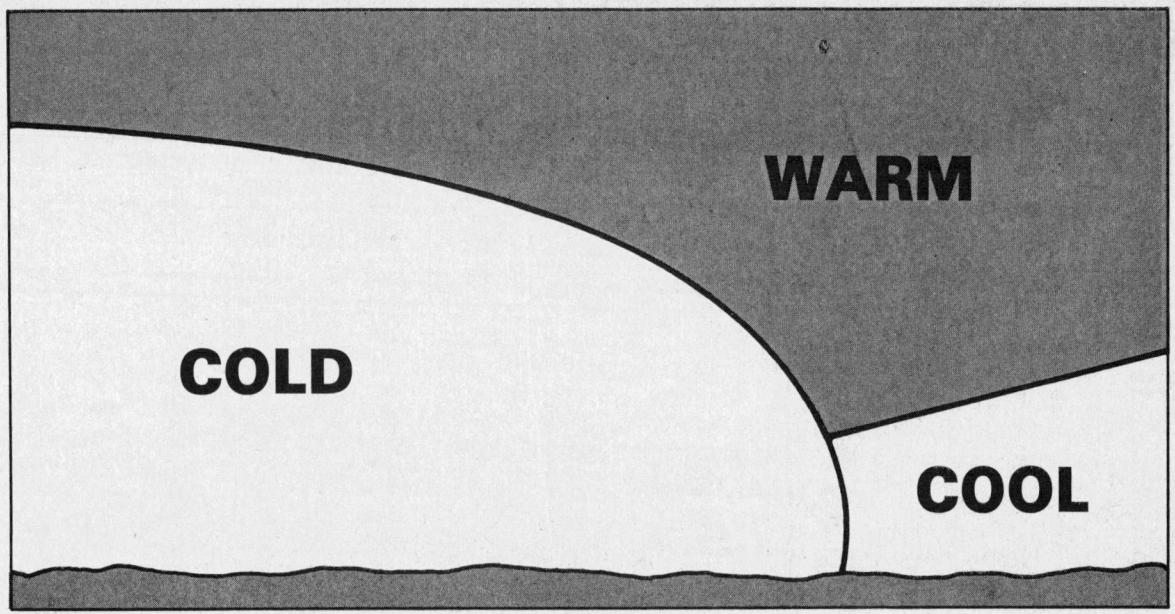

FIGURE 64. Cross section of a cold-front occlusion. Its weather map symbol is the same as for a warm-front occlusion shown in Figure 63. In the cold-front occlusion, the coldest air is under the cold front. When it overtakes the warm front, it lifts the warm front aloft; and cold air replaces cool air at the surface.

ally during the warmer months. Small nonfrontal lows over the western mountains are common as is the semistationary thermal low in extreme Southwestern United States. Tropical lows are also nonfrontal.

FRONTOLYSIS AND FRONTOGENESIS

As adjacent air masses modify and as temperature and pressure differences equalize across a front, the front dissipates. This process, frontolysis, is illustrated in figure 65. Frontogenesis is the generation of a front. It occurs when a relatively sharp zone of transition develops over an area between two air masses which have densities gradually becoming more and more in contrast with each other. The necessary wind flow pattern develops at the same time. Figure 66 shows an example of frontogenesis with the symbol.

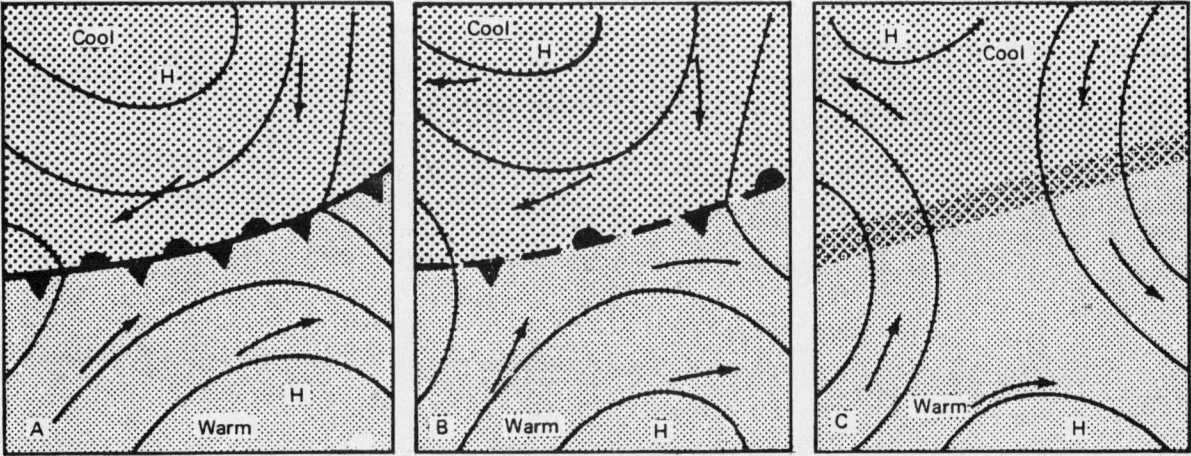

FIGURE 65. Frontolysis of a stationary front.

71

Air Masses and Fronts

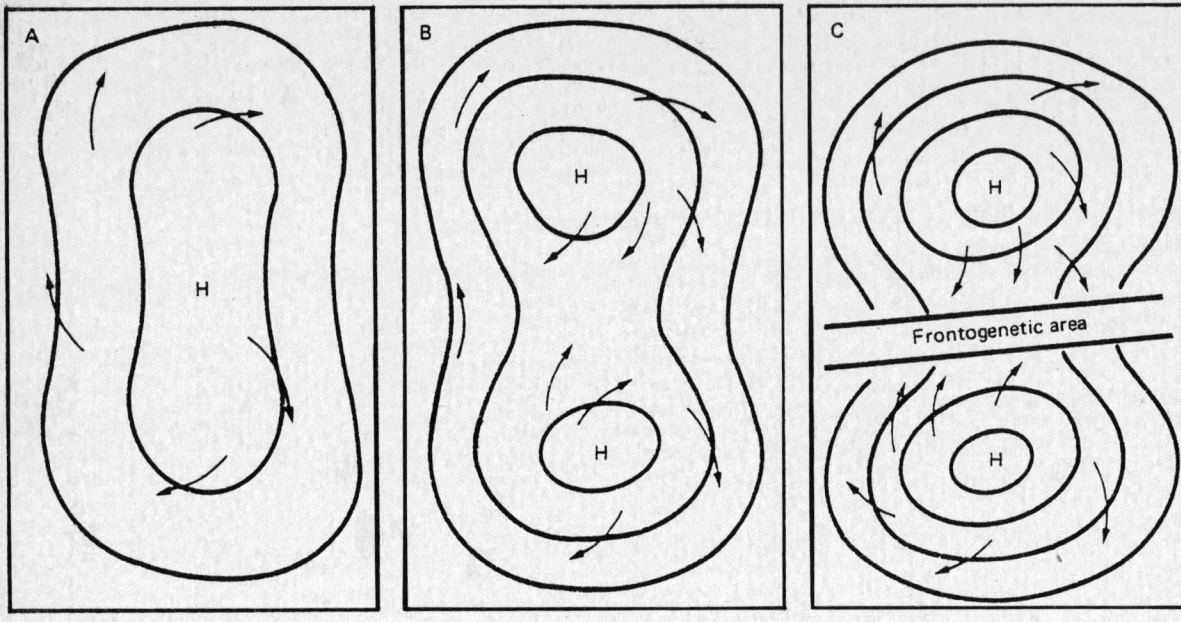

FIGURE 66. Frontogenesis of a stationary front.

FRONTAL WEATHER

In fronts, flying weather varies from virtually clear skies to extreme hazards including hail, turbulence, icing, low clouds, and poor visibility. Weather occurring with a front depends on (1) the amount of moisture available, (2) the degree of stability of the air that is forced upward, (3) the slope of the front, (4) the speed of frontal movement, and (5) the upper wind flow.

Sufficient moisture must be available for clouds to form, or there will be no clouds. As an inactive front comes into an area of moisture, clouds and precipitation may develop rapidly. A good example of this is a cold front moving eastward from the dry slopes of the Rocky Mountains into a tongue of moist air from the Gulf of Mexico over the Plains States. Thunderstorms may build rapidly and catch a pilot unaware.

The degree of stability of the lifted air determines whether cloudiness will be predominately stratiform or cumuliform. If the warm air overriding the front is stable, stratiform clouds develop. If the warm air is unstable, cumuliform clouds develop. Precipitation from stratiform clouds is usually steady as illustrated in figure 67 and there is little or no turbulence. Precipitation from cumuliform clouds is of a shower type as in figure 68, and the clouds are turbulent.

Shallow frontal surfaces tend to give extensive cloudiness with large precipitation areas (figure 69). Widespread precipitation associated with a gradual sloping front often causes low stratus and fog. In this case, the rain raises the humidity of the cold air to saturation. This and related effects may produce low ceiling and poor visibility over thousands of square miles. If temperature of the cold air near the surface is below freezing but the warmer air aloft is above freezing, precipitation falls as freezing rain or ice pellets; however, if temperature of the warmer air aloft is well below freezing, precipitation forms as snow.

When the warm air overriding a shallow front is moist and unstable, the usual widespread cloud mass forms; but embedded in the cloud mass are altocumulus, cumulus, and even thunderstorms as in figures 70 and 71. These embedded storms are more common with warm and stationary fronts but may occur with a slow moving, shallow cold front. A good preflight briefing helps you to foresee the presence of these hidden thunderstorms. Radar also helps in this situation and is discussed in chapter 11.

A fast moving, steep cold front forces upward motion of the warm air along its leading edge. If the warm air is moist, precipitation occurs immediately along the surface position of the front as shown in figure 72.

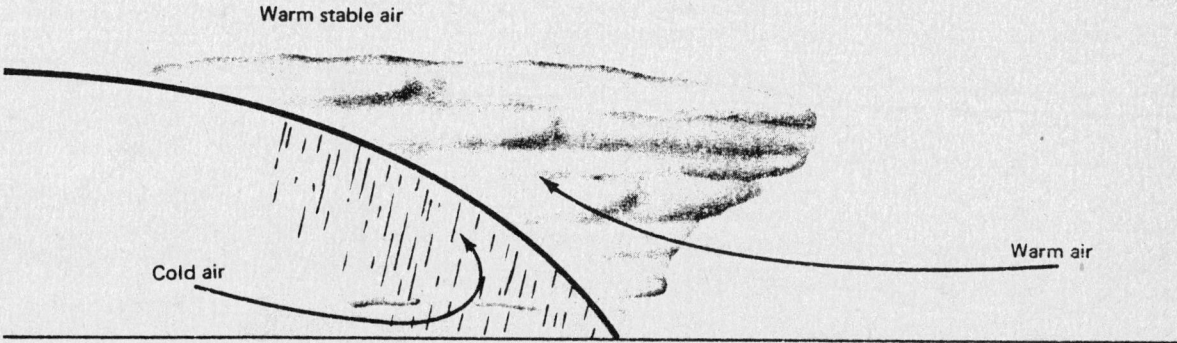

FIGURE 67. A cold front underrunning warm, moist, stable air. Clouds are stratified and precipitation continuous. Precipitation induces stratus in the cold air.

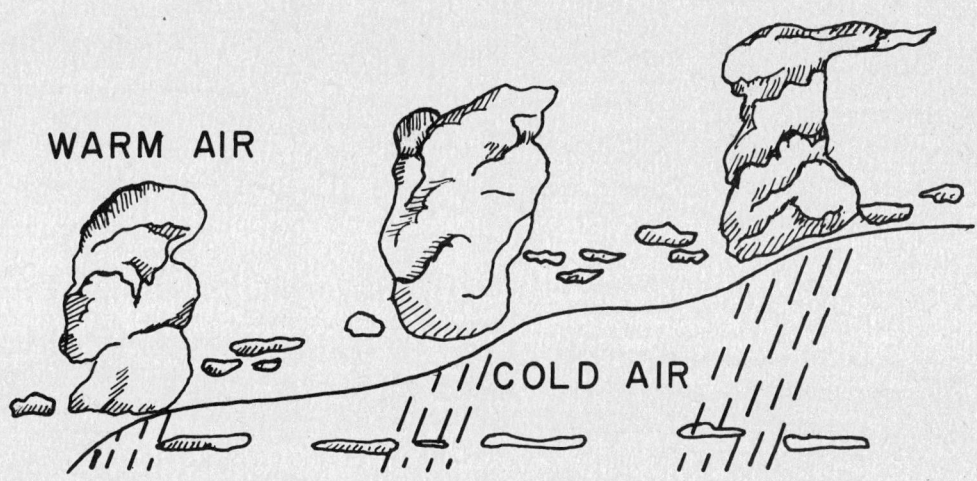

FIGURE 68. A cold front underrunning warm, moist, unstable air. Clouds are cumuliform with possible showers or thunderstorms near the surface position of the front. Convective clouds often develop in the warm air ahead of the front. The warm, wet ground behind the front generates low-level convection and fair weather cumulus in the cold air.

Air Masses and Fronts

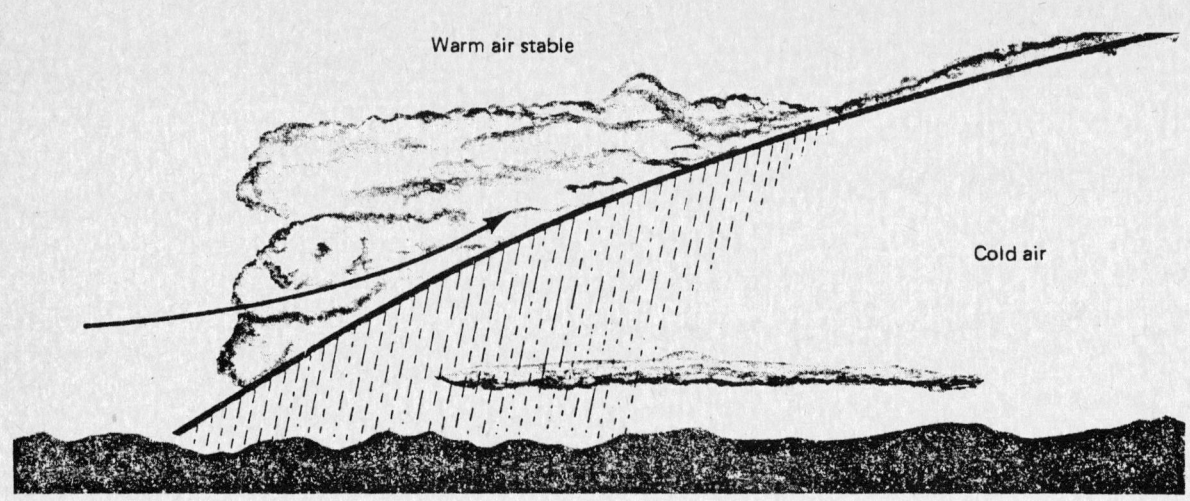

FIGURE 69. A warm front with overrunning moist, stable air. Clouds are stratiform and widespread over the shallow front. Precipitation is continuous and induces widespread stratus in the cold air.

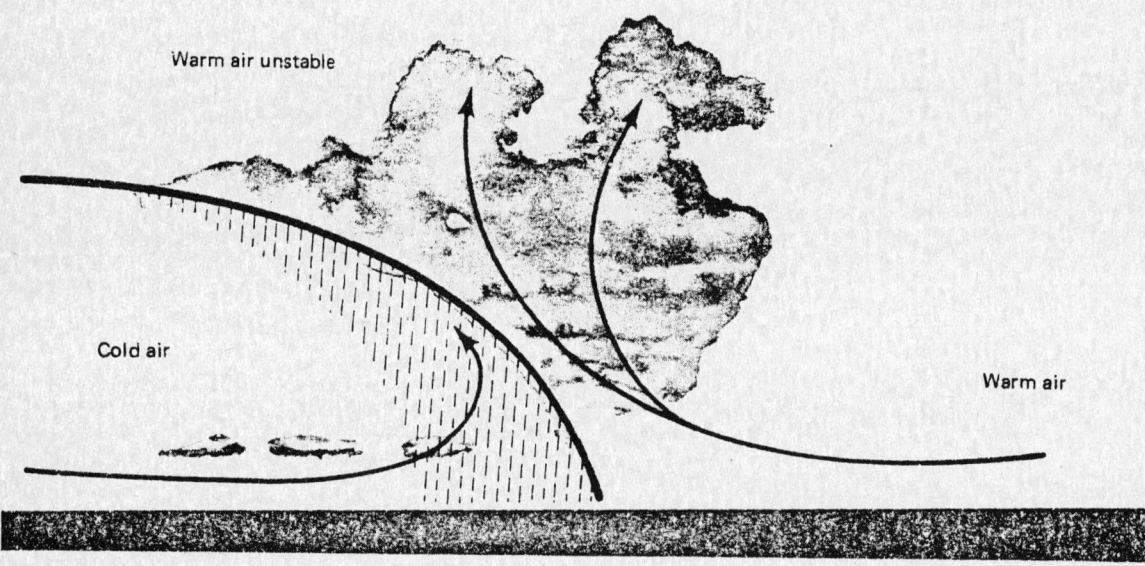

FIGURE 70. A slow-moving cold front underrunning warm, moist, unstable air. Note that the front is more shallow than the fast-moving front shown in figure 68. Clouds are stratified with embedded cumulonimbus and thunderstorms. This type of frontal weather is especially hazardous since the individual thunderstorms are hidden and cannot be avoided unless the aircraft is equipped with airborne radar.

Since an occluded front develops when a cold front overtakes a warm front, weather with an occluded front is a combination of both warm and cold frontal weather. Figures 73 and 74 show warm and cold occlusions and associated weather.

A front may have little or no cloudiness associated with it. Dry fronts occur when the warm air aloft is flowing down the frontal slope or the air is so dry that any cloudiness that occurs is at high levels.

Air Masses and Fronts

FIGURE 71. A warm front with overrunning warm, moist, unstable air. Weather, clouds, and hazards are similar to those described in figure 70 except that they generally are more widespread.

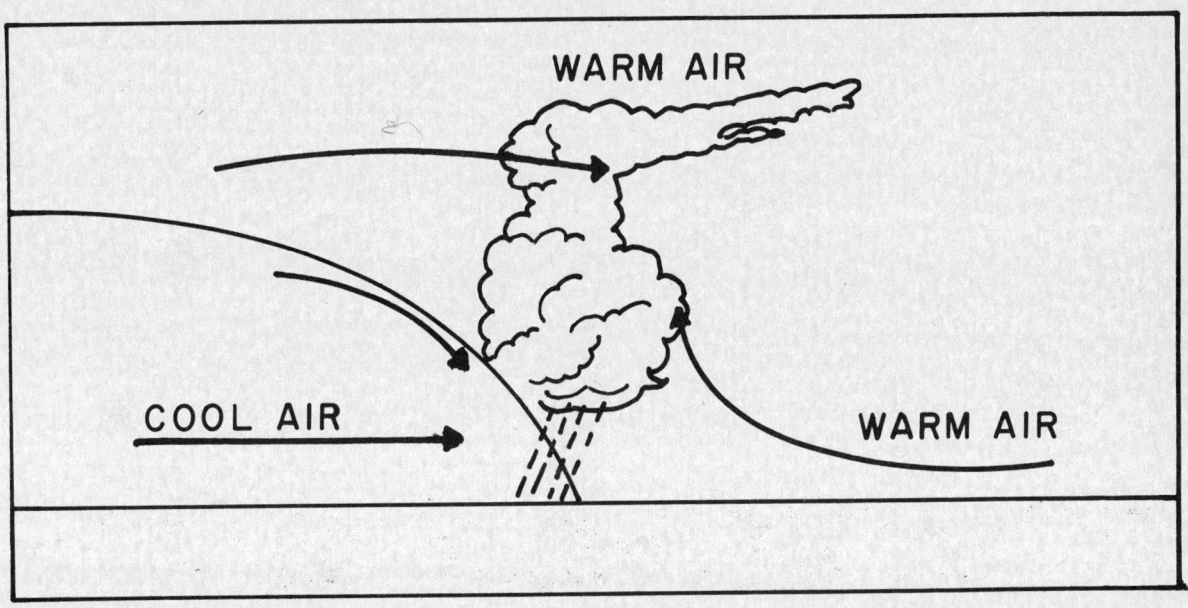

FIGURE 72. A fast moving cold front underrunning warm, moist, unstable air. Showers and thunderstorms develop along the surface position of the front.

The upper wind flow dictates to a great extent the amount of cloudiness and rain accompanying a frontal system as well as movement of the front itself. Remember in chapter 4 we said that systems tend to move with the upper winds. When winds aloft blow across a front, it tends to move with the wind. When winds aloft parallel a front, the front moves slowly if at all. A deep, slow moving trough aloft forms extensive cloudiness and precipitation, while a rapid moving minor trough more often restricts weather to a rather narrow band. However, the latter often breeds severe, fast moving, turbulent spring weather.

INSTABILITY LINE

An instability line is a narrow, nonfrontal line or band of convective activity. If the activity is fully developed thunderstorms, figure 75, the line is a

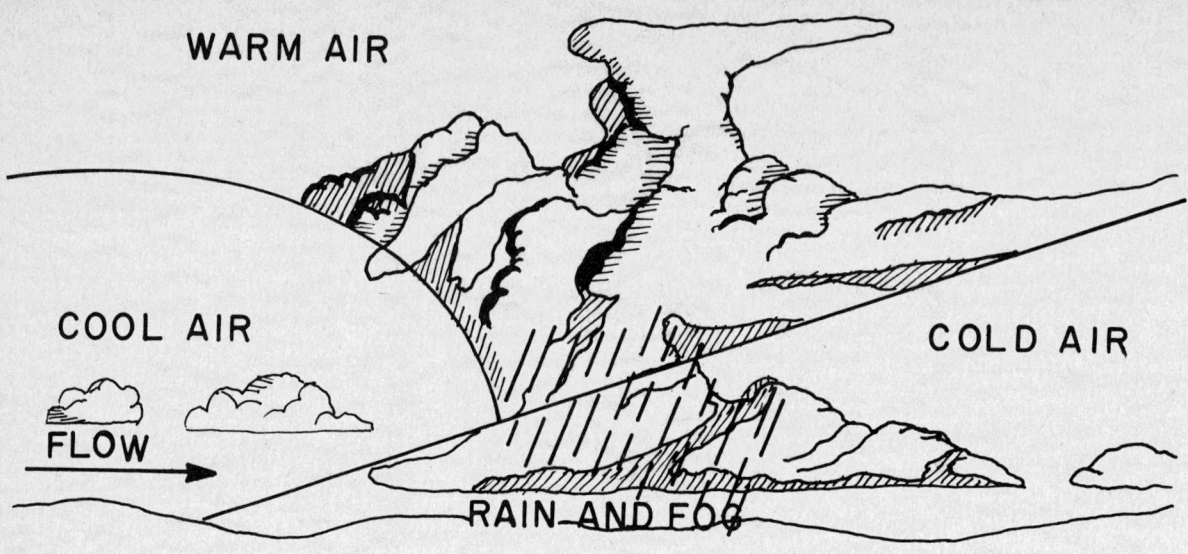

FIGURE 73. A warm front occlusion lifting warm, moist, unstable air. Note that the associated weather is complex and encompasses all types of weather associated with both the warm and cold fronts when air is moist and unstable.

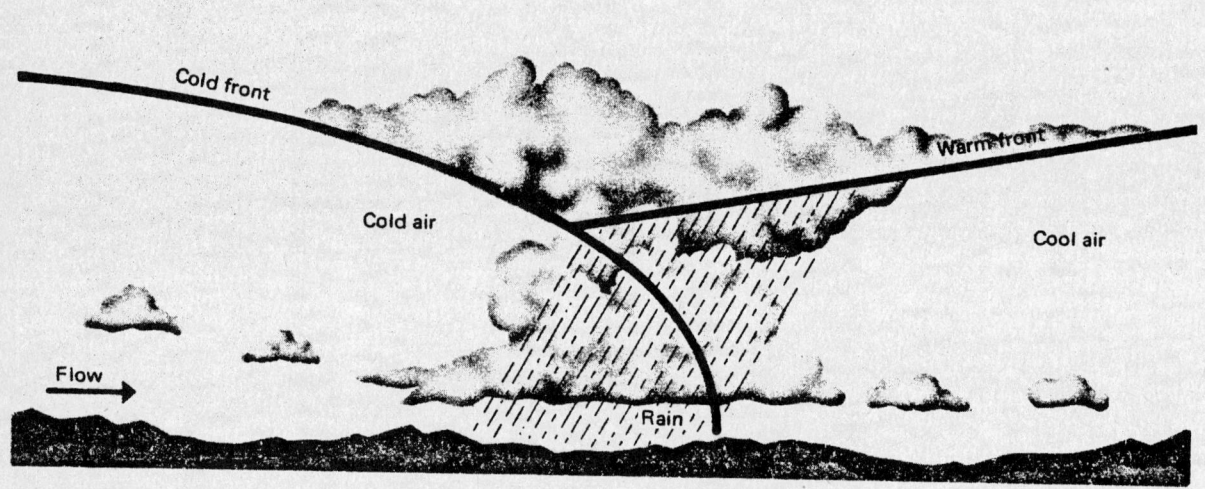

FIGURE 74. A cold front occlusion lifting warm, moist, stable air. Associated weather encompasses types of weather associated with both warm and cold fronts when air is moist and stable.

squall line (chapter 11, "Thunderstorms"). Instability lines form in moist unstable air. An instability line may develop far from any front. More often, it develops ahead of a cold front, and sometimes a series of these lines move out ahead of the front. A favored location for instability lines which frequently erupt into severe thunderstorms is a dew point front or dry line.

DEW POINT FRONT OR DRY LINE

During a considerable part of the year, dew point fronts are common in Western Texas and New Mexico northward over the Plains States. Moist air flowing north from the Gulf of Mexico abuts the dryer and therefore slightly denser air flowing from the southwest. Except for moisture differences, there is seldom any significant air mass contrast across this "Front"; and therefore, it is commonly called a "dry line." Nighttime and early morning fog and low-level clouds often prevail on the moist side of the line while generally clear skies mark the dry side. In spring and early summer over Texas, Oklahoma, and Kansas, and for some distance eastward, the dry line is a favored spawning area for squall lines and tornadoes.

FIGURE 75. An aerial view of a portion of a squall line.

FRONTS AND FLIGHT PLANNING

Surface weather charts pictorially portray fronts and, in conjunction with other forecast charts and special analyses, aid you in determining expected weather conditions along your proposed route. Knowing the locations of fronts and associated weather helps you determine if you can proceed as planned. Often you can change your route to avoid adverse weather.

Frontal weather may change rapidly. For example, there may be only cloudiness associated with a cold front over northern Illinois during the morning but with a strong squall line forecast by afternoon. Skies may be partly cloudy during the afternoon over Atlanta in advance of a warm front, but by sunset drizzle and dense fog are forecast. A cold front in Kansas is producing turbulent thunderstorms, but by midnight the upper flow is expected to dissipate the thunderstorms and weaken the front. A Pacific front is approaching Seattle and is expected to produce heavy rain by midnight.

A mental picture of what is happening and what is forecast should greatly aid you in avoiding adverse weather conditions: If unexpected adverse weather develops en route, your mental picture aids you in planning the best diversion. *If possible, always obtain a good preflight weather briefing.*

We suggest you again look at figures 67 through 75 and review weather conditions associated with different types of fronts and stability conditions. These are only a few of many possibilities, but they should give some help during preflight planning or inflight diversion.

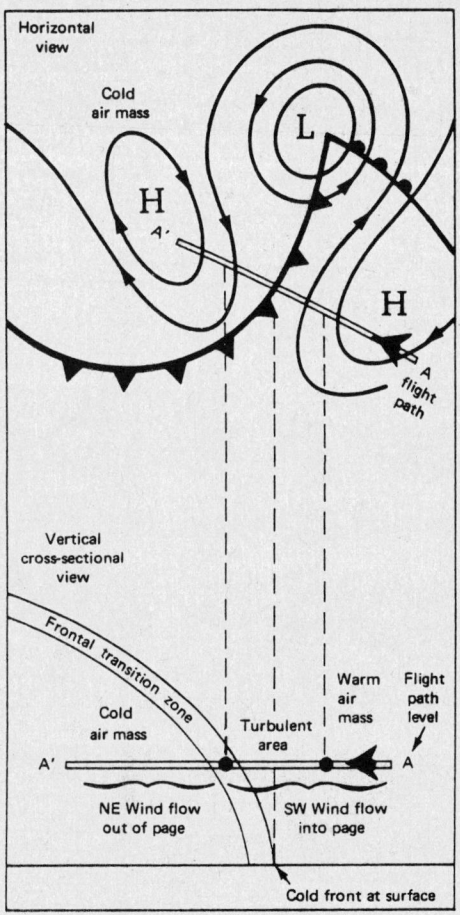

Chapter 9
TURBULENCE

Everyone who flies encounters turbulence at some time or other. A turbulent atmosphere is one in which air currents vary greatly over short distances. These currents range from rather mild eddies to strong currents of relatively large dimensions. As an aircraft moves through these currents, it undergoes changing accelerations which jostle it from its smooth flight path. This jostling is turbulence. Turbulence ranges from bumpiness which can annoy crew and passengers to severe jolts which can structurally damage the aircraft or injure its passengers.

Aircraft reaction to turbulence varies with the difference in windspeed in adjacent currents, size of the aircraft, wing loading, airspeed, and aircraft attitude. When an aircraft travels rapidly from one current to another, it undergoes abrupt changes in acceleration. Obviously, if the aircraft moved more slowly, the changes in acceleration would be more gradual. The first rule in flying turbulence is to reduce airspeed. Your aircraft manual most likely lists recommended airspeed for penetrating turbulence.

Knowing where to expect turbulence helps a

Turbulence

pilot avoid or minimize turbulence discomfort and hazards. The main causes of turbulence are (1) convective currents, (2) obstructions to wind flow, and (3) wind shear. Turbulence also occurs in the wake of moving aircraft whenever the airfoils exert lift—wake turbulence. Any combination of causes may occur at one time.

CONVECTIVE CURRENTS

Convective currents are a common cause of turbulence, especially at low altitudes. These currents are localized vertical air movements, both *ascending* and *descending*. For every rising current, there

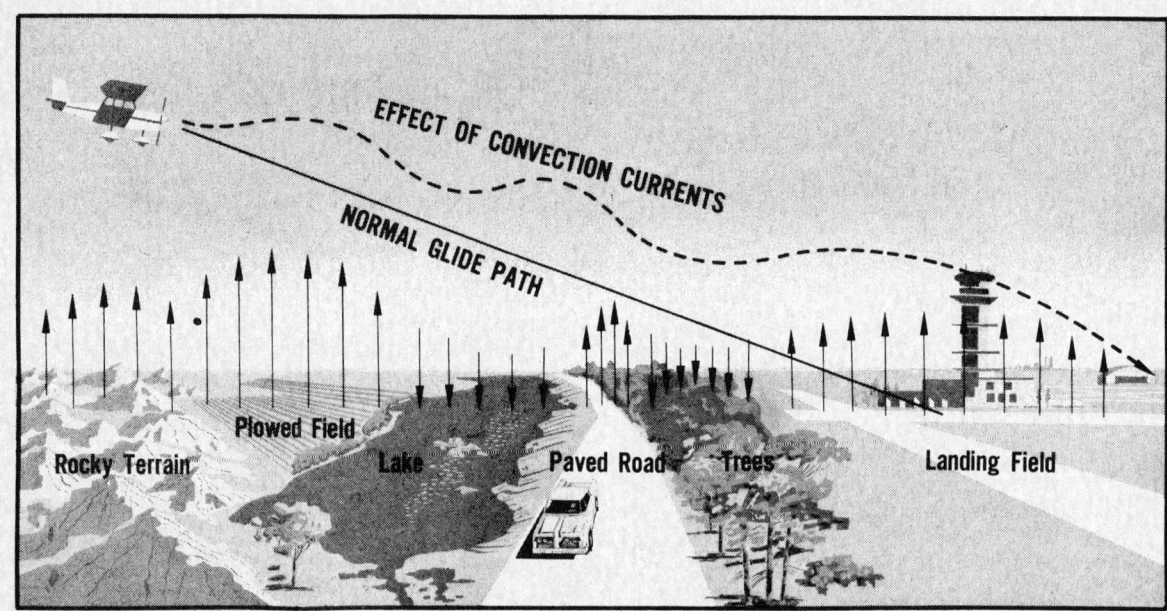

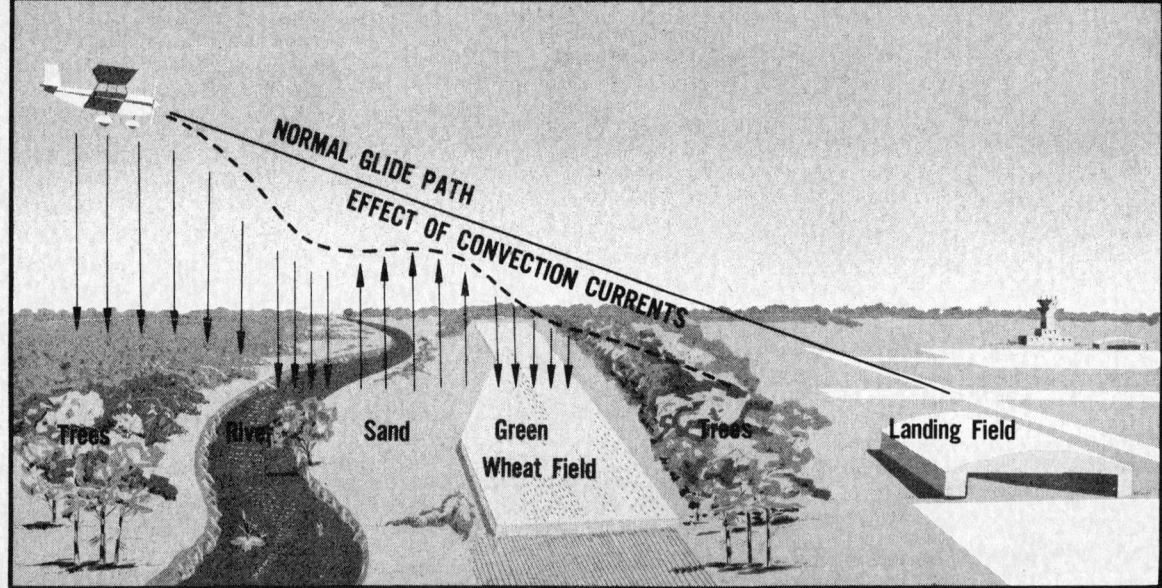

FIGURE 76. Effect of convective currents on final approach. Predominantly upward currents (top) tend to cause the aircraft to overshoot. Predominantly downward currents (bottom) tend to cause the craft to undershoot.

is a compensating downward current. The downward currents frequently occur over broader areas than do the upward currents, and therefore, they have a slower vertical speed than do the rising currents.

Convective currents are most active on warm summer afternoons when winds are light. Heated air at the surface creates a shallow, unstable layer, and the warm air is forced upward. Convection increases in strength and to greater heights as surface heating increases. Barren surfaces such as sandy or rocky wastelands and plowed fields become hotter than open water or ground covered by vegetation. Thus, air at and near the surface heats unevenly. Because of uneven heating, the strength of convective currents can vary considerably within short distances.

When cold air moves over a warm surface, it becomes unstable in lower levels. Convective currents extend several thousand feet above the surface resulting in rough, choppy turbulence when flying in the cold air. This condition often occurs in any season after the passage of a cold front.

Figure 76 illustrates the effect of low-level convective turbulence on aircraft approaching to land. Turbulence on approach can cause abrupt changes in airspeed and may even result in a stall at a dangerously low altitude. To prevent the danger, increase airspeed slightly over normal approach speed. This procedure may appear to conflict with the rule of reducing airspeed for turbulence penetration; but remember, the approach speed for your aircraft is well below the recommended turbulence penetration speed.

As air moves upward, it cools by expansion. A convective current continues upward until it reaches a level where its temperature cools to the same as that of the surrounding air. If it cools to saturation, a cloud forms. Billowy fair weather cumulus clouds, usually seen on sunny afternoons, are signposts in the sky indicating convective turbulence. The cloud top usually marks the approximate upper limit of the convective current. A pilot can expect to encounter turbulence beneath or in the clouds, while above the clouds, air generally is smooth. You will find most comfortable flight above the cumulus as illustrated in figure 77.

When convection extends to greater heights, it develops larger towering cumulus clouds and cumulonimbus with anvil-like tops. The cumulonimbus gives visual warning of violent convective turbulence discussed in more detail in chapter 11.

The pilot should also know that when air is too dry for cumulus to form, convective currents still can be active. He has little indication of their presence until he encounters turbulence.

FIGURE 77. Avoiding turbulence by flying above convective clouds.

Turbulence

OBSTRUCTIONS TO WIND FLOW

Obstructions such as buildings, trees, and rough terrain disrupt smooth wind flow into a complex snarl of eddies as diagrammed in figure 78. An aircraft flying through these eddies experiences turbulence. This turbulence we classify as "mechanical" since it results from mechanical disruption of the ambient wind flow.

The degree of mechanical turbulence depends on wind speed and roughness of the obstructions. The higher the speed and/or the rougher the surface, the greater is the turbulence. The wind carries the turbulent eddies downstream—how far depends on wind speed and stability of the air. Unstable air allows larger eddies to form than those that form in stable air; but the instability breaks up the eddies quickly, while in stable air they dissipate slowly.

Mechanical turbulence can also cause cloudiness near the top of the mechanically disturbed layer. However, the type of cloudiness tells you whether

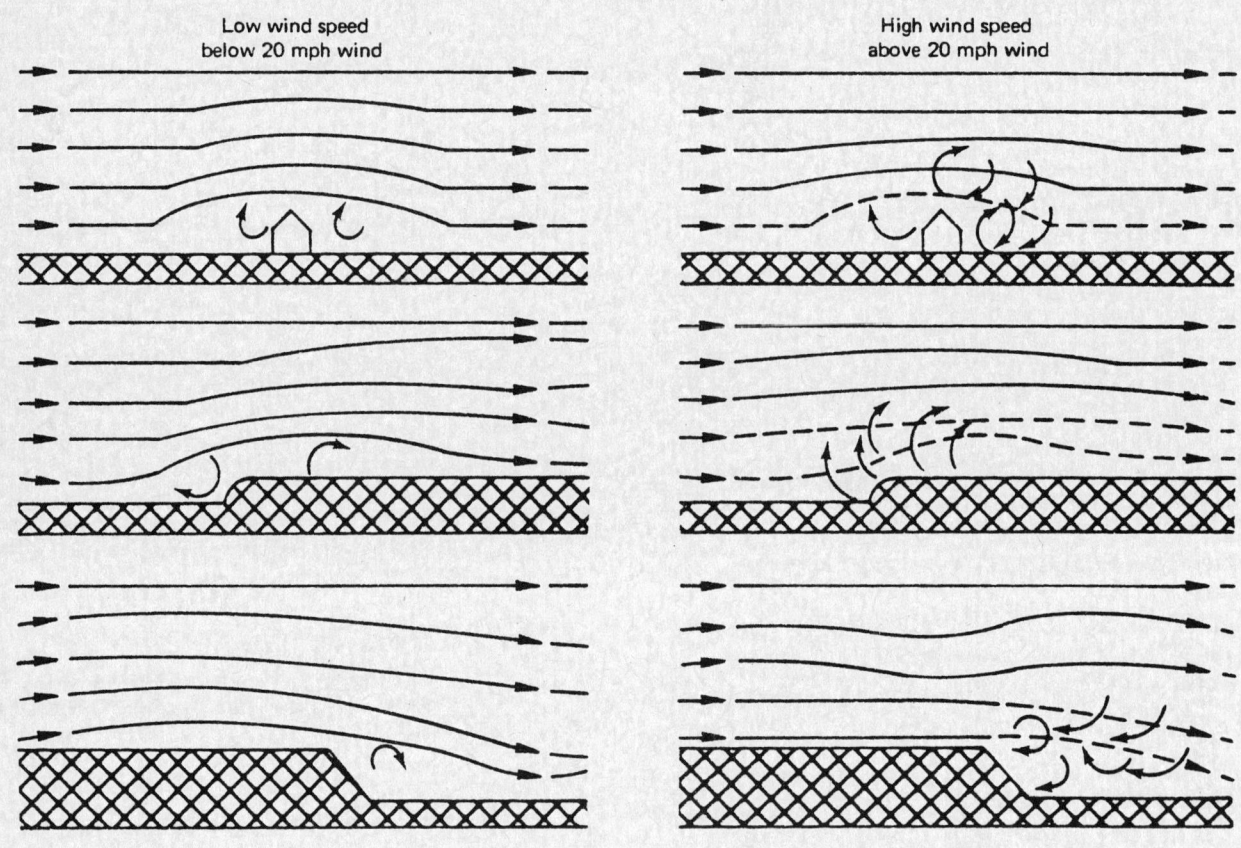

FIGURE 78. Eddy currents formed by wind blowing over uneven ground or over obstructions.

it is from mechanical or convective mixing. Mechanical mixing produces stratocumulus clouds in rows or bands, while convective clouds form a random pattern. The cloud rows developed by mechanical mixing may be parallel to or perpendicular to the wind depending on meteorological factors which we do not discuss here.

The airport area is especially vulnerable to mechanical turbulence which invariably causes gusty surface winds. When an aircraft is in a low-level approach or a climb, airspeed fluctuates in the gusts, and the aircraft may even stall. During extremely gusty conditions, maintain a margin of airspeed above normal approach or climb speed to allow for changes in airspeed. When landing with a gusty crosswind as illustrated in figure 79, be alert for mechanical turbulence and control problems caused by airport structures upwind. Surface gusts also create taxi problems.

Mechanical turbulence can affect low-level cross-country flight about anywhere. Mountains can generate turbulence to altitudes much higher than the mountains themselves.

When flying over rolling hills, you may experience mechanical turbulence. Generally, such turbulence is not hazardous, but it may be annoying or uncomfortable. A climb to higher altitude should reduce the turbulence.

When flying over rugged hills or mountains, however, you may have some real turbulence problems. Again, we cannot discuss mechanical turbulence without considering wind speed and stability. When wind speed across mountains exceeds about 40 knots, you can anticipate turbulence. Where and to what extent depends largely on stability.

If the air crossing the mountains is unstable, turbulence on the windward side is almost certain. If sufficient moisture is present, convective clouds form intensifying the turbulence. Convective clouds over a mountain or along a ridge are a sure sign of unstable air and turbulence on the windward side and over the mountain crest.

As the unstable air crosses the barrier, it spills down the leeward slope often as a violent downdraft. Sometimes the downward speed exceeds the maximum climb rate for your aircraft and may drive the craft into the mountainside as shown in figure 80. In the process of crossing the mountains, mixing reduces the instability to some extent. Therefore, hazardous turbulence in unstable air generally does not extend a great distance downwind from the barrier.

MOUNTAIN WAVE

When stable air crosses a mountain barrier, the turbulent situation is somewhat reversed. Air flowing up the windward side is relatively smooth. Wind flow across the barrier is laminar—that is, it tends to flow in layers. The barrier may set up waves in these layers much as waves develop on

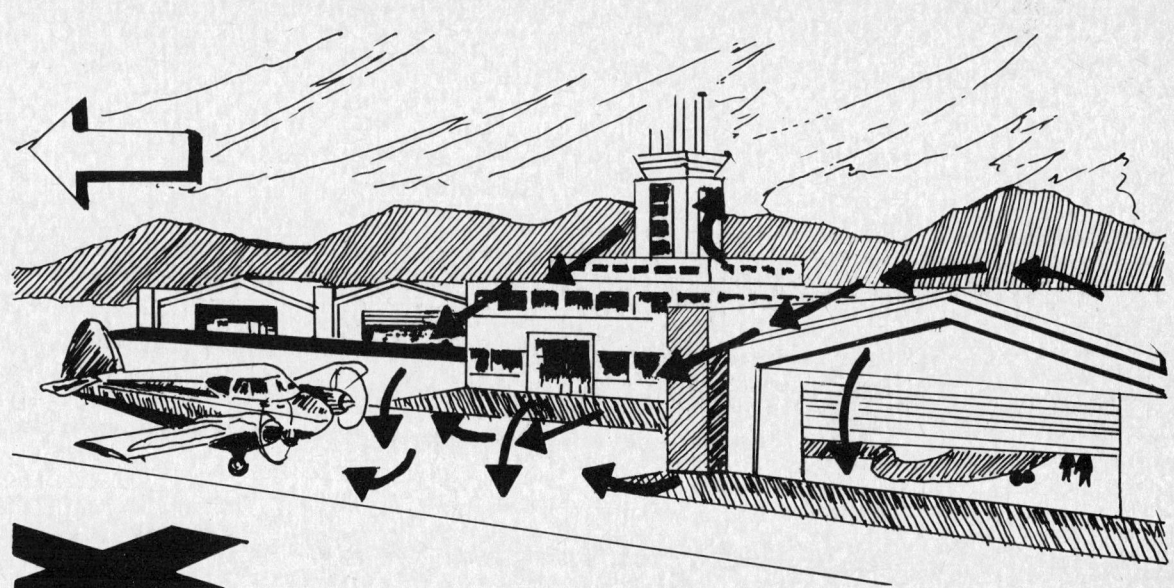

FIGURE 79. Turbulent air in the landing area.

Turbulence

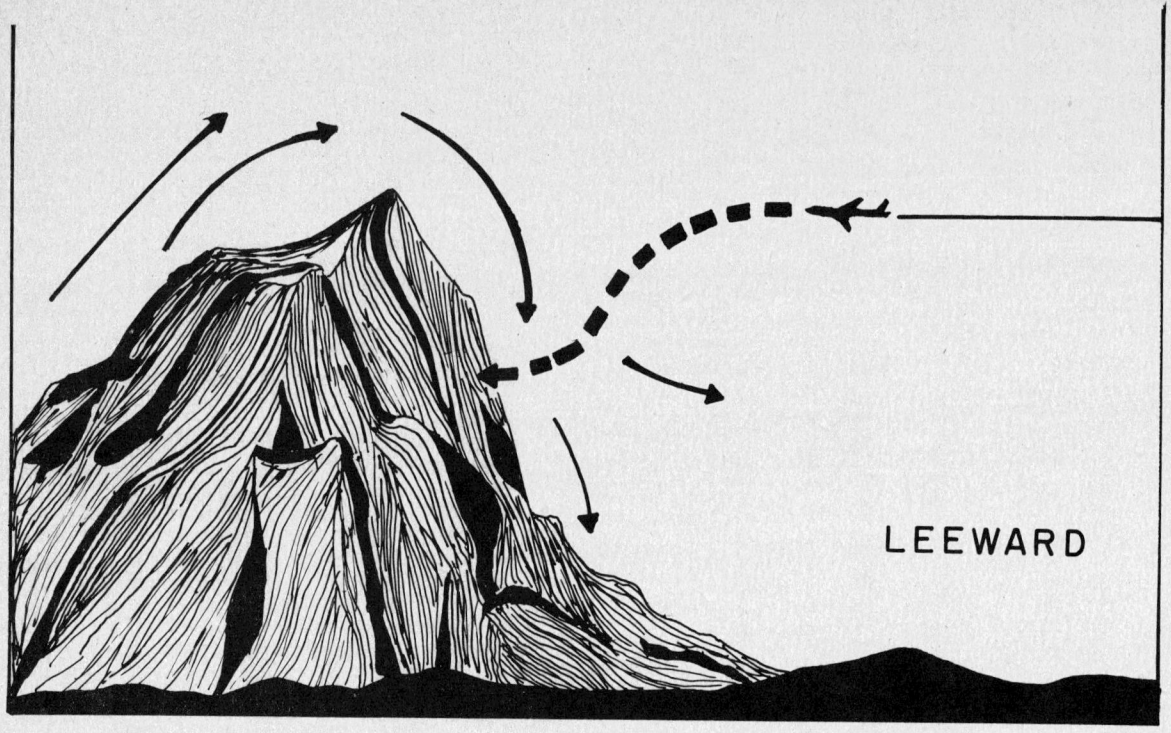

FIGURE 80. Wind flow in mountain areas. Dangerous downdrafts may be encountered on the lee side.

a disturbed water surface. The waves remain nearly stationary while the wind blows rapidly through them. The wave pattern, diagrammed in figure 81, is a "standing" or "mountain" wave, so named because it remains essentially stationary and is associated with the mountain. The wave pattern may extend 100 miles or more downwind from the barrier.

Wave crests extend well above the highest mountains, sometimes into the lower stratosphere. Under

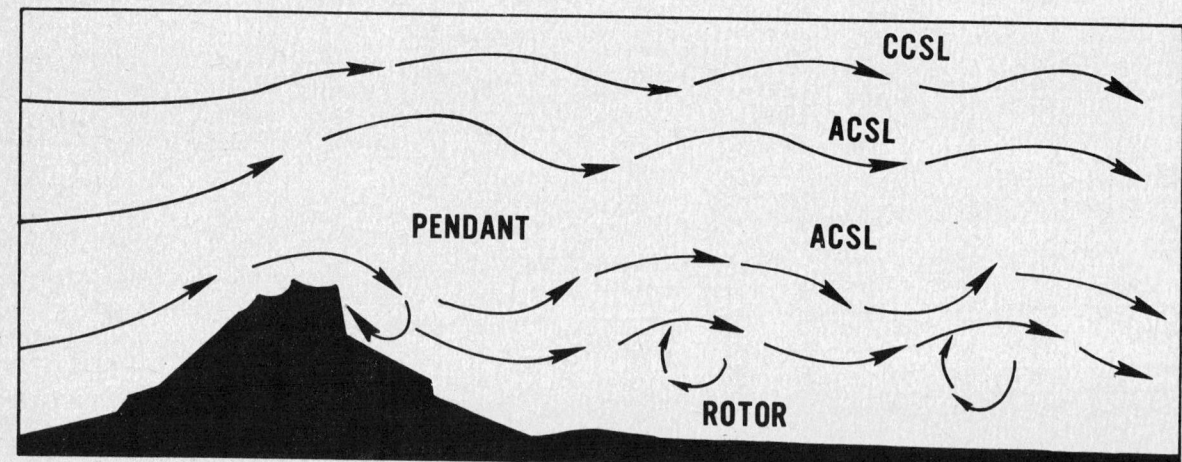

FIGURE 81. Schematic cross section of a mountain wave. Note the standing wave pattern downwind from the mountain. Note also the rotary circulation below the wave crests. When the air contains sufficient moisture, characteristic clouds form.

84

each wave crest is a rotary circulation also diagrammed in figure 81. The "rotor" forms below the elevation of the mountain peaks. Turbulence can be violent in the overturning rotor. Updrafts and downdrafts in the waves can also create violent turbulence.

Figure 81 further illustrates clouds often associated with a mountain wave. When moisture is sufficient to produce clouds on the windward side, they are stratified. Crests of the standing waves may be marked by stationary, lens-shaped clouds known as "standing lenticular" clouds. Figure 82 is a photograph of standing lenticular clouds. They form in the updraft and dissipate in the downdraft, so they do not move as the wind blows through them. The rotor may also be marked by a "rotor" cloud. Figure 83 is a photograph of a series of rotor clouds, each under the crest of a wave. But remember, clouds are not always present to mark the mountain wave. Sometimes, the air is too dry. Always anticipate possible mountain wave turbulence when strong winds of 40 knots or greater blow across a mountain or ridge and the air is stable.

You should not be surprised at any degree of turbulence in a mountain wave. Reports of turbulence range from none to turbulence violent enough to damage the aircraft, but most reports show something in between.

MOUNTAIN FLYING

When planning a flight over mountainous terrain, gather as much preflight information as possible on cloud reports, wind direction, wind speed, and stability of air. Satellites often help locate mountain waves. Figures 84 and 85 are photographs of mountain wave clouds taken from spacecraft. Adequate information may not always be available, so remain alert for signposts in the sky. What should you look for both during preflight planning and during your inflight observations?

Wind at mountain top level in excess of 25 knots suggests some turbulence. Wind in excess of 40 knots across a mountain barrier dictates caution. Stratified clouds mean stable air. Standing lentic-

FIGURE 82. Standing lenticular clouds associated with a mountain wave.

FIGURE 83. Standing wave rotor clouds marking the rotary circulation beneath mountain waves.

ular and/or rotor clouds suggest a mountain wave; expect turbulence many miles to the lee of mountains and relative smooth flight on the windward side. Convective clouds on the windward side of mountains mean unstable air; expect turbulence in close proximity to and on either side of the mountain.

When approaching mountains from the leeward side during strong winds, begin your climb well away from the mountains—100 miles in a mountain wave and 30 to 50 miles otherwise. Climb to an altitude 3,000 to 5,000 feet above mountain tops before attempting to cross. The best procedure is to approach a ridge at a 45° angle to enable a rapid retreat to calmer air. If unable to make good on your first attempt and you have higher altitude capabilities, you may back off and make another attempt at higher altitude. Sometimes you may have to choose between turning back or detouring the area.

Flying mountain passes and valleys is not a safe procedure during high winds. The mountains funnel the wind into passes and valleys thus increasing wind speed and intensifying turbulence. If winds at mountain top level are strong, go high, or go around.

Surface wind may be relatively calm in a valley surrounded by mountains when wind aloft is strong. If taking off in the valley, climb above mountain top level before leaving the valley. Maintain lateral clearance from the mountains sufficient to allow recovery if caught in a downdraft.

WIND SHEAR

As discussed in chapter 4, wind shear generates eddies between two wind currents of differing velocities. The differences may be in wind speed, wind direction, or in both. Wind shear may be associated with either a wind shift or a wind speed gradient at any level in the atmosphere. Three conditions are of special interest—(1) wind shear with a low-level temperature inversion, (2) wind shear

FIGURE 84. Mountain wave clouds over the Tibetan Plateau photographed from a manned spacecraft.

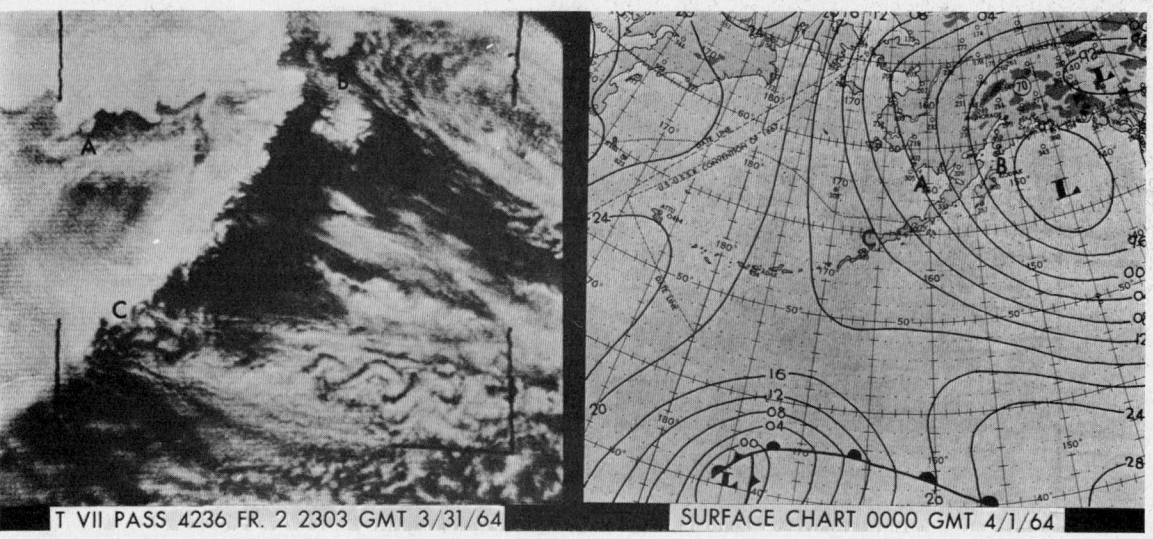

FIGURE 85. Satellite photograph of a mountain wave and the surface weather map for approximately the same time. A single mountain in the Aleutian chain generates the wave. Note how it spirals downwind from the source. Without the satellite, the turbulent wave would have gone undetected unless some aircraft had flown into it.

in a frontal zone, and (3) clear air turbulence (CAT) at high levels associated with a jet stream or strong circulation. High-level clear air turbulence is discussed in detail in chapter 13, "High Altitude Weather."

WIND SHEAR WITH A LOW-LEVEL TEMPERATURE INVERSION

A temperature inversion forms near the surface on a clear night with calm or light surface wind as discussed in chapter 2. Wind just above the inversion may be relatively strong. As illustrated in figure 86, a wind shear zone develops between the calm and the stronger winds above. Eddies in the shear zone cause airspeed fluctuations as an aircraft climbs or descends through the inversion. An aircraft most likely is either climbing from takeoff or approaching to land when passing through the inversion; therefore, airspeed is slow—only a few knots greater than stall speed. The fluctuation in airspeed can induce a stall precariously close to the ground.

Since surface wind is calm or very light, takeoff or landing can be in any direction. Takeoff may be in the direction of the wind above the inversion. If so, the aircraft encounters a sudden tailwind and a corresponding loss of airspeed when climbing through the inversion. Stall is possible. If approach is into the wind above the inversion, the headwind is suddenly lost when descending through the inversion. Again, a sudden loss in airspeed may induce a stall.

When taking off or landing in calm wind under clear skies within a few hours before or after sunrise, be prepared for a temperature inversion near the ground. You can be relatively certain of a shear zone in the inversion if you know the wind at 2,000 to 4,000 feet is 25 knots or more. Allow a margin of airspeed above normal climb or approach speed to alleviate danger of stall in event of turbulence or sudden change in wind velocity.

WIND SHEAR IN A FRONTAL ZONE

As you have learned in chapter 8, a front can contain many hazards. However, a front can be between two dry stable airmasses and can be devoid of clouds. Even so, wind changes abruptly in the frontal zone and can induce wind shear turbulence. The degree of turbulence depends on the magnitude of the wind shear. When turbulence is expected in a frontal zone, follow turbulence penetration procedures recommended in your aircraft manual.

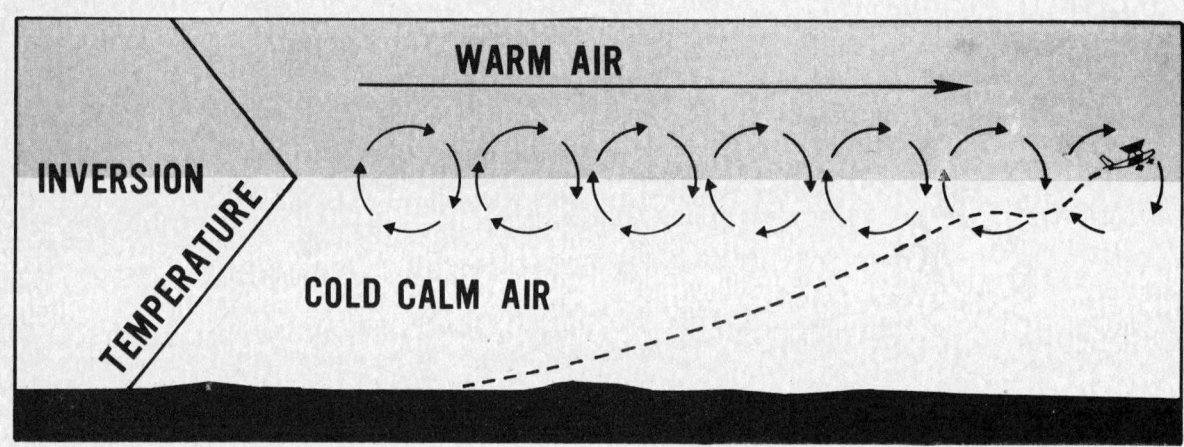

FIGURE 86. Wind shear in a zone between relatively calm wind below an inversion and strong wind above the inversion. This condition is most common at night or in early morning. It can cause an abrupt turbulence encounter at low altitude.

WAKE TURBULENCE

An aircraft receives its lift by accelerating a mass of air downward. Thus, whenever the wings are providing lift, air is forced downward under the wings generating rotary motions or vortices off the wing tips. When the landing gear bears the entire weight of the aircraft, no wing tip vortices develop. But the instant the pilot "hauls back" on the controls, these vortices begin. Figure 87 illustrates how they might appear if visible behind the plane as it breaks ground. These vortices continue throughout the flight and until the craft again settles firmly on its landing gear.

Turbulence

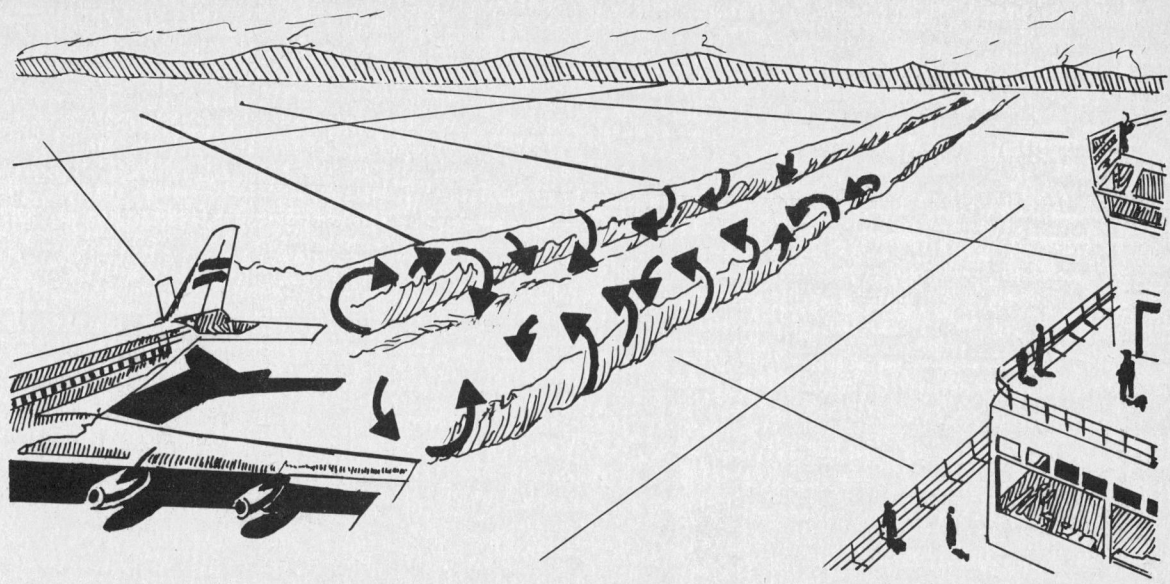

FIGURE 87. Wake turbulence wing tip vortices developing as aircraft breaks ground. These vortices develop when the aircraft is rotated into a flying attitude and the wings begin developing lift.

These vortices spread downward and outward from the flight path. They also drift with the wind. Strength of the vortices is proportional to the weight of the aircraft as well as other factors. Therefore, wake turbulence is more intense behind large, transport category aircraft than behind small aircraft. Generally, it is a problem only when following the larger aircraft.

The turbulence persists several minutes and may linger after the aircraft is out of sight. At controlled airports, the controller generally warns pilots in the vicinity of possible wake turbulence. When left to your own resources, you could use a few pointers. Most jets when taking off lift the nose wheel about midpoint in the takeoff roll; therefore, vortices begin about the middle of the takeoff roll. Vortices behind propeller aircraft begin only a short distance behind lift-off. Following a landing of either type of aircraft, vortices end at about the point where the nose wheel touches down. Avoid flying through these vortices. More specifically, when using the same runway as a heavier aircraft:

(1) if landing behind another aircraft, keep your approach above his approach and keep your touchdown beyond the point where his nose wheel touched the runway (figure 88 (A));

(2) if landing behind a departing aircraft, land only if you can complete your landing roll before reaching the midpoint of his takeoff roll (figure 88 (B));

(3) if departing behind another departing aircraft, take off only if you can become airborne before reaching the midpoint of his takeoff roll and only if you can climb fast enough to stay above his flight path (figure 88 (C)); and

(4) if departing behind a landing aircraft, don't unless you can taxi onto the runway beyond the point at which his nose wheel touched down and have sufficient runway left for safe takeoff (figure 88 (D)).

If parallel runways are available and the heavier aircraft takes off with a crosswind on the downwind runway, you may safely use the upwind runway. Never land or take off downwind from the heavier aircraft. When using a runway crossing his runway, you may safely use the upwind portion of your runway. You may cross behind a departing aircraft behind the midpoint of his takeoff roll. You may cross ahead of a landing aircraft ahead of the point at which his nose wheel touches down. If none of these procedures is possible, wait 5 minutes or so for the vortices to dissipate or to blow off the runway.

The foregoing procedures are elementary. The problem of wake turbulence is more operational than meteorological. The FAA issues periodic ad-

Turbulence

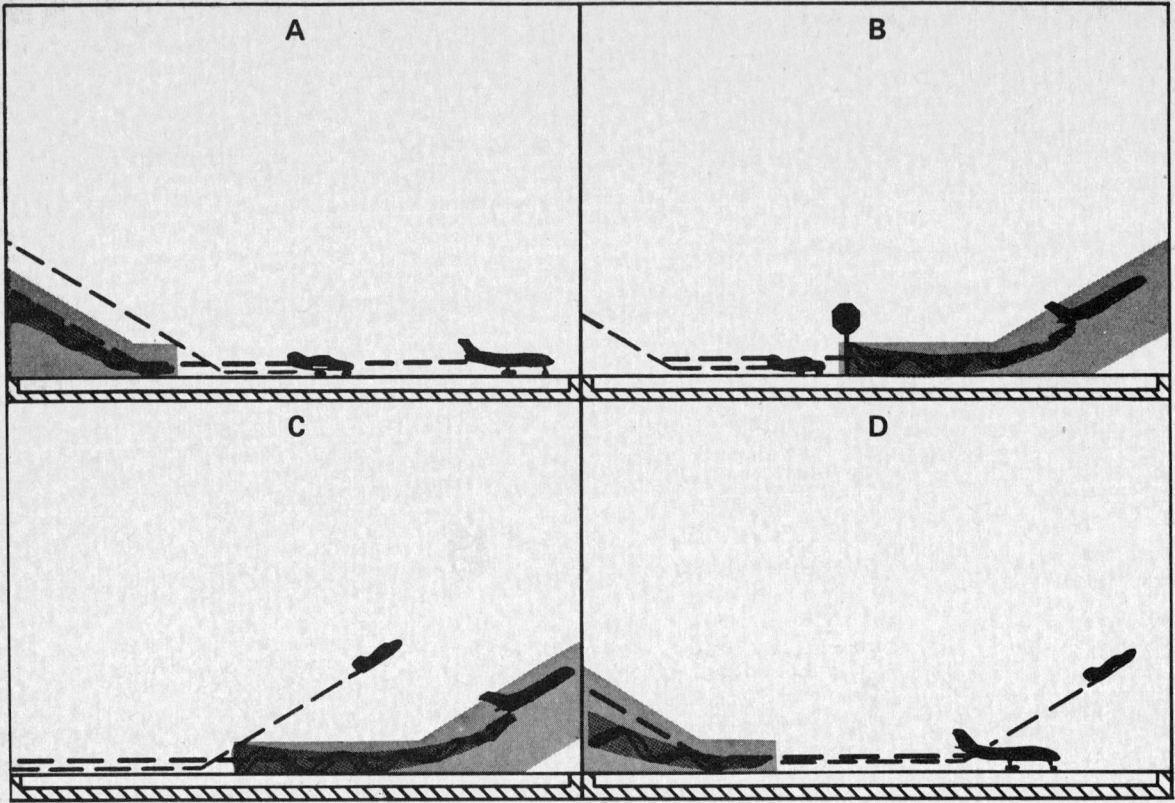

FIGURE 88. Planning landing or takeoff to avoid heavy aircraft wake turbulence.

visory circulars of operational problems. If you plan to operate out of airports used routinely by air carriers, we highly recommend you read the latest advisory circulars on wake turbulence. Titles of these circulars are listed in the FAA "Advisory Circular Checklist and Status of Regulations."

IN CLOSING

We have discussed causes of turbulence, classified it into types, and offered some flight procedures to avoid it or minimize its hazards. Occurrences of turbulence, however, are local in extent and transient in character. A forecast of turbulence specifies a volume of airspace that is small when compared to useable airspace but relatively large compared to the localized extent of the hazard. Although general forecasts of turbulence are quite good, forecasting precise locations is at present impossible.

Generally, when a pilot receives a forecast, he plans his flight to avoid areas of *most probable turbulence*. Yet the best laid plans can go astray and he may encounter turbulence. Since no instruments are currently available for directly observing turbulence, the man on the ground can only confirm its existence or absence via pilot reports. **HELP YOUR FELLOW PILOT AND THE WEATHER SERVICE—SEND PILOT REPORTS.**

To make reports and forecasts meaningful, turbulence is classified into intensities based on the effects it has on the aircraft and passengers. Section 16 of AVIATION WEATHER SERVICES (AC 00-45) lists and describes these intensities. Use this guide in reporting your turbulence encounters.

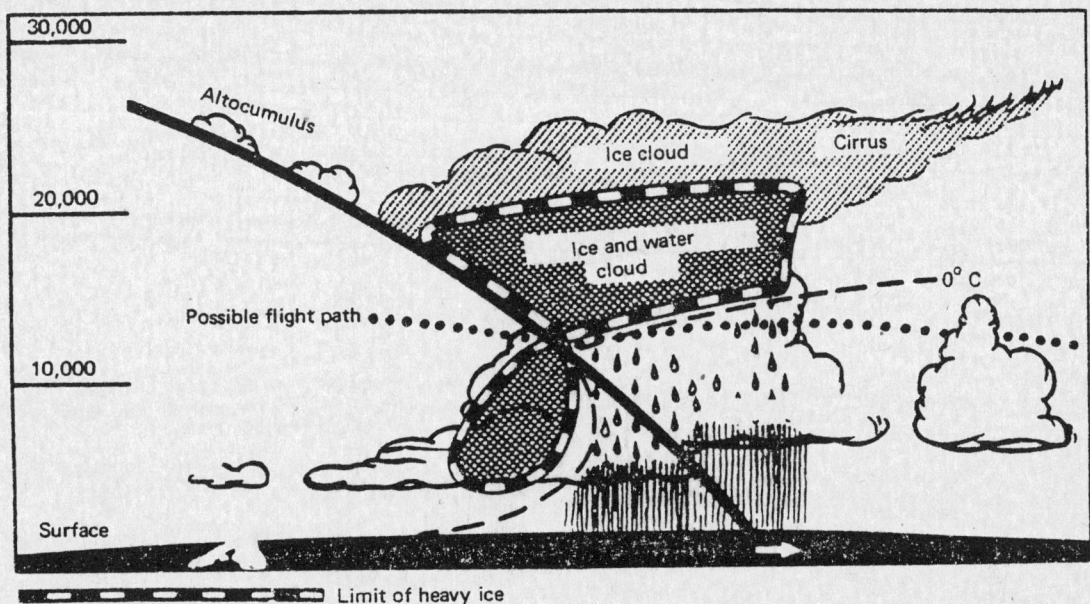

Chapter 10
ICING

Aircraft icing is one of the major weather hazards to aviation. Icing is a cumulative hazard. It reduces aircraft efficiency by increasing weight, reducing lift, decreasing thrust, and increasing drag. As shown in figure 89, each effect tends to either slow the aircraft or force it downward. Icing also seriously impairs aircraft engine performance. Other icing effects include false indications on flight instruments, loss of radio communications, and loss of operation of control surfaces, brakes, and landing gear.

In this chapter we discuss the principles of structural, induction system, and instrument icing and relate icing to cloud types and other factors. Although ground icing and frost are structural icing, we discuss them separately because of their different effect on an aircraft. And we wind up the chapter with a few operational pointers.

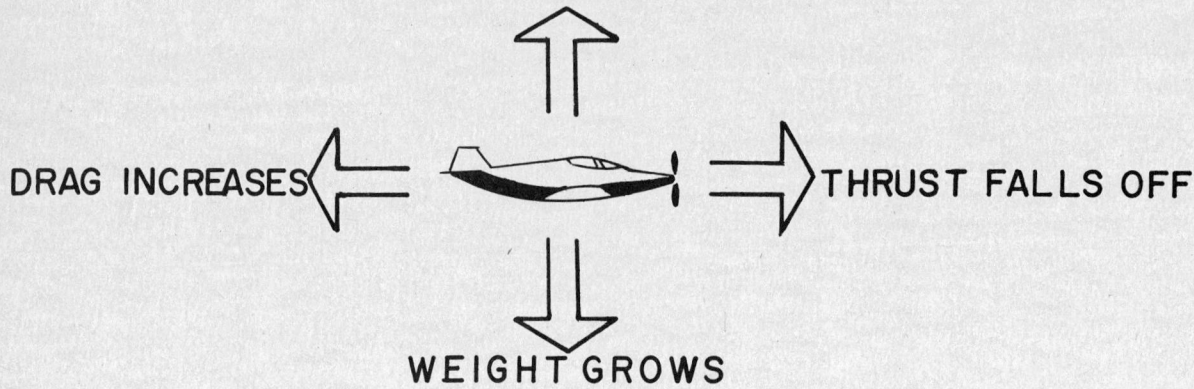

Figure 89. Effects of structural icing.

STRUCTURAL ICING

Two conditions are necessary for structural icing in flight: (1) the aircraft must be flying through visible water such as rain or cloud droplets, and (2) the temperature at the point where the moisture strikes the aircraft must be 0° C or colder. Aerodynamic cooling can lower temperature of an airfoil to 0° C even though the ambient temperature is a few degrees warmer.

Supercooled water increases the rate of icing and is essential to rapid accretion. Supercooled water is in an unstable liquid state; when an aircraft strikes a supercooled drop, part of the drop freezes instantaneously. The latent heat of fusion released by the freezing portion raises the temperature of the remaining portion to the melting point. Aerodynamic effects may cause the remaining portion to freeze. The way in which the remaining portion freezes determines the type of icing. The types of structural icing are clear, rime, and a mixture of the two. Each type has its identifying features.

CLEAR ICE

Clear ice forms when, after initial impact, the remaining liquid portion of the drop flows out over the aircraft surface gradually freezing as a smooth sheet of solid ice. This type forms when drops are large as in rain or in cumuliform clouds.

Figure 90 illustrates ice on the cross section of an airfoil, clear ice shown at the top. Figures 91 and 92 are photographs of clear structural icing. Clear ice is hard, heavy, and tenacious. Its removal by deicing equipment is especially difficult.

RIME ICE

Rime ice forms when drops are small, such as those in stratified clouds or light drizzle. The liquid portion remaining after initial impact freezes rapidly before the drop has time to spread over the aircraft surface. The small frozen droplets trap air between them giving the ice a white appearance as shown at the center of figure 90. Figure 93 is a photograph of rime.

Rime ice is lighter in weight than clear ice and its weight is of little significance. However, its irregular shape and rough surface make it very effective in decreasing aerodynamic efficiency of airfoils, thus reducing lift and increasing drag. Rime ice is brittle and more easily removed than clear ice.

MIXED CLEAR AND RIME ICING

Mixed ice forms when drops vary in size or when liquid drops are intermingled with snow or ice particles. It can form rapidly. Ice particles become imbedded in clear ice, building a very rough accumulation sometimes in a mushroom shape on leading edges as shown at the bottom of figure 90. Figure 94 is a photo of mixed icing built up on a pitot tube.

ICING INTENSITIES

By mutual agreement and for standardization the FAA, National Weather Service, the military aviation weather services, and aircraft operating organizations have classified aircraft structural icing into intensity categories.

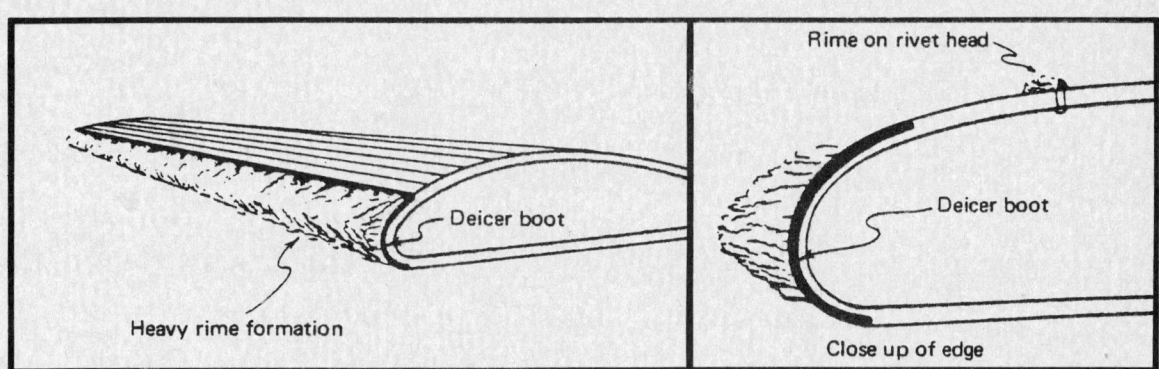

FIGURE 90.

Icing

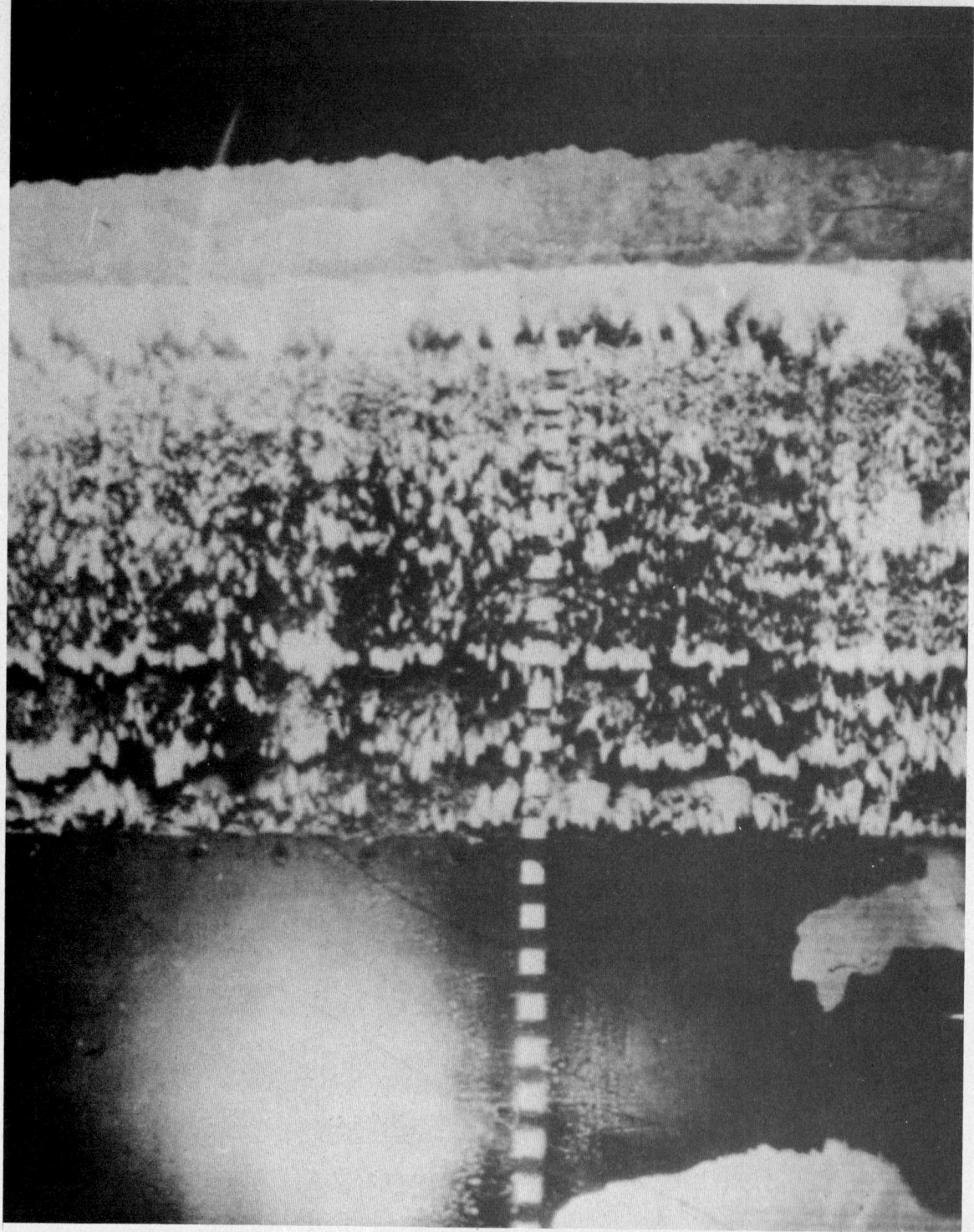

FIGURE 91. Clear wing icing (leading edge and underside). (Courtesy Dean T. Bowden, General Dynamics/Convair.)

FIGURE 92. Propeller icing. Ice may form on propellers just as on any airfoil. It reduces propeller efficiency and may induce severe vibrations.

FIGURE 93. Rime icing on the nose of a Mooney "Mark 21" aircraft. (Photo by Norman Hoffman, Mooney Aircraft, Inc., courtesy the A.O.P.A. Pilot Magazine.)

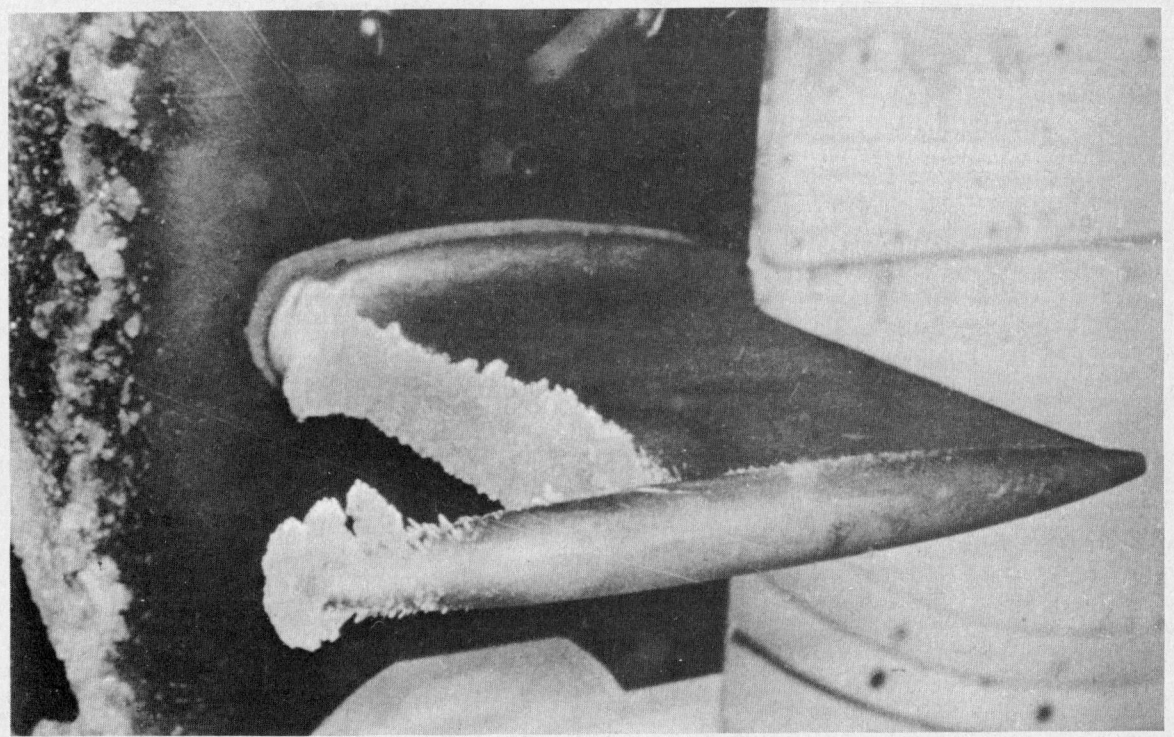

FIGURE 94. External icing on a pitot tube.

INDUCTION SYSTEM ICING

Ice frequently forms in the air intake of an engine robbing the engine of air to support combustion. This type icing occurs with both piston and jet engines, and almost everyone in the aviation community is familiar with carburetor icing. The downward moving piston in a piston engine or the compressor in a jet engine forms a partial vacuum in the intake. Adiabatic expansion in the partial vacuum cools the air. Ice forms when the temperature drops below freezing and sufficient moisture is present for sublimation. In piston engines, fuel evaporation produces additional cooling. Induction icing always lowers engine performance and can even reduce intake flow below that necessary for the engine to operate. Figure 95 illustrates carburetor icing.

Induction icing potential varies greatly among different aircraft and occurs under a wide range of meteorological conditions. It is primarily an engineering and operating problem rather than meteorological.

Icing

FIGURE 95. Carburetor icing. Expansional cooling of air and vaporization of fuel can induce freezing and cause ice to clog the carburetor intake.

INSTRUMENT ICING

Icing of the pitot tube as seen in figure 96 reduces ram air pressure on the airspeed indicator and renders the instrument unreliable. Most modern aircraft also have an outside static pressure port as part of the pitot-static system. Icing of the static pressure port reduces reliability of all instruments on the system—the airspeed, rate-of-climb, and the altimeter.

Ice forming on the radio antenna distorts its shape, increases drag, and imposes vibrations that may result in failure in the communications system of the aircraft. The severity of this icing depends upon the shape, location, and orientation of the antenna. Figure 97 is a photograph of clear ice on an antenna mast.

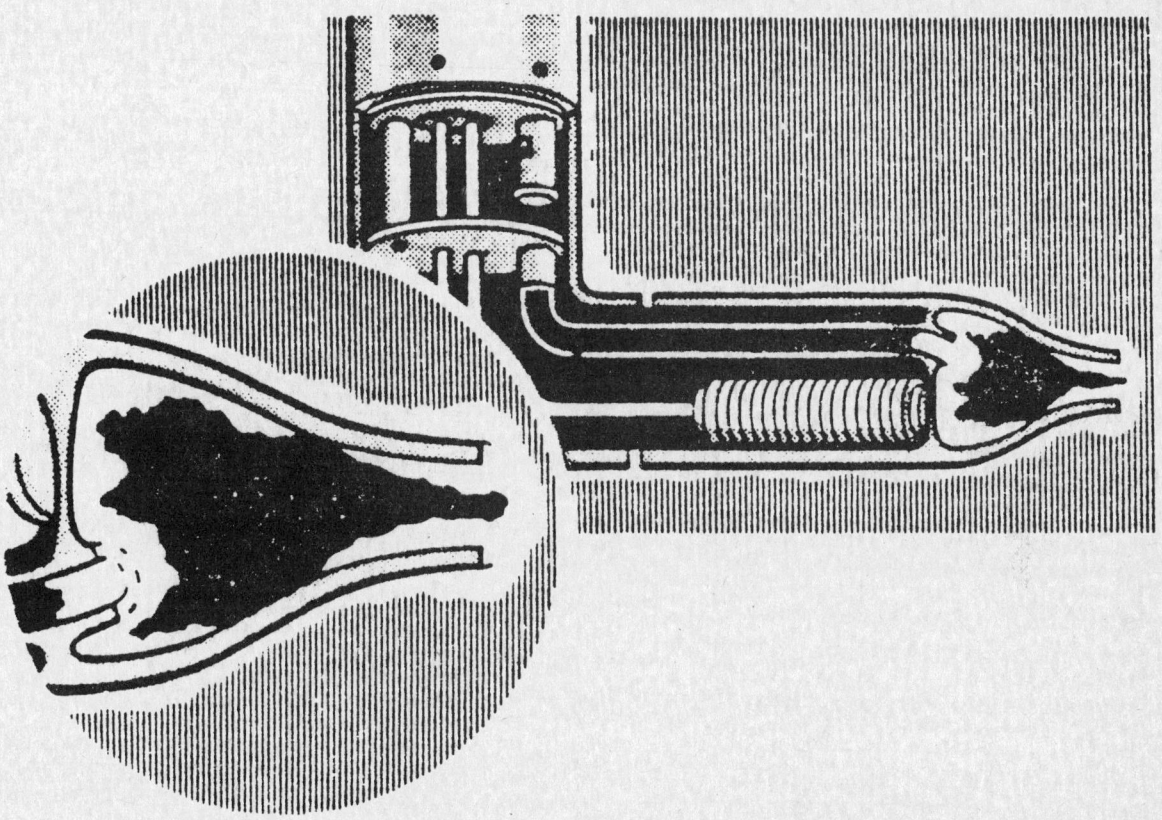

FIGURE 96. Internal pitot tube icing. It renders airspeed indicator unreliable.

ICING AND CLOUD TYPES

Basically, all clouds at subfreezing temperatures have icing potential. However, drop size, drop distribution, and aerodynamic effects of the aircraft influence ice formation. Ice may not form even though the potential exists.

The condition most favorable for very hazardous icing is the presence of many large, supercooled water drops. Conversely, an equal or lesser number of smaller droplets favors a slower rate of icing.

Small water droplets occur most often in fog and low-level clouds. Drizzle or very light rain is evidence of the presence of small drops in such clouds; but in many cases there is no precipitation at all. The most common type of icing found in lower-level stratus clouds is rime.

On the other hand, thick extensive stratified clouds that produce continuous rain such as altostratus and nimbostratus usually have an abundance of liquid water because of the relatively larger drop size and number. Such cloud systems in winter may cover thousands of square miles and present very serious icing conditions for protracted flights. Particularly in thick stratified clouds, concentrations of liquid water normally are greater with warmer temperatures. Thus, heaviest icing usually will be found at or slightly above the freezing level where temperature is never more than a few degrees below freezing. In layer type clouds, continuous icing conditions are rarely found to be more than 5,000 feet above the freezing level, and usually are two or three thousand feet thick.

The upward currents in cumuliform clouds are

Icing

FIGURE 97. Clear ice on an aircraft antenna mast.

favorable for the formation and support of many large water drops. The size of raindrops and rainfall intensity normally experienced from showers and thunderstorms confirm this. When an aircraft enters the heavy water concentrations found in cumuliform clouds, the large drops break and spread rapidly over the leading edge of the airfoil forming a film of water. If temperatures are freezing or colder, the water freezes quickly to form a solid sheet of clear ice. Pilots usually avoid cumuliform clouds when possible. Consequently, icing reports from such clouds are rare and do not indicate the frequency with which it can occur.

The updrafts in cumuliform clouds carry large amounts of liquid water far above the freezing level. On rare occasions icing has been encountered in thunderstorm clouds at altitudes of 30,000 to 40,000 feet where the free air temperature was colder than minus 40° C.

While an upper limit of critical icing potential cannot be specified in cumuliform clouds, the cellular distribution of such clouds usually limits the horizontal extent of icing conditions. An exception, of course, may be found in a protracted flight through a broad zone of thunderstorms or heavy showers.

OTHER FACTORS IN ICING

In addition to the above, other factors also enter into icing. Some of the more important ones are discussed below.

FRONTS

A condition favorable for rapid accumulation of clear icing is freezing rain below a frontal surface. Rain forms above the frontal surface at temperatures warmer than freezing. Subsequently, it falls through air at temperatures below freezing and becomes supercooled. The supercooled drops freeze on impact with an aircraft surface. Figure 98 diagrams this type of icing. It may occur with either a warm front (top) or a cold front. The icing can be critical because of the large amount of supercooled water. Icing can also become serious in cumulonimbus clouds along a surface cold front, along a squall line, or embedded in the cloud shield of a warm front.

TERRAIN

Air blowing upslope is cooled adiabatically. When the air is cooled below the freezing point, the water becomes supercooled. In stable air blowing up a gradual slope, the cloud drops generally remain comparatively small since larger drops fall out as rain. Ice accumulation is rather slow and you should have ample time to get out of it before the accumulation becomes extremely dangerous. When air is unstable, convective clouds develop a more serious hazard as described in "Icing and Cloud Types."

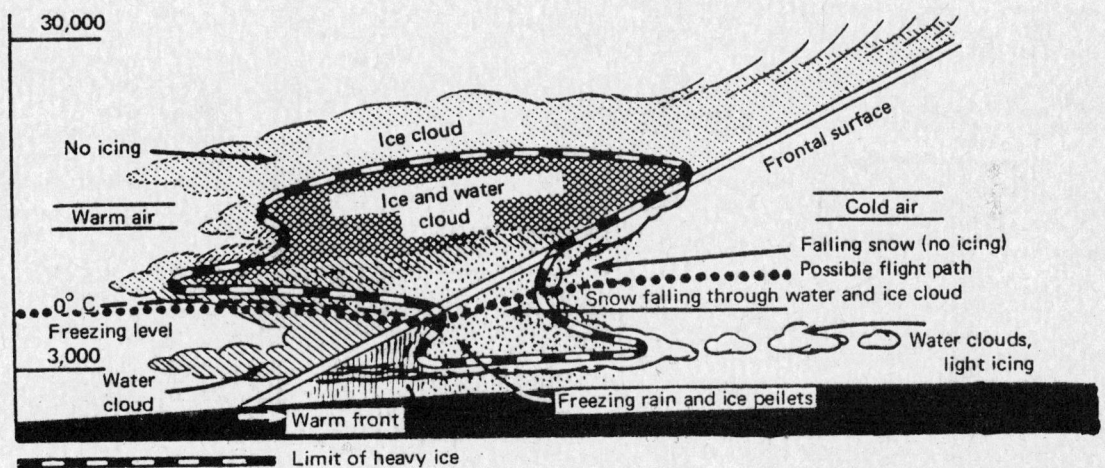

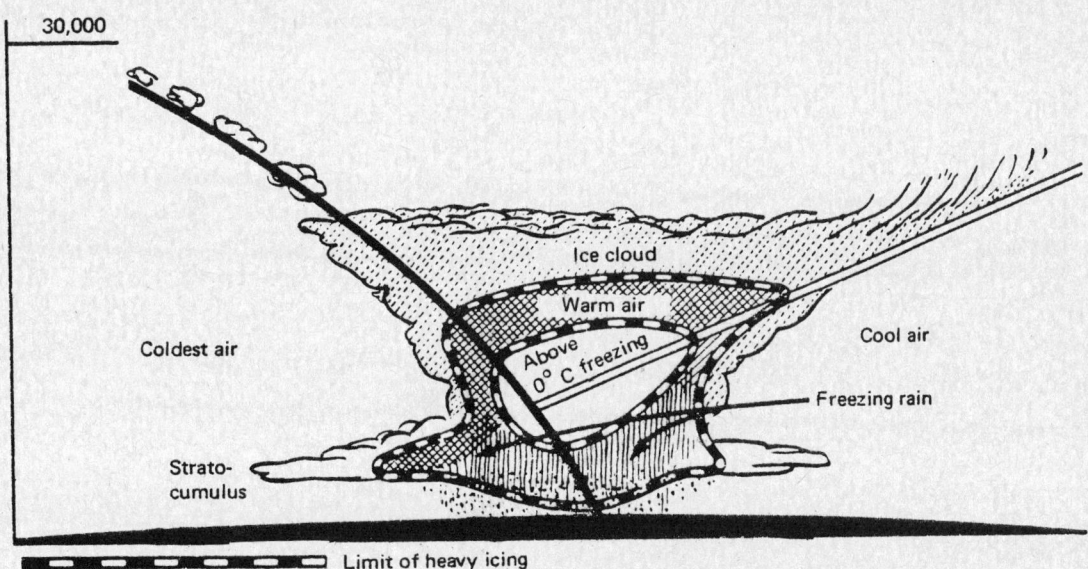

FIGURE 98. Freezing rain with a warm front (top) and a cold front (bottom). Rainfall through warm air aloft into subfreezing cold air near the ground. The rain becomes supercooled and freezes on impact.

Icing

Icing is more probable and more hazardous in mountainous regions than over other terrain. Mountain ranges cause rapid upward air motions on the windward side, and these vertical currents support large water drops. The movement of a frontal system across a mountain range often combines the normal frontal lift with the upslope effect of the mountains to create extremely hazardous icing zones.

Each mountainous region has preferred areas of icing depending upon the orientation of mountain ranges to the wind flow. The most dangerous icing takes place above the crests and to the windward side of the ridges. This zone usually extends about 5,000 feet above the tops of the mountains; but when clouds are cumuliform, the zone may extend much higher.

SEASONS

Icing may occur during any season of the year; but in temperate climates such as cover most of the contiguous United States, icing is more frequent in winter. The freezing level is nearer the ground in winter than in summer leaving a smaller low-level layer of airspace free of icing conditions. Cyclonic storms also are more frequent in winter, and the resulting cloud systems are more extensive. Polar regions have the most dangerous icing conditions in spring and fall. During the winter the air is normally too cold in the polar regions to contain heavy concentrations of moisture necessary for icing, and most cloud systems are stratiform and are composed of ice crystals.

GROUND ICING

Frost, ice pellets, frozen rain, or snow may accumulate on parked aircraft. You should remove all ice prior to takeoff, for it reduces flying efficiency of the aircraft. Water blown by propellers or splashed by wheels of an airplane as it taxis or runs through pools of water or mud may result in serious aircraft icing. Ice may form in wheel wells, brake mechanisms, flap hinges, etc., and prevent proper operation of these parts. Ice on runways and taxiways create traction and braking problems.

FROST

Frost is a hazard to flying long recognized in the aviation community. Experienced pilots have learned to remove all frost from airfoils prior to takeoff. Frost forms near the surface primarily in clear, stable air and with light winds—conditions which in all other respects make weather ideal for flying. Because of this, the real hazard is often minimized. Thin metal airfoils are especially vulnerable surfaces on which frost will form. Figure 99 is a photograph of frost on an airfoil.

Frost does not change the basic aerodynamic shape of the wing, but the roughness of its surface spoils the smooth flow of air thus causing a slowing of the airflow. This slowing of the air causes early air flow separation over the affected airfoil resulting in a loss of lift. A heavy coat of hard frost will cause a 5 to 10 percent increase in stall speed. Even a small amount of frost on airfoils may prevent an aircraft from becoming airborne at normal takeoff speed. Also possible is that, once airborne, an aircraft could have insufficient margin of airspeed above stall so that moderate gusts or turning flight could produce incipient or complete stalling.

Frost formation in flight offers a more complicated problem. The extent to which it will form is still a matter of conjecture. At most, it is comparatively rare.

IN CLOSING

Icing is where you find it. As with turbulence, icing may be local in extent and transient in character. Forecasters can identify regions in which icing is possible. However, they cannot define the precise small pockets in which it occurs. You should plan your flight to avoid those areas where icing probably will be heavier than your aircraft can handle. And you must be prepared to avoid or to escape the hazard when encountered en route.

Icing

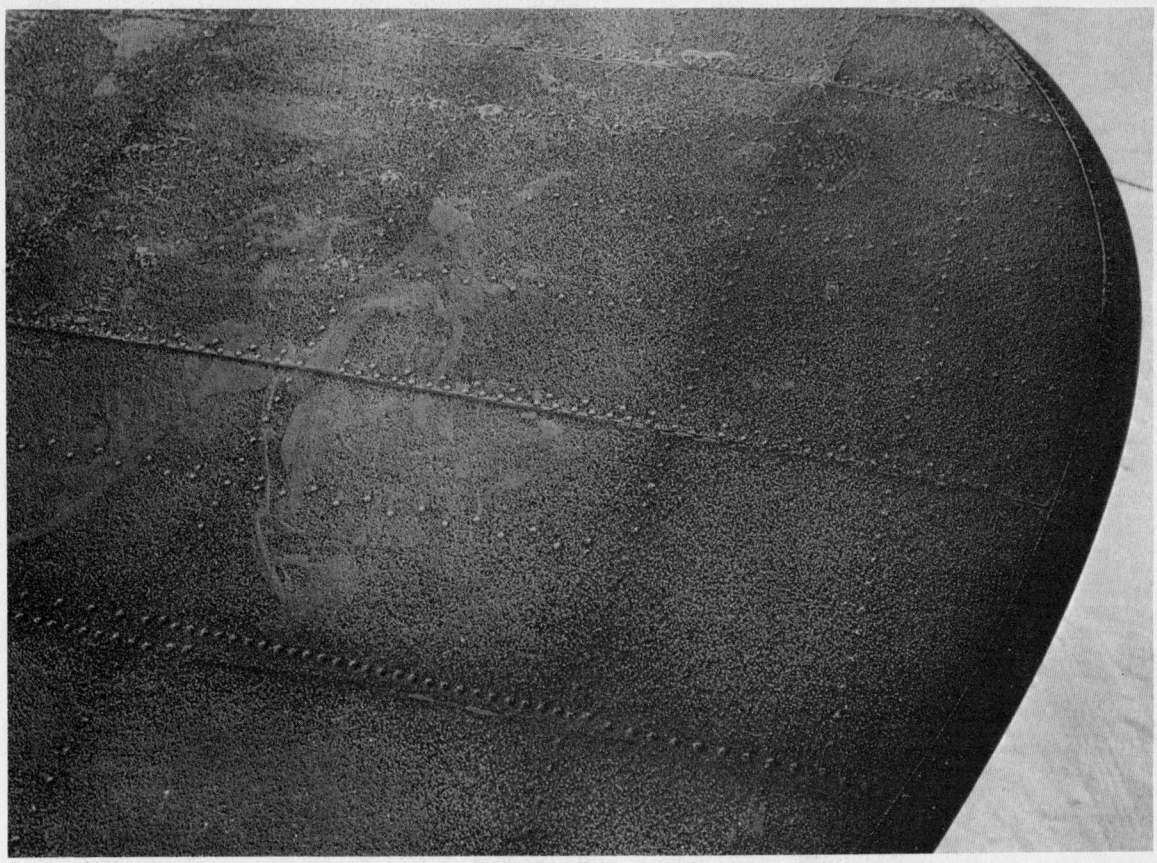

FIGURE 99. Frost on an aircraft. Always remove ice or frost before attempting takeoff.

Here are a few specific points to remember:
1. Before takeoff, check weather for possible icing areas along your planned route. Check for pilot reports, and if possible talk to other pilots who have flown along your proposed route.
2. If your aircraft is not equipped with deicing or anti-icing equipment, avoid areas of icing. Water (clouds or precipitation) must be visible and outside air temperature must be near 0° C or colder for structural ice to form.
3. Always remove ice or frost from airfoils before attempting takeoff.
4. In cold weather, avoid, when possible, taxiing or taking off through mud, water, or slush. If you have taxied through any of these, make a preflight check to ensure freedom of controls.
5. When climbing out through an icing layer, climb at an airspeed a little faster than normal to avoid a stall.
6. Use deicing or anti-icing equipment when accumulations of ice are not too great. When such equipment becomes less than totally effective, change course or altitude to get out of the icing as rapidly as possible.
7. If your aircraft is not equipped with a pitot-static system deicer, be alert for erroneous readings from your airspeed indicator, rate-of-climb indicator, and altimeter.
8. In stratiform clouds, you can likely alleviate icing by changing to a flight level and above-freezing temperatures or to one colder than $-10°$ C. An altitude change also may take you out of clouds. Rime icing in stratiform clouds can be very extensive horizontally.
9. In frontal freezing rain, you may be able to climb or descend to a layer warmer than freezing. Temperature is always warmer than freezing at some higher altitude. If you are going to climb, move quickly; procrastination may leave you with too much

Icing

ice. If you are going to descend, you must know the temperature and terrain below.
10. Avoid cumuliform clouds if at all possible. Clear ice may be encountered anywhere above the freezing level. Most rapid accumulations are usually at temperatures from 0° C to −15° C.
11. Avoid abrupt maneuvers when your aircraft is heavily coated with ice since the aircraft has lost some of its aerodynamic efficiency.
12. When "iced up," fly your landing approach with power.

The man on the ground has no way of observing actual icing conditions. His only confirmation of the existence or absence of icing comes from pilots. Help your fellow pilot and the weather service by sending pilot reports when you encounter icing or when icing is forecast but none encountered. Use the table in Section 16 of AVIATION WEATHER SERVICES as a guide in reporting intensities.

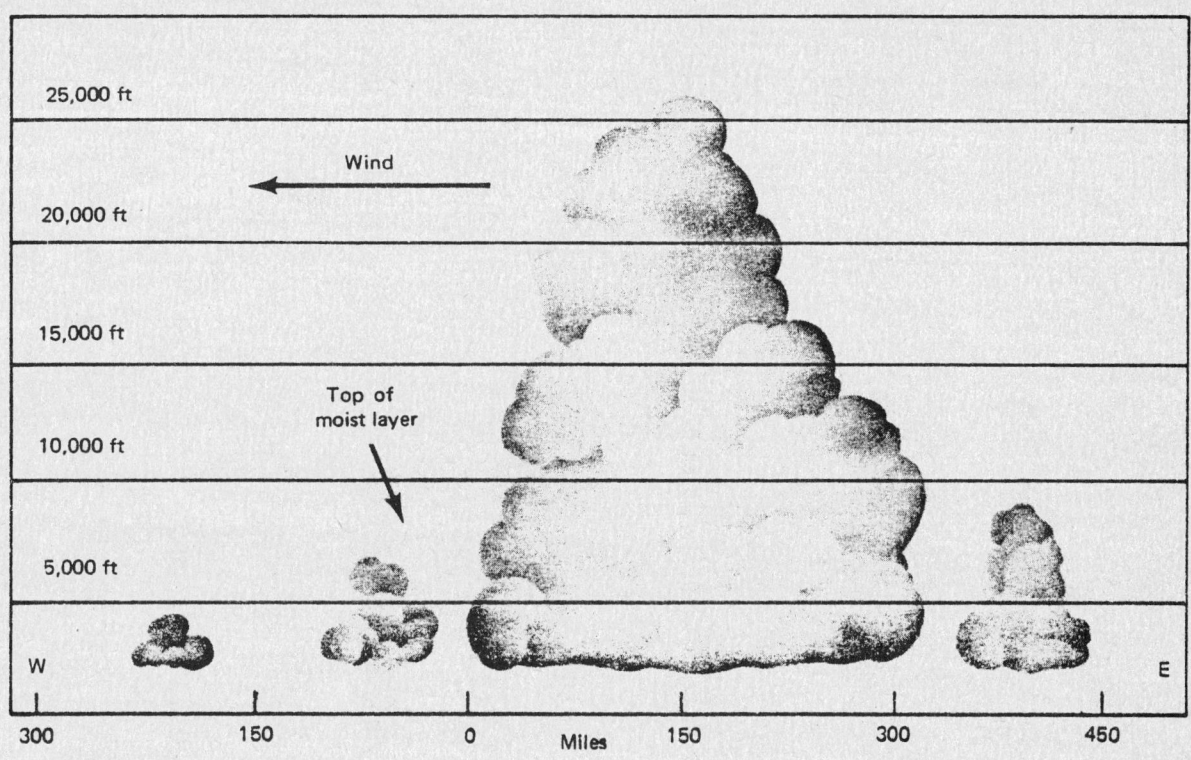

Chapter 11
THUNDERSTORMS

Many times you have to make decisions involving thunderstorms and flying. This chapter looks at where and when thunderstorms occur most frequently, explains what creates a storm, and looks inside the storm at what goes on and what it can do to an aircraft. The chapter also describes how you can use radar and suggests some do's and don'ts of thunderstorm flying.

WHERE AND WHEN?

In some tropical regions, thunderstorms occur year-round. In midlatitudes, they develop most frequently in spring, summer, and fall. Arctic regions occasionally experience thunderstorms during summer.

Figure 100 shows the average number of thunderstorms each year in the adjoining 48 States. Note the frequent occurrences in the south-central and southeastern States. The number of days on which thunderstorms occur varies widely from season to season as shown in figures 101 through 104. In general, thunderstorms are most frequent during July and August and least frequent in December and January.

Thunderstorms

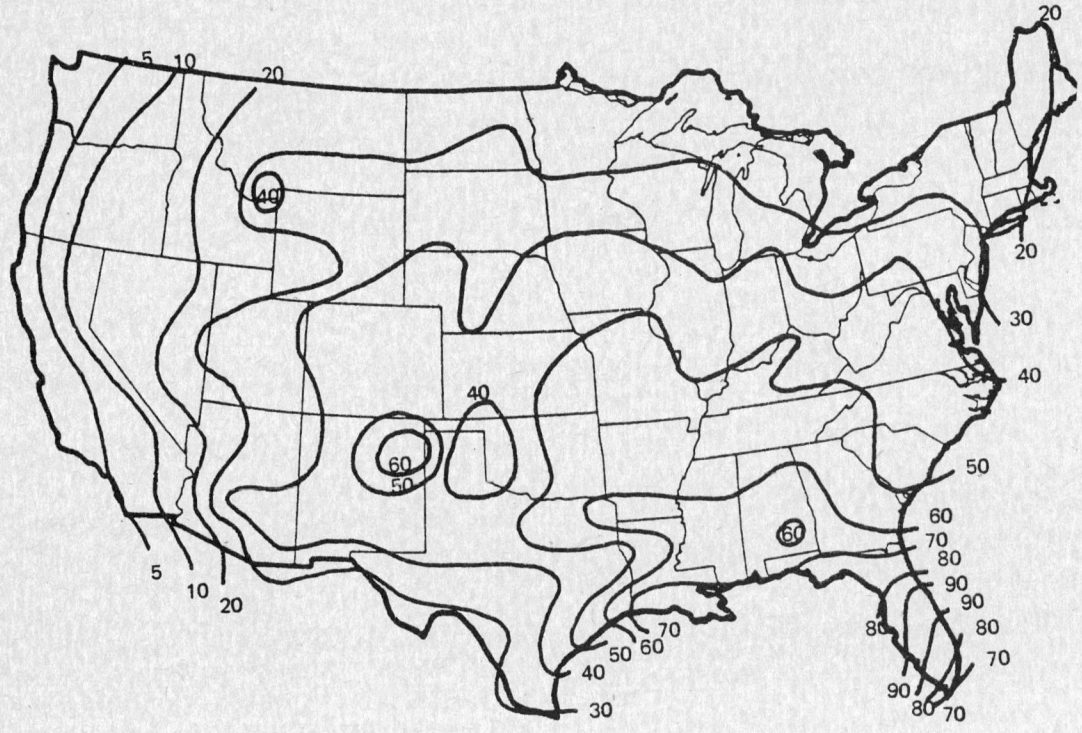

FIGURE 100. The average number of thunderstorms each year.

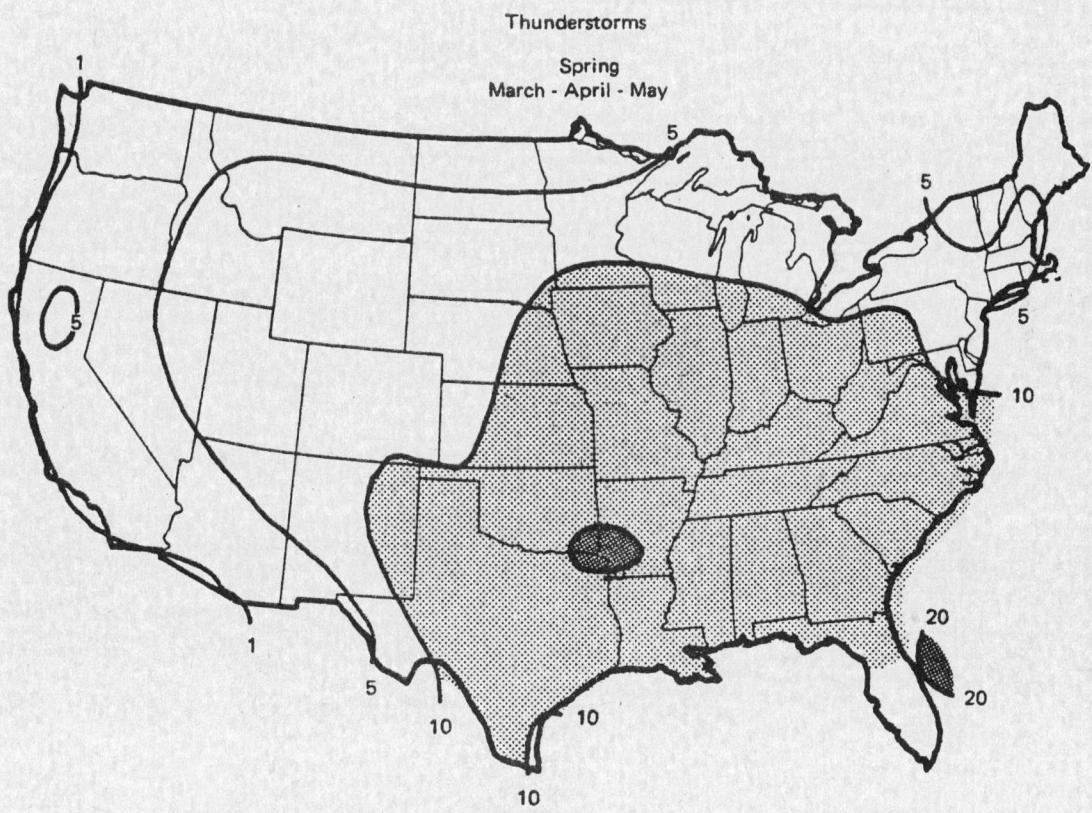

Figure 101. The average number of days with thunderstorms during spring.

Thunderstorms

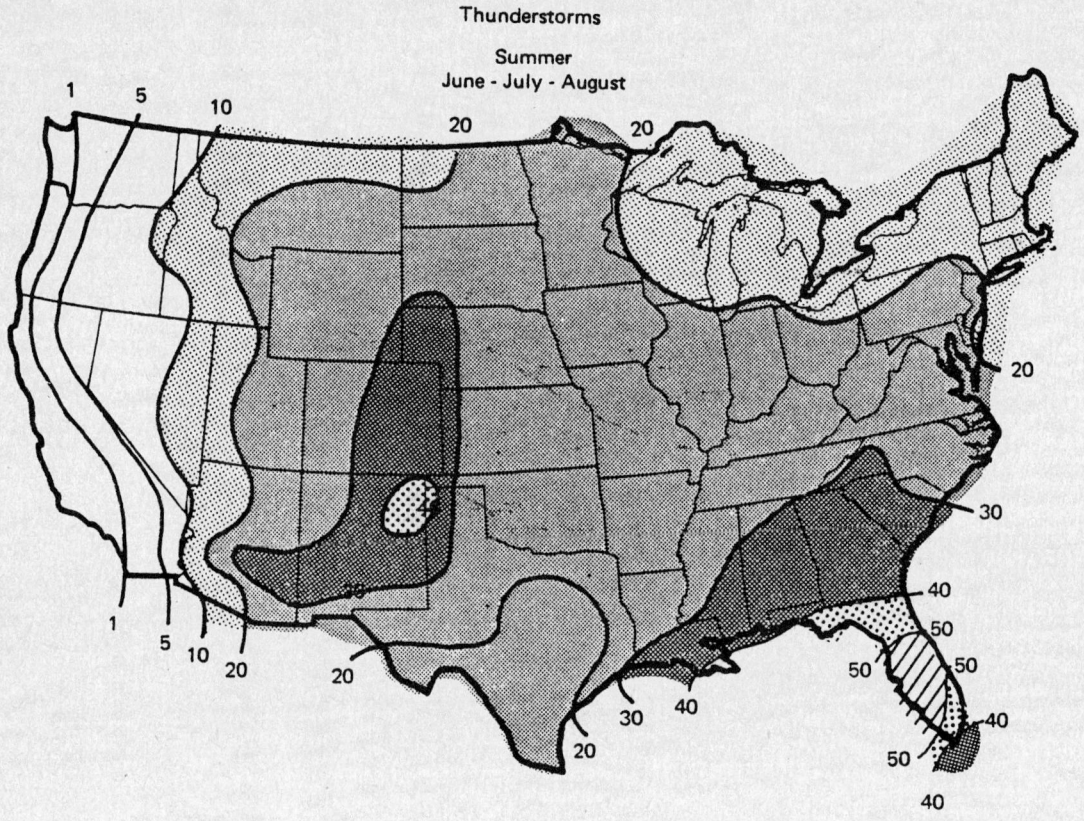

FIGURE 102. The average number of days with thunderstorms during summer.

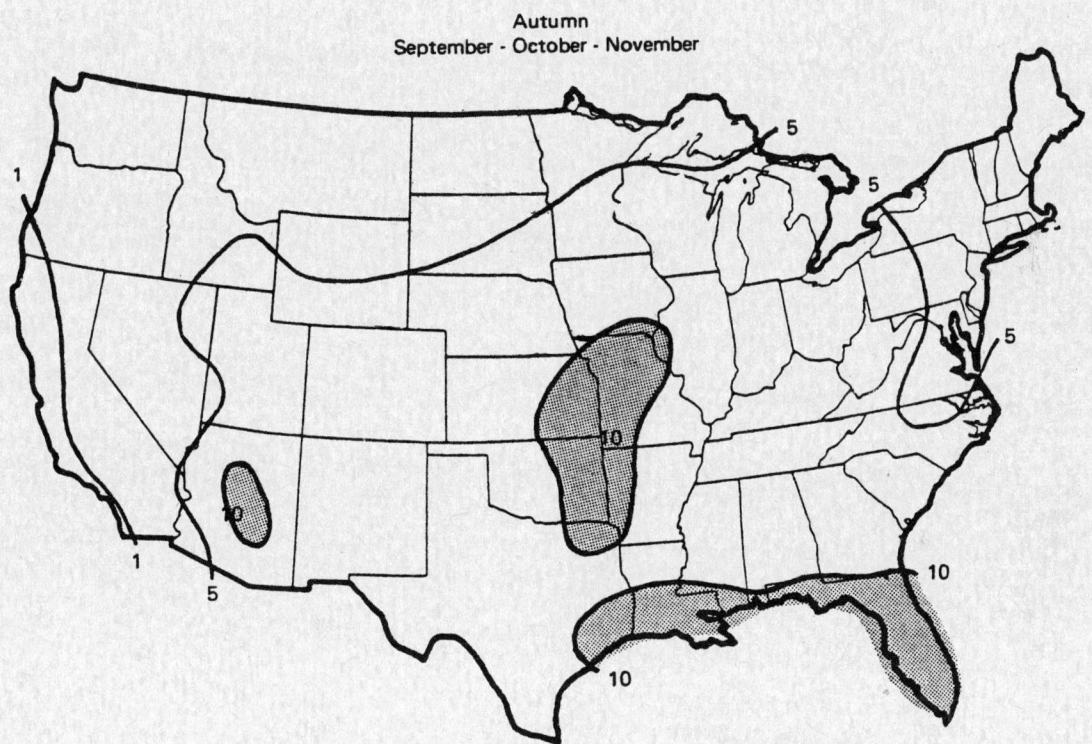

FIGURE 103. The average number of days with thunderstorms during fall.

Thunderstorms

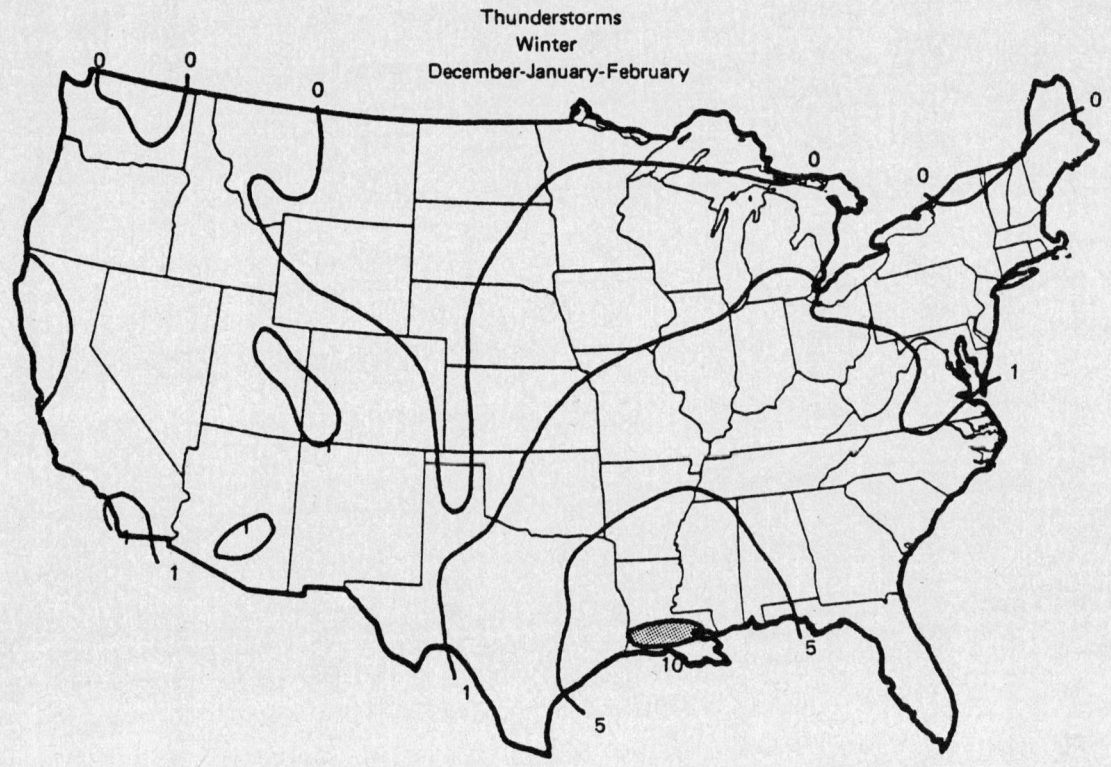

FIGURE 104. The average number of days with thunderstorms during winter.

Thunderstorms

THEY DON'T JUST HAPPEN

For a thunderstorm to form, the air must have (1) sufficient water vapor, (2) an unstable lapse rate, and (3) an initial upward boost (lifting) to start the storm process in motion. We discussed water vapor in chapter 5 and stability in chapter 6; but, what about lifting? Surface heating, converging winds, sloping terrain, a frontal surface, or any combination of these can provide the lift.

Thunderstorms have been a subject of considerable investigation for many years as they are today. Figuratively speaking, let's look inside a thunderstorm.

THE INSIDE STORY

Forced upward motion creates an initial updraft. Cooling in the updraft results in condensation and the beginning of a cumulus cloud. Condensation releases latent heat which partially offsets cooling in the saturated updraft and increases buoyancy within the cloud. This increased buoyancy drives the updraft still faster drawing more water vapor into the cloud; and, for awhile, the updraft becomes self-sustaining. All thunderstorms progress through a life cycle from their initial development through maturity and into degeneration.

LIFE CYCLE

A thunderstorm cell during its life cycle progresses through three stages—(1) the cumulus, (2) the mature, and (3) the dissipating. It is virtually impossible to visually detect the transition from one stage to another; the transition is subtle and by no means abrupt. Furthermore, a thunderstorm may be a cluster of cells in different stages of the life cycle.

The Cumulus Stage

Although most cumulus clouds do not grow into thunderstorms, every thunderstorm begins as a cumulus. The key feature of the cumulus stage is an updraft as illustrated in figure 105(A). The updraft varies in strength and extends from very near the surface to the cloud top. Growth rate of the cloud may exceed 3,000 feet per minute, so it is inadvisable to attempt to climb over rapidly building cumulus clouds.

Early during the cumulus stage, water droplets are quite small but grow to raindrop size as the cloud grows. The upwelling air carries the liquid water above the freezing level creating an icing hazard. As the raindrops grow still heavier, they fall. The cold rain drags air with it creating a cold downdraft coexisting with the updraft; the cell has reached the mature stage.

The Mature Stage

Precipitation beginning to fall from the cloud base is your signal that a downdraft has developed and a cell has entered the mature stage. Cold rain in the downdraft retards compressional heating, and the downdraft remains cooler than surrounding air. Therefore, its downward speed is accelerated and may exceed 2,500 feet per minute. The downrushing air spreads outward at the surface as shown in figure 105(B) producing strong, gusty surface winds, a sharp temperature drop, and a rapid rise in pressure. The surface wind surge is a "plow wind" and its leading edge is the "first gust."

Meanwhile, updrafts reach a maximum with speeds possibly exceeding 6,000 feet per minute. Updrafts and downdrafts in close proximity create strong vertical shear and a very turbulent environment. All thunderstorm hazards reach their greatest intensity during the mature stage.

The Dissipating Stage

Downdrafts characterize the dissipating stage of the thunderstorm cell as shown in figure 105(C) and the storm dies rapidly. When rain has ended and downdrafts have abated, the dissipating stage is complete. When all cells of the thunderstorm have completed this stage, only harmless cloud remnants remain.

HOW BIG?

Individual thunderstorms measure from less than 5 miles to more than 30 miles in diameter. Cloud bases range from a few hundred feet in very moist climates to 10,000 feet or higher in drier regions. Tops generally range from 25,000 to 45,000 feet but occasionally extend above 65,000 feet.

Thunderstorms

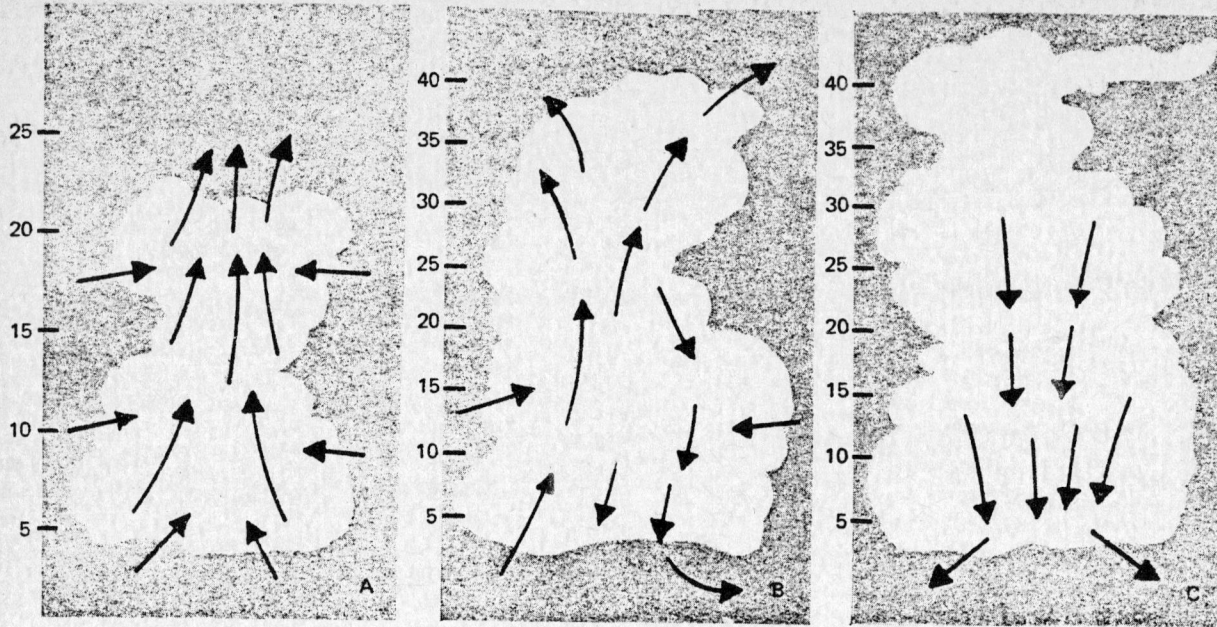

FIGURE 105. The stages of a thunderstorm. (A) is the cumulus stage; (B), the mature stage; and (C), the dissipating stage. Arrows depict air flow.

ROUGH AND ROUGHER

Duration of the mature stage is closely related to severity of the thunderstorm. Some storms occur at random in unstable air, last for only an hour or two, and produce only moderate gusts and rainfall. These are the "air mass" type, but even they are dangerously rough to fly through. Other thunderstorms form in lines, last for several hours, dump heavy rain and possibly hail, and produce strong, gusty winds and possibly tornadoes. These storms are the "steady state" type, usually are rougher than air mass storms, and virtually defy flight through them.

AIR MASS THUNDERSTORMS

Air mass thunderstorms most often result from surface heating. When the storm reaches the mature stage, rain falls through or immediately beside the updraft. Falling precipitation induces frictional drag, retards the updraft and reverses it to a downdraft. The storm is self-destructive. The downdraft and cool precipitation cool the lower portion of the storm and the underlying surface. Thus, it cuts off the inflow of water vapor; the storm runs out of energy and dies. A self-destructive cell usually has a life cycle of 20 minutes to 1½ hours.

Since air mass thunderstorms generally result from surface heating, they reach maximum intensity and frequency over land during middle and late afternoon. Off shore, they reach a maximum during late hours of darkness when land temperature is coolest and cool air flows off the land over the relatively warm water.

STEADY STATE THUNDERSTORMS

Steady state thunderstorms usually are associated with weather systems. Fronts, converging winds, and troughs aloft force upward motion spawning these storms which often form into squall lines. Afternoon heating intensifies them.

In a steady state storm, precipitation falls outside the updraft as shown in figure 106 allowing the updraft to continue unabated. Thus, the mature stage updrafts become stronger and last much longer than in air mass storms—hence, the name, "steady state." A steady state cell may persist for several hours.

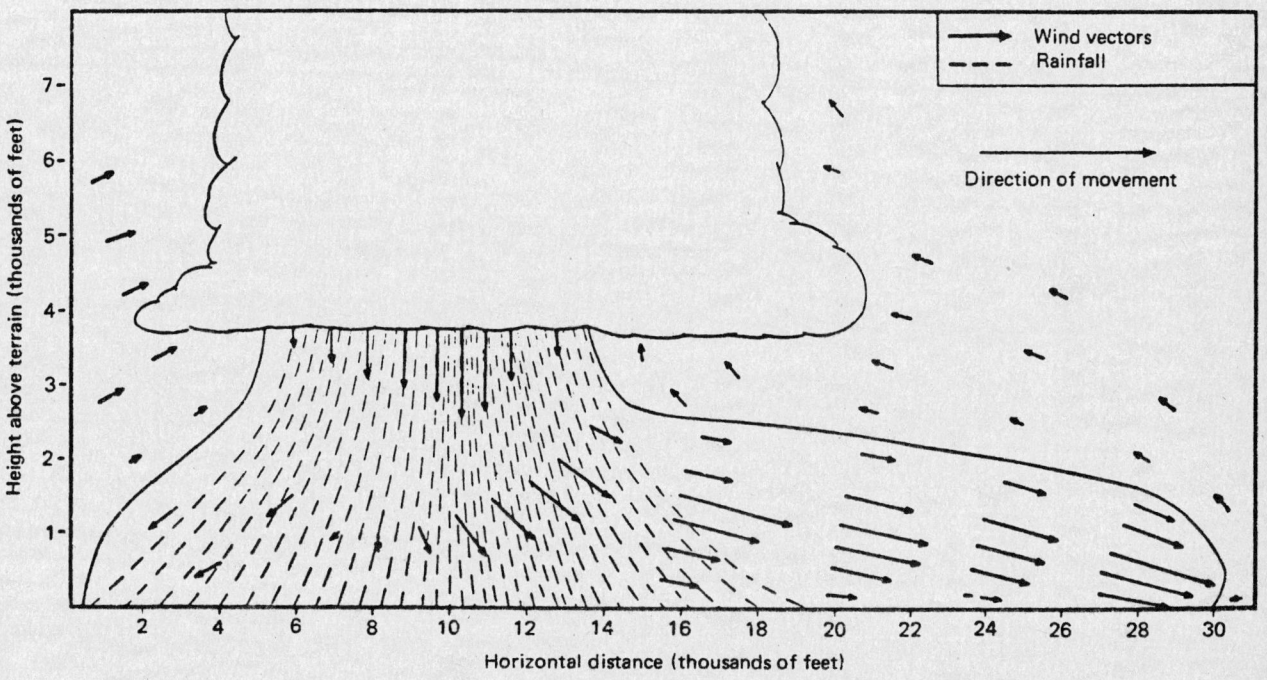

FIGURE 106. Schematic of the mature stage of a steady state thunderstorm cell showing a sloping updraft with the downdraft and precipitation outside the updraft not impeding it. The steady state mature cell may continue for many hours and deliver the most violent thunderstorm hazards.

HAZARDS

A thunderstorm packs just about every weather hazard known to aviation into one vicious bundle. Although the hazards occur in numerous combinations, let's separate them and examine each individually.

TORNADOES

The most violent thunderstorms draw air into their cloud bases with great vigor. If the incoming air has any initial rotating motion, it often forms an extremely concentrated vortex from the surface well into the cloud. Meteorologists have estimated that wind in such a vortex can exceed 200 knots; pressure inside the vortex is quite low. The strong winds gather dust and debris, and the low pressure generates a funnel-shaped cloud extending downward from the cumulonimbus base. If the cloud does not reach the surface, it is a "funnel cloud," figure 109; if it touches a land surface, it is a "tornado," figure 107; if it touches water, it is a "waterspout," figure 108.

Tornadoes occur with isolated thunderstorms at times, but much more frequently, they form with steady state thunderstorms associated with cold

Thunderstorms

FIGURE 107. A tornado.

fronts or squall lines. Reports or forecasts of tornadoes indicate that atmospheric conditions are favorable for violent turbulence.

An aircraft entering a tornado vortex is almost certain to suffer structural damage. Since the vortex extends well into the cloud, any pilot inadvertently caught on instruments in a severe thunderstorm could encounter a hidden vortex.

Families of tornadoes have been observed as appendages of the main cloud extending several miles outward from the area of lightning and precipitation. Thus, any cloud connected to a severe thunderstorm carries a threat of violence.

Frequently, cumulonimbus mamma clouds occur in connection with violent thunderstorms and tornadoes. The cloud displays rounded, irregular pockets or festoons from its base and is a signpost of violent turbulence. Figure 110 is a photograph of a cumulonimbus mamma cloud. Surface aviation reports specifically mention this and other especially hazardous clouds.

Tornadoes occur most frequently in the Great Plains States east of the Rocky Mountains. Figure 111 shows, however, that they have occurred in every State.

FIGURE 108. A waterspout.

SQUALL LINES

A *squall line* is a non-frontal, narrow band of active thunderstorms. Often it develops ahead of a cold front in moist, unstable air, but it may develop in unstable air far removed from any front. The line may be too long to easily detour and too wide and severe to penetrate. It often contains severe steady-state thunderstorms and presents the single most intense weather hazard to aircraft. It usually forms rapidly, generally reaching maximum intensity during the late afternoon and the first few hours of darkness. Figure 112 is a photograph of an advancing squall line.

TURBULENCE

Hazardous turbulence is present in **all** thunderstorms; and in a severe thunderstorm, it can damage an airframe. Strongest turbulence within the cloud occurs with shear between updrafts and

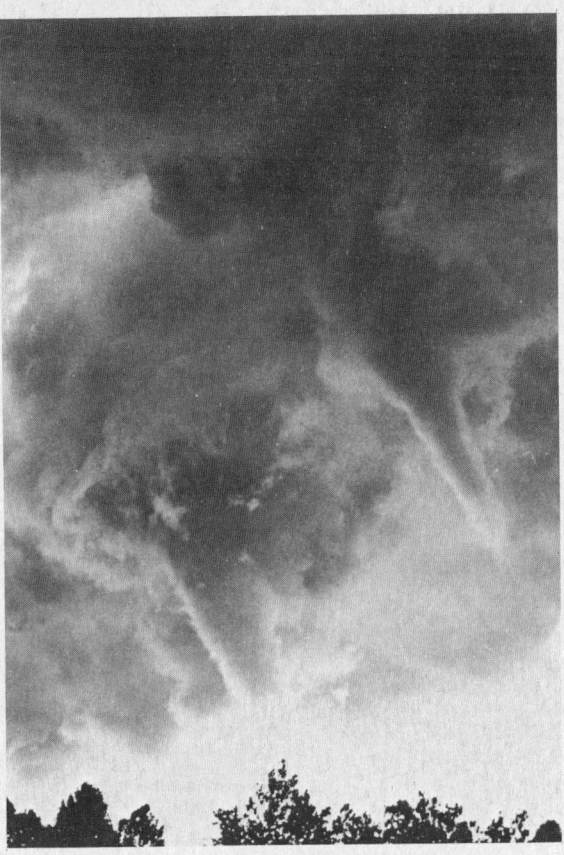

FIGURE 109. Funnel clouds.
(Photograph by Paul Hexter, NWS.)

downdrafts. Outside the cloud, shear turbulence has been encountered several thousand feet above and 20 miles laterally from a severe storm. A low level turbulent area is the shear zone between the plow wind and surrounding air. Often, a "roll cloud" on the leading edge of a storm marks the eddies in this shear. The roll cloud is most prevalent with cold frontal or squall line thunderstorms and signifies an extremely turbulent zone. The first gust causes a rapid and sometimes drastic change in surface wind ahead of an approaching storm. Figure 113 shows a schematic cross section of a thunderstorm with areas outside the cloud where turbulence may be encountered.

It is almost impossible to hold a constant altitude in a thunderstorm, and maneuvering in an attempt to do so greatly increases stresses on the aircraft. Stresses will be least if the aircraft is held in a constant *attitude* and allowed to "ride the waves." To date, we have no sure way to pick "soft spots" in a thunderstorm.

ICING

Updrafts in a thunderstorm support abundant liquid water; and when carried above the freezing level, the water becomes supercooled. When temperature in the upward current cools to about $-15°$ C, much of the remaining water vapor sublimates as ice crystals; and above this level, the amount of supercooled water decreases.

Supercooled water freezes on impact with an aircraft (see chapter 10). Clear icing can occur at any altitude above the freezing level; but at high levels, icing may be rime or mixed rime and clear. The abundance of supercooled water makes clear icing very rapid between $0°$ C and $-15°$ C, and encounters can be frequent in a cluster of cells. Thunderstorm icing can be extremely hazardous.

HAIL

Hail competes with turbulence as the greatest thunderstorm hazard to aircraft. Supercooled drops above the freezing level begin to freeze. Once a drop has frozen, other drops latch on and freeze to it, so the hailstone grows—sometimes into a huge iceball. Large hail occurs with severe thunderstorms usually built to great heights. Eventually the hailstones fall, possibly some distance from the storm core. Hail has been observed in clear air several miles from the parent thunderstorm.

As hailstones fall through the melting level, they begin to melt, and precipitation may reach the ground as either hail or rain. Rain at the surface does not mean the absence of hail aloft. You should anticipate possible hail with *any* thunderstorm, especially beneath the anvil of a large cumulonimbus. Hailstones larger than one-half inch in diameter can significantly damage an aircraft in a few seconds. Figure 114 is a photograph of an aircraft flown through a "hail" of a thunderstorm.

LOW CEILING AND VISIBILITY

Visibility generally is near zero within a thunderstorm cloud. Ceiling and visibility also can become restricted in precipitation and dust between the cloud base and the ground. The restrictions create the same problem as all ceiling and visibility restrictions; but the hazards are increased many fold when associated with the other thunderstorm hazards of turbulence, hail, and lightning which make precision instrument flying virtually impossible.

FIGURE 110. Cumulonimbus Mamma clouds, associated with cumulonimbus clouds, indicate extreme instability.

EFFECT ON ALTIMETERS

Pressure usually falls rapidly with the approach of a thunderstorm, then rises sharply with the onset of the first gust and arrival of the cold downdraft and heavy rain showers, falling back to normal as the storm moves on. This cycle of pressure change may occur in 15 minutes. If the altimeter setting is not corrected, the indicated altitude may be in error by over 100 feet.

THUNDERSTORM ELECTRICITY

Electricity generated by thunderstorms is rarely a great hazard to aircraft, but it may cause damage and is annoying to flight crews. Lightning is the most spectacular of the electrical discharges.

Lightning

A lightning strike can puncture the skin of an aircraft and can damage communication and electronic navigational equipment. Lightning has been suspected of igniting fuel vapors causing explosion; however, serious accidents due to lightning strikes are extremely rare. Nearby lightning can blind the pilot rendering him momentarily unable to navigate either by instrument or by visual reference. Nearby lightning can also induce permanent errors in the magnetic compass. Lightning discharges, even distant ones, can disrupt radio communications on low and medium frequencies.

A few pointers on lightning:
1. The more frequent the lightning, the more severe the thunderstorm.

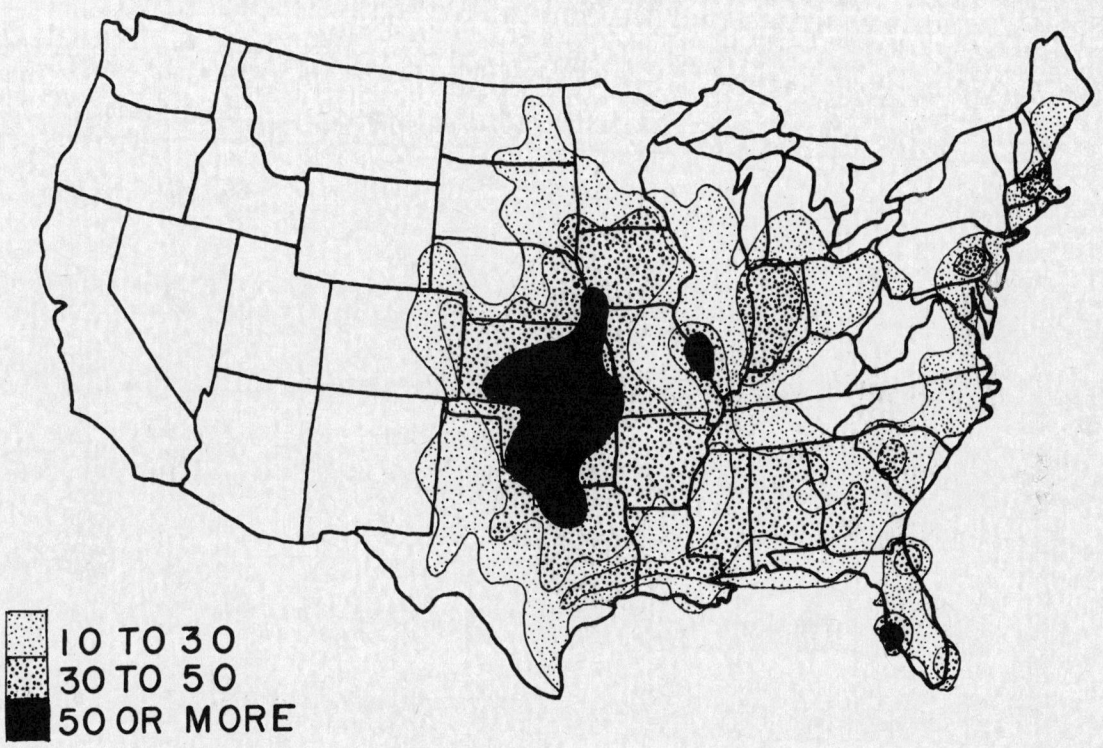

FIGURE 111. Tornado incidence by State and area.

2. Increasing frequency of lightning indicates a growing thunderstorm.
3. Decreasing lightning indicates a storm nearing the dissipating stage.
4. At night, frequent distant flashes playing along a large sector of the horizon suggest a probable squall line.

Precipitation Static

Precipitation static, a steady, high level of noise in radio receivers is caused by intense corona discharges from sharp metallic points and edges of flying aircraft. It is encountered often in the vicinity of thunderstorms. When an aircraft flies through clouds, precipitation, or a concentration of solid particles (ice, sand, dust, etc.), it accumulates a charge of static electricity. The electricity discharges onto a nearby surface or into the air causing a noisy disturbance at lower frequencies.

The corona discharge is weakly luminous and may be seen at night. Although it has a rather eerie appearance, it is harmless. It was named "St. Elmo's Fire" by Mediterranean sailors, who saw the brushy discharge at the top of ship masts.

FIGURE 112. Squall line thunderstorms.

Thunderstorms

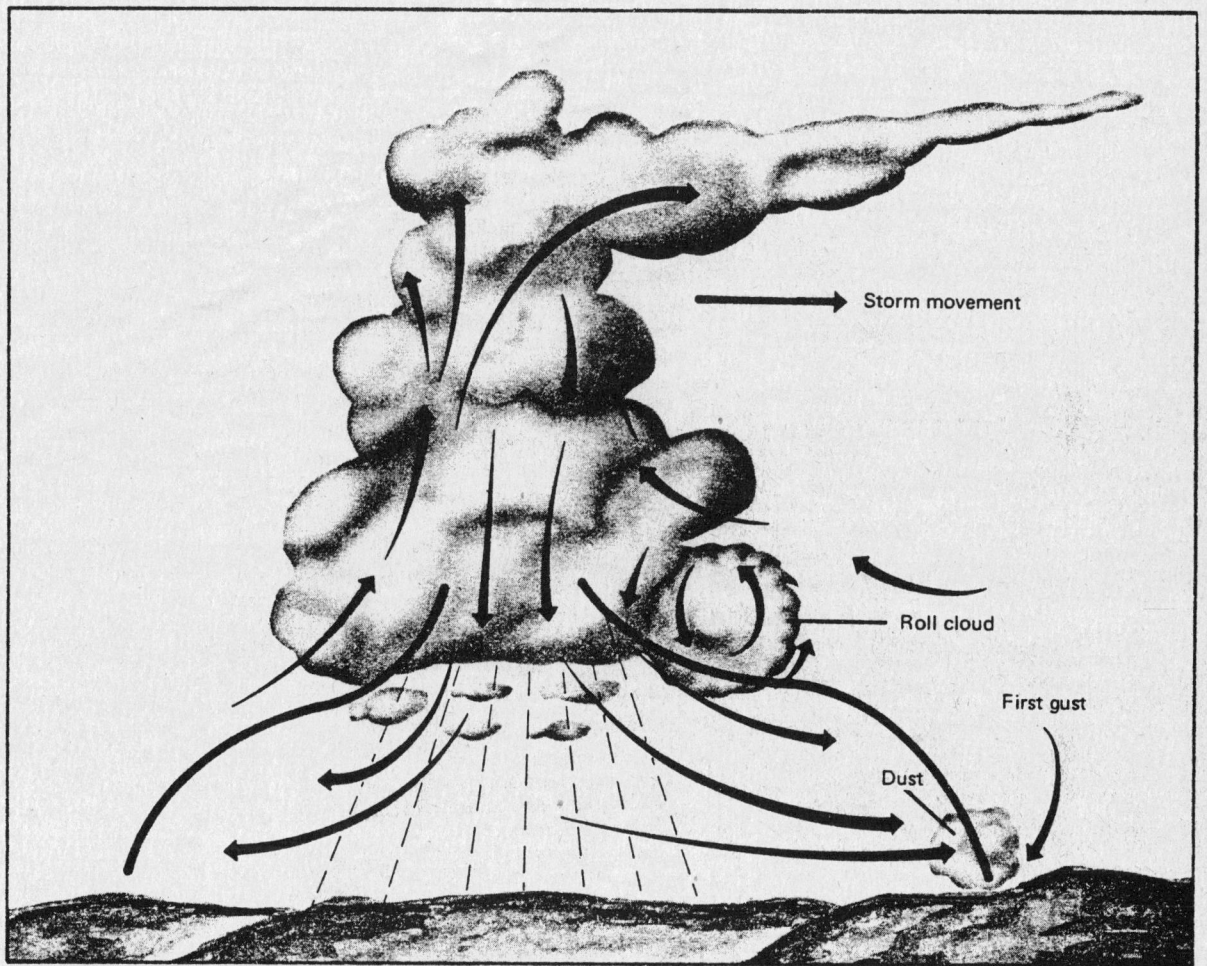

Figure 113. Schematic cross section of a thunderstorm. Note areas outside the main cloud where turbulence may be encountered.

Thunderstorms

FIGURE 114. Hail damage to an aircraft.

THUNDERSTORMS AND RADAR

Weather radar detects droplets of precipitation size. Strength of the radar return (echo) depends on drop size and number. The greater the number of drops, the stronger is the echo; and the larger the drops, the stronger is the echo. Drop size determines echo intensity to a much greater extent than does drop number.

Meteorologists have shown that drop size is almost directly proportional to rainfall rate; and the greatest rainfall rate is in thunderstorms. Therefore, the strongest echoes are thunderstorms. Hailstones usually are covered with a film of water and, therefore, act as huge water droplets giving the strongest of all echoes. Showers show less intense echoes; and gentle rain and snow return the weakest of all echoes. Figure 115 is a photograph of a ground based radar scope.

Since the strongest echoes identify thunderstorms, they also mark the areas of greatest hazards. Radar information can be valuable both from ground based radar for preflight planning and from airborne radar for severe weather avoidance.

Thunderstorms build and dissipate rapidly, and they also may move rapidly. Therefore, **do not attempt to preflight plan a course between echoes.** The best use of ground radar information is to isolate general areas and coverage of echoes. You must evade individual storms from inflight observations either by visual sighting or by airborne radar.

Airborne weather avoidance radar is, as its name implies, for avoiding severe weather—not for penetrating it. Whether to fly into an area of radar echoes depends on echo intensity, spacing between the echoes, and the capabilities of you and your

Thunderstorms

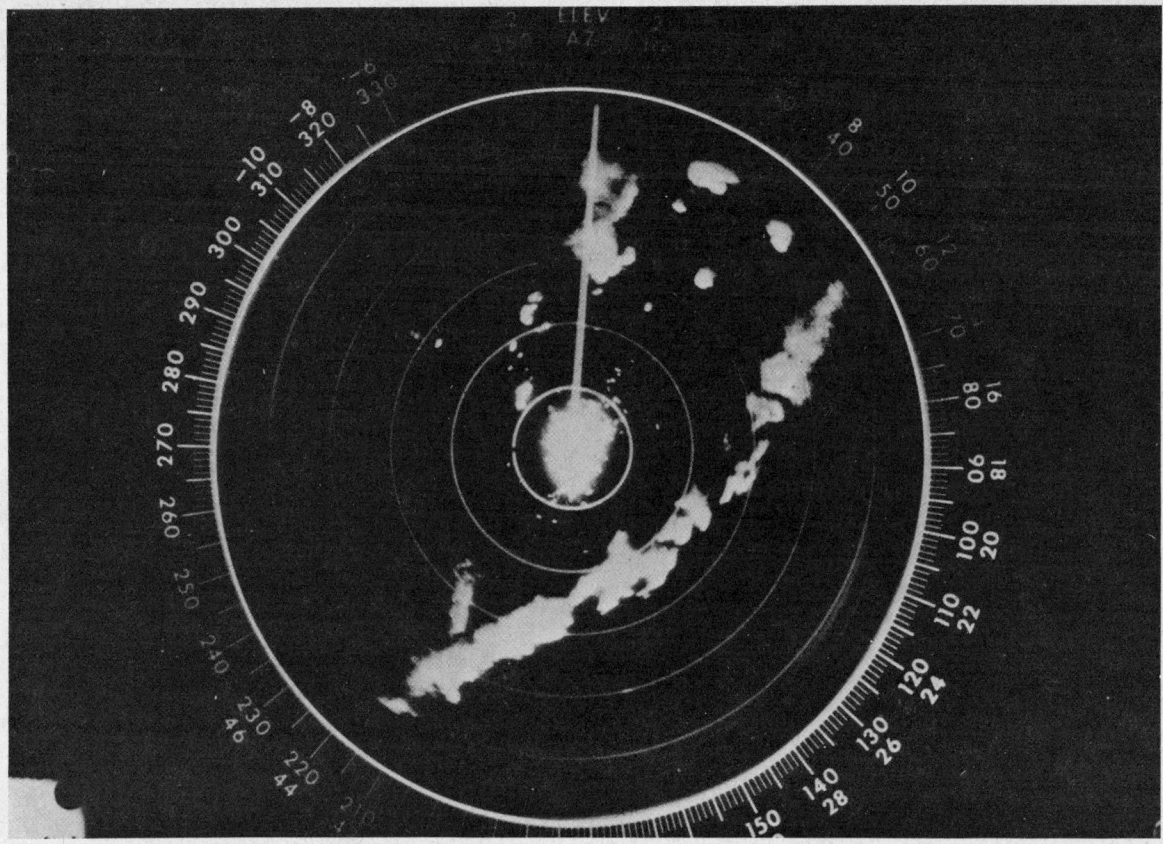

FIGURE 115. Radar photograph of a line of thunderstorms.

aircraft. Remember that weather radar detects only precipitation drops; it does not detect minute cloud droplets. Therefore, *the radar scope provides no assurance of avoiding instrument weather in clouds and fog.* Your scope may be clear between intense echoes; this clear area does not necessarily mean you can fly between the storms and maintain visual sighting of them.

The most intense echoes are severe thunderstorms. Remember that hail may fall several miles from the cloud, and hazardous turbulence may extend as much as 20 miles from the cloud. Avoid the most intense echoes by at least 20 miles; that is, echoes should be separated by at least 40 miles before you fly between them. As echoes diminish in intensity, you can reduce the distance by which you avoid them. Figure 116 illustrates use of airborne radar in avoiding thunderstorms.

DO'S AND DON'TS OF THUNDERSTORM FLYING

Above all, remember this: *never regard any thunderstorm as "light"* even when radar observers report the echoes are of light intensity. *Avoiding thunderstorms is the best policy.* Following are some Do's and Don'ts of thunderstorm *avoidance:*

1. Don't land or take off in the face of an approaching thunderstorm. A sudden wind shift or low level turbulence could cause loss of control.
2. Don't attempt to fly under a thunderstorm even if you can see through to the other side. Turbulence under the storm could be disastrous.
3. Don't try to circumnavigate thunderstorms covering 6/10 of an area or more either visually or by airborne radar.
4. Don't fly without airborne radar into a cloud mass containing scattered embedded

Thunderstorms

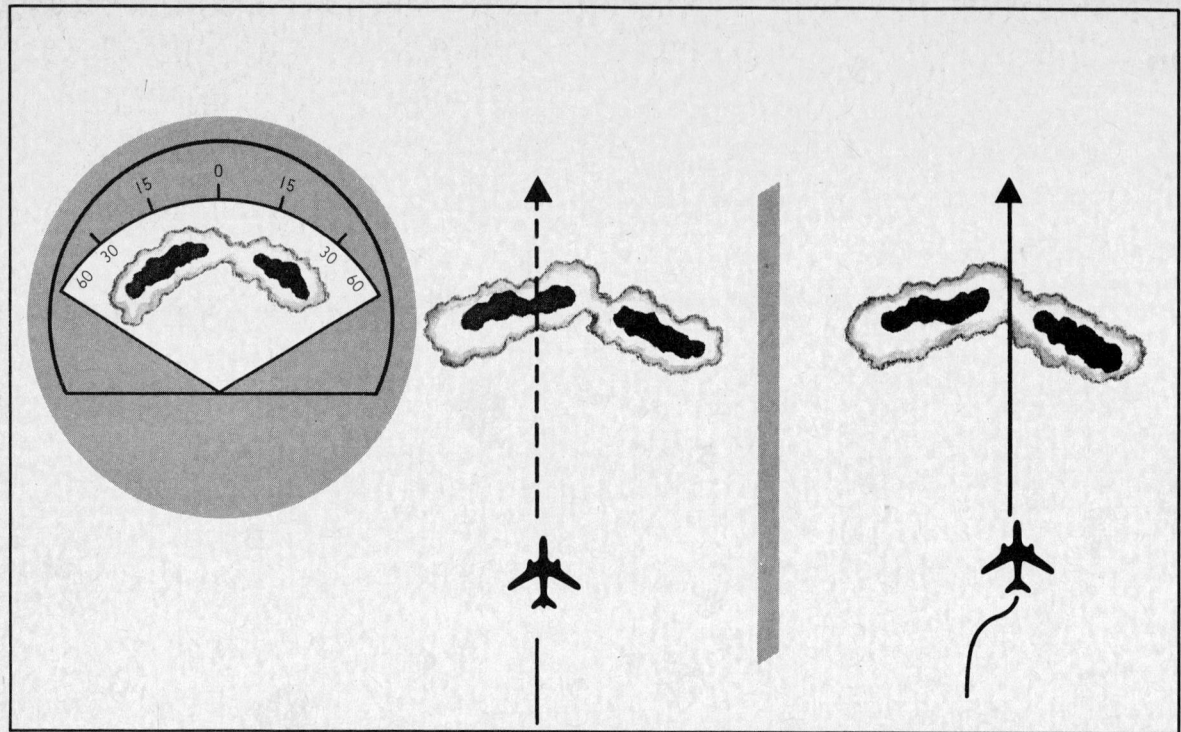

FIGURE 116. Use of airborne radar to avoid heavy precipitation and turbulence. When echoes are extremely intense, avoid the most intense echoes by at least 20 miles. You should avoid flying between these very intense echoes unless they are separated by at least 40 miles. Hazardous turbulence and hail often extend several miles from the storm centers.

thunderstorms. Scattered thunderstorms not embedded usually can be visually circumnavigated.

5. Do avoid by at least 20 miles any thunderstorm identified as severe or giving an intense radar echo. This is especially true under the anvil of a large cumulonimbus.
6. Do clear the top of a known or suspected severe thunderstorm by at least 1,000 feet altitude for each 10 knots of wind speed at the cloud top. This would exceed the altitude capability of most aircraft.
7. Do remember that vivid and frequent lightning indicates a severe thunderstorm.
8. Do regard as severe any thunderstorm with tops 35,000 feet or higher whether the top is visually sighted or determined by radar.

If you *cannot* avoid penetrating a thunderstorm, following are some Do's **Before** entering the storm:

1. Tighten your safety belt, put on your shoulder harness if you have one, and secure all loose objects.
2. Plan your course to take you through the storm in a minimum time and *hold* it.
3. To avoid the most critical icing, establish a penetration altitude below the freezing level or above the level of $-15°$ C.
4. Turn on pitot heat and carburetor or jet inlet heat. Icing can be rapid at any altitude and cause almost instantaneous power failure or loss of airspeed indication.
5. Establish power settings for reduced turbulence penetration airspeed recommended in your aircraft manual. Reduced airspeed lessens the structural stresses on the aircraft.
6. Turn up cockpit lights to highest intensity to lessen danger of temporary blindness from lightning.
7. If using automatic pilot, disengage altitude hold mode and speed hold mode. The automatic altitude and speed controls will increase maneuvers of the aircraft thus increasing structural stresses.
8. If using airborne radar, tilt your antenna

up and down occasionally. Tilting it up may detect a hail shaft that will reach a point on your course by the time you do. Tilting it down may detect a growing thunderstorm cell that may reach your altitude.

Following are some Do's and Don'ts *During* thunderstorm penetration:

1. Do keep your eyes on your instruments. Looking outside the cockpit can increase danger of temporary blindness from lightning.
2. Don't change power settings; maintain settings for reduced airspeed.
3. Do maintain a constant *attitude;* let the aircraft "ride the waves." Maneuvers in trying to maintain constant altitude increase stresses on the aircraft.
4. Don't turn back once you are in the thunderstorm. A straight course through the storm most likely will get you out of the hazards most quickly. In addition, turning maneuvers increase stresses on the aircraft.

Chapter 12
COMMON IFR PRODUCERS

Most aircraft accidents related to low ceilings and visibilities involve pilots who are not instrument qualified. These pilots attempt flight by visual reference into weather that is suitable at best only for instrument flight. When you lose sight of the visual horizon, your senses deceive you; you lose sense of direction—you can't tell up from down. You may doubt that *you* will lose your sense of direction, but one good scare has changed the thinking of many a pilot. *"Continued VFR into adverse weather" is the cause of about 25 percent of all fatal general aviation accidents.*

Minimum values of ceiling and visibility determine Visual Flight Rules. Lower ceiling and/or visibility require instrument flight. Ceiling is the maximum height from which a pilot can maintain VFR in reference to the ground. Visibility is how far he can see. AVIATION WEATHER SERVICES (AC 00-45) contains details of ceiling and visibility reports.

Don't let yourself be caught in the statistics of "continued VFR into adverse weather." IFR producers are fog, low clouds, haze, smoke, blowing obstructions to vision, and precipitation. Fog and low stratus restrict navigation by visual reference more often than all other weather parameters.

FOG

Fog is a surface based cloud composed of either water droplets or ice crystals. Fog is the most frequent cause of surface visibility below 3 miles, and is one of the most common and persistent weather hazards encountered in aviation. The rapidity with which fog can form makes it especially hazardous. It is not unusual for visibility to drop from VFR to less than a mile in a few minutes. It is primarily a hazard during takeoff and landing, but it is also important to VFR pilots who must maintain visual reference to the ground.

Small temperature-dew point spread is essential for fog to form. Therefore, fog is prevalent in coastal areas where moisture is abundant. However, fog can occur anywhere. Abundant condensation nuclei enhances the formation of fog. Thus, fog is prevalent in industrial areas where byproducts of combustion provide a high concentration of these nuclei. Fog occurs most frequently in the colder months, but the season and frequency of occurrence vary from one area to another.

Fog may form (1) by cooling air to its dew point, or (2) by adding moisture to air near the ground. Fog is classified by the way it forms. Formation may involve more than one process.

RADIATION FOG

Radiation fog is relatively shallow fog. It may be dense enough to hide the entire sky or may conceal only part of the sky. "Ground fog" is a form of radiation fog. As viewed by a pilot in flight, dense radiation fog may obliterate the entire surface below him; a less dense fog may permit his observation of a small portion of the surface directly below him. Tall objects such as buildings, hills, and towers may protrude upward through ground fog giving the pilot fixed references for VFR flight. Figure 117 illustrates ground fog as seen from the air.

Conditions favorable for radiation fog are clear sky, little or no wind, and small temperature-dew point spread (high relative humidity). The fog forms almost exclusively at night or near daybreak. Terrestrial radiation cools the ground; in turn, the cool ground cools the air in contact with it. When the air is cooled to its dew point, fog forms. When rain soaks the ground, followed by clearing skies,

FIGURE 117. Ground fog as seen from the air.

Common IFR Producers

radiation fog is not uncommon the following morning.

Radiation fog is restricted to land because water surfaces cool little from nighttime radiation. It is shallow when wind is calm. Winds up to about 5 knots mix the air slightly and tend to deepen the fog by spreading the cooling through a deeper layer. Stronger winds disperse the fog or mix the air through a still deeper layer with stratus clouds forming at the top of the mixing layer.

Ground fog usually "burns off" rather rapidly after sunrise. Other radiation fog generally clears before noon unless clouds move in over the fog.

ADVECTION FOG

Advection fog forms when moist air moves over colder ground or water. It is most common along coastal areas but often develops deep in continental areas. At sea it is called "sea fog." Advection fog deepens as wind speed increases up to about 15 knots. Wind much stronger than 15 knots lifts the fog into a layer of low stratus or stratocumulus.

The west coast of the United States is quite vulnerable to advection fog. This fog frequently forms offshore as a result of cold water as shown in figure 118 and then is carried inland by the wind. During the winter, advection fog over the central and eastern United States results when moist air from the Gulf of Mexico spreads northward over cold ground as shown in figure 119. The fog may extend as far north as the Great Lakes. Water areas in northern latitudes have frequent dense sea fog in summer as a result of warm, moist, tropical air flowing northward over colder Arctic waters.

A pilot will notice little difference between flying over advection fog and over radiation fog except that skies may be cloudy above the advection fog. Also, advection fog is usually more extensive and much more persistent than radiation fog. Advection fog can move in rapidly regardless of the time of day or night.

UPSLOPE FOG

Upslope fog forms as a result of moist, stable air being cooled adiabatically as it moves up sloping terrain. Once the upslope wind ceases, the fog dissipates. Unlike radiation fog, it can form under cloudy skies. Upslope fog is common along the

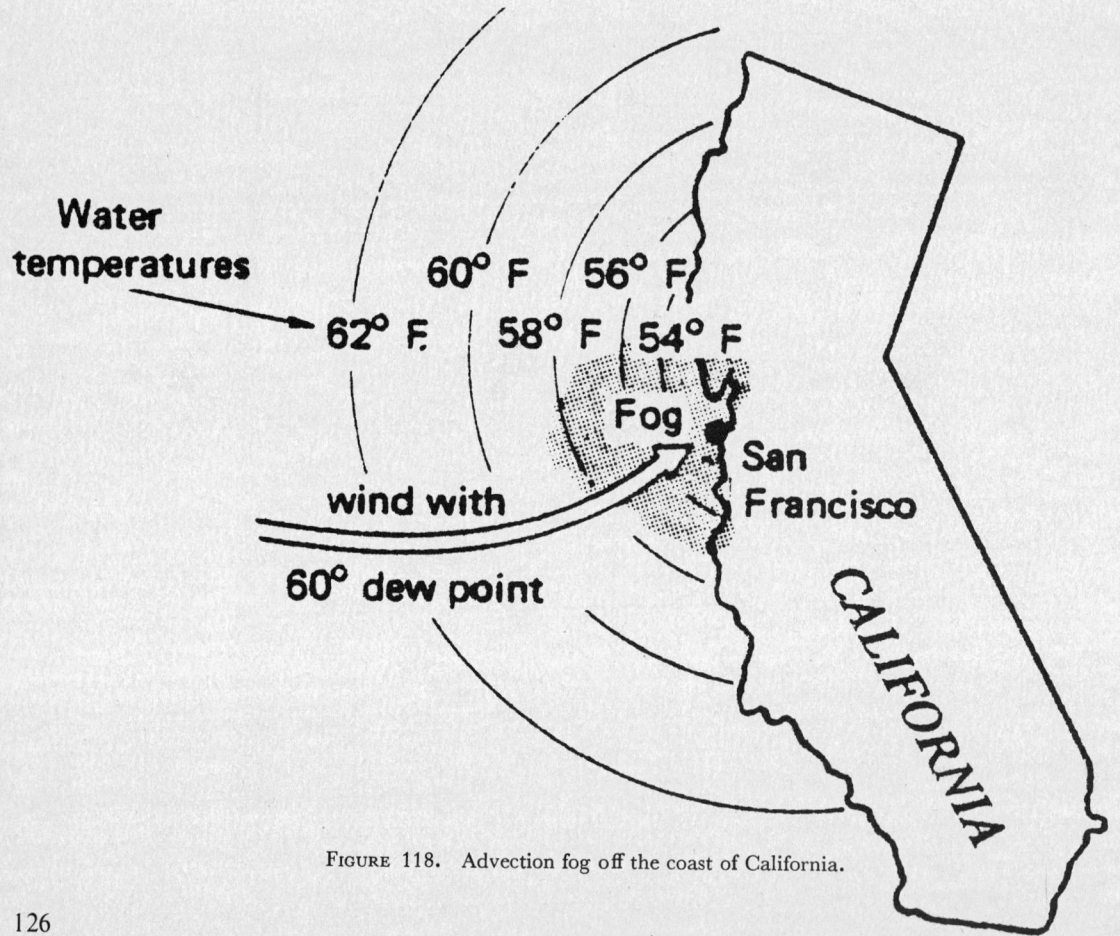

FIGURE 118. Advection fog off the coast of California.

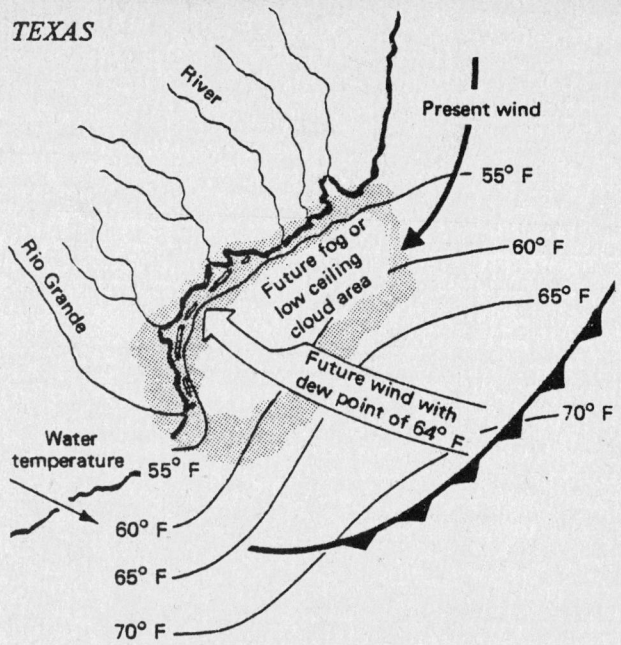

Low water temperatures in the gulf caused by considerable inland cold front precipitation draining into the gulf and previous and present polar air traversing across the gulf cooling the shallow water offshore.

FIGURE 119. Advection fog over the southeastern United States and Gulf Coast. The fog often may spread to the Great Lakes and northern Appalachians.

eastern slopes of the Rockies and somewhat less frequent east of the Appalachians. Upslope fog often is quite dense and extends to high altitudes.

PRECIPITATION-INDUCED FOG

When relatively warm rain or drizzle falls through cool air, evaporation from the precipitation saturates the cool air and forms fog. Precipitation-induced fog can become quite dense and continue for an extended period of time. This fog may extend over large areas, completely suspending air operations. It is most commonly associated with warm fronts, but can occur with slow moving cold fronts and with stationary fronts.

Fog induced by precipitation is in itself hazardous as is any fog. It is especially critical, however, because it occurs in the proximity of precipitation and other possible hazards such as icing, turbulence, and thunderstorms.

ICE FOG

Ice fog occurs in cold weather when the temperature is much below freezing and water vapor sublimates directly as ice crystals. Conditions favorable for its formation are the same as for radiation fog except for cold temperature, usually —25° F or colder. It occurs mostly in the Arctic regions, but is not unknown in middle latitudes during the cold season. Ice fog can be quite blinding to someone flying into the sun.

LOW STRATUS CLOUDS

Stratus clouds, like fog, are composed of extremely small water droplets or ice crystals suspended in air. An observer on a mountain in a stratus layer would call it fog. Stratus and fog frequently exist together. In many cases there is no real line of distinction between the fog and stratus; rather, one gradually merges into the other. Flight visibility may approach zero in stratus clouds. Stratus tends to be lowest during night and early morning, lifting or dissipating due to solar heating during the late morning or afternoon. Low stratus clouds often occur when moist air mixes with a colder air mass or in any situation where temperature-dew point spread is small.

HAZE AND SMOKE

Haze is a concentration of salt particles or other dry particles not readily classified as dust or other phenomenon. It occurs in stable air, is usually only a few thousand feet thick, but sometimes may extend as high as 15,000 feet. Haze layers often have definite tops above which horizontal visibility is good. However, downward visibility from above a haze layer is poor, especially on a slant. Visibility in haze varies greatly depending upon whether the pilot is facing the sun. Landing an aircraft into the sun is often hazardous if haze is present.

Smoke concentrations form primarily in industrial areas when air is stable. It is most prevalent at night or early morning under a temperature inversion but it can persist throughout the day. Figure 120 illustrates smoke trapped under a temperature inversion.

When skies are clear above haze or smoke, visibility generally improves during the day; however, the improvement is slower than the clearing of fog. Fog evaporates, but haze or smoke must be dispersed by movement of air. Haze or smoke may be blown away; or heating during the day may cause convective mixing spreading the smoke or haze to a higher altitude, decreasing the concentration near the surface. At night or early morning, radiation fog or stratus clouds often combine with haze or smoke. The fog and stratus may clear rather rapidly during the day but the haze and smoke will linger. A heavy cloud cover above haze or smoke may block sunlight preventing dissipation; visibility will improve little, if any, during the day.

BLOWING RESTRICTIONS TO VISIBILITY

Strong wind lifts blowing dust in both stable and unstable air. When air is unstable, dust is lifted to great heights (as much as 15,000 feet) and may be spread over wide areas by upper winds. Visibility is restricted both at the surface and aloft. When air is stable, dust does not extend to as great a height as in unstable air and usually is not as widespread.

Dust, once airborne, may remain suspended and restrict visibility for several hours after the wind subsides. Figure 121 is a photograph of a dust storm moving in with an approaching cold front.

Blowing sand is more local than blowing dust; the sand is seldom lifted above 50 feet. However, visibilities within it may be near zero. Blowing sand

FIGURE 120. Smoke trapped in stagnant air under an inversion.

may occur in any dry area where loose sand is exposed to strong wind.

Blowing snow can be troublesome. Visibility at ground level often will be near zero and the sky may become obscured when the particles are raised to great heights.

FIGURE 121. Aerial photograph of blowing dust approaching with a cold front. The dust cloud outlines the leading surface of the advancing cold air.

PRECIPITATION

Rain, drizzle, and snow are the forms of precipitation which most commonly present ceiling and/or visibility problems. Drizzle or snow restricts visibility to a greater degree than rain. Drizzle falls in stable air and, therefore, often accompanies fog, haze, or smoke, frequently resulting in extremely poor visibility. Visibility may be reduced to zero in heavy snow. Rain seldom reduces surface visibility below 1 mile except in brief, heavy showers, but rain does limit cockpit visibility. When rain streams over the aircraft windshield, freezes on it, or fogs over the inside surface, the pilot's visibility to the outside is greatly reduced.

OBSCURED OR PARTIALLY OBSCURED SKY

To be classified as obscuring phenomena, smoke, haze, fog, precipitation, or other visibility restricting phenomena must extend upward from the surface. When the sky is totally hidden by the surface based phenomena, the ceiling is the vertical visibility from the ground upward into the obscuration. If clouds or part of the sky can be seen above the obscuring phenomena, the condition is defined as a partial obscuration; a partial obscuration does not define a ceiling. However, a cloud layer above a partial obscuration may constitute a ceiling.

An obscured ceiling differs from a cloud ceiling. With a cloud ceiling you normally can see the ground and runway once you descend below the cloud base. However, with an obscured ceiling, the obscuring phenomena restricts visibility between your altitude and the ground, and you have restricted slant visibility. Thus, you cannot always clearly see the runway or approach lights even after penetrating the level of the obscuration ceiling as shown in figure 122.

Partial obscurations also present a visibility problem for the pilot approaching to land but usually to a lesser degree than the total obscuration. However, be especially aware of erratic visibility reduction in the partial obscuration. Visibility along the runway or on the approach can instantaneously become zero. This abrupt and unexpected reduction in visibility can be extremely hazardous especially on touchdown.

IN CLOSING

In your preflight preparation, be aware of or alert for phenomena that may produce IFR or marginal VFR flight conditions. Current charts and special analyses along with forecast and prognostic charts are your best sources of information.

You may get your preflight weather from a briefer; or, you may rely on recorded briefings; and you always have your own inflight observations. No weather observation is more current or more accurate than the one you make through your cockpit

Common IFR Producers

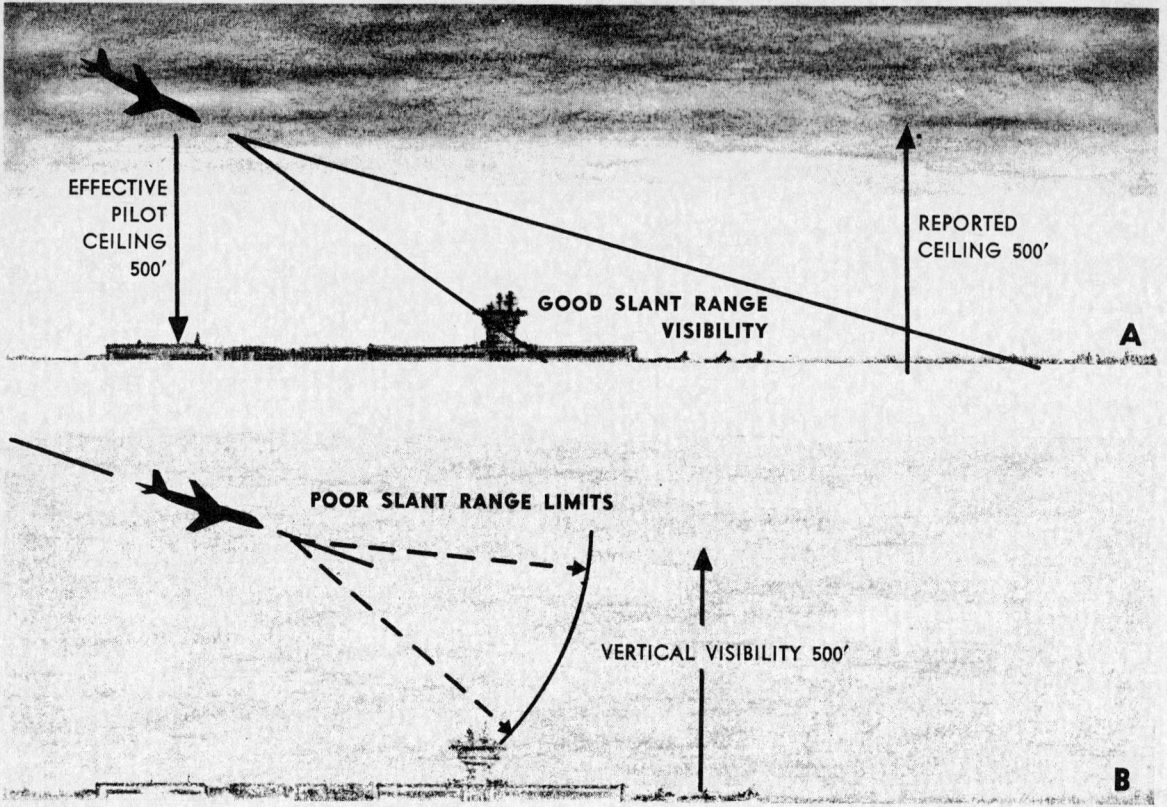

FIGURE 122. Difference between the ceiling caused by a surface-based obscuration (B) and the ceiling caused by a layer aloft (A). When visibility is not restricted, slant range vision is good upon breaking out of the base of a layer aloft.

window. In any event, your understanding of IFR producers will help you make better preflight and inflight decisions.

Do not fly VFR in weather suitable only for IFR. If you do, you endanger not only your own life but the lives of others both in the air and on the ground. Remember, the single cause of the greatest number of general aviation fatal accidents is "continued VFR into adverse weather." The most common cause is vertigo, but you also run the risk of flying into unseen obstructions. Furthermore, pilots who attempt to fly VFR under conditions below VFR minimums are violating Federal Aviation Regulations.

The threat of flying VFR into adverse weather is far greater than many pilots might realize. A pilot may press onward into lowering ceiling and visibility complacent in thinking that better weather still lies behind him. Eventually, conditions are too low to proceed; he no longer can see a horizon ahead. But when he attempts to turn around, he finds so little difference in conditions that he cannot re-establish a visual horizon. He continued too far into adverse weather; he is a prime candidate for vertigo.

Don't let an overwhelming desire to reach your destination entice you into taking the chance of flying too far into adverse weather. The IFR pilot may think it easier to "sneak" through rather than go through the rigors of getting an IFR clearance. The VFR pilot may think, "if I can only make it a little farther." If you can go IFR, get a clearance *before* you lose your horizon. If you must stay VFR, do a 180 while you still have a horizon. The 180 is not the maneuver of cowards. *Any pilot knows how to make a 180; a good pilot knows when.*

Be especially alert for development of:

1. Fog the following morning when at dusk temperature–dew point spread is 15° F or less, skies are clear, and winds are light.
2. Fog when moist air is flowing from a relatively warm surface to a colder surface.

3. Fog when temperature-dew point spread is 5° F or less and decreasing.
4. Fog or low stratus when a moderate or stronger moist wind is blowing over an extended upslope. (Temperature and dew point converge at about 4° F for every 1,000 feet the air is lifted.)
5. Steam fog when air is blowing from a cold surface (either land or water) over warmer water.
6. Fog when rain or drizzle falls through cool air. This is especially prevalent during winter ahead of a warm front and behind a stationary front or stagnating cold front.
7. Low stratus clouds whenever there is an influx of low level moisture overriding a shallow cold air mass.
8. Low visibilities from haze and smoke when a high pressure area stagnates over an industrial area.
9. Low visibilities due to blowing dust or sand over semiarid or arid regions when winds are strong and the atmosphere is unstable. This is especially prevalent in spring. If the dust extends upward to moderate or greater heights, it can be carried many miles beyond its source.
10. Low visibility due to snow or drizzle.
11. An undercast when you must make a VFR descent.

Expect little if any improvement in visibility when:
1. Fog exists below heavily overcast skies.
2. Fog occurs with rain or drizzle and precipitation is forecast to continue.
3. Dust extends to high levels and no frontal passage or precipitation is forecast.
4. Smoke or haze exists under heavily overcast skies.
5. A stationary high persists over industrial areas.

Part II
Aviation Weather Services

Chapter 13
THE AVIATION WEATHER SERVICE PROGRAM

Weather service to aviation is a joint effort of the National Weather Service (NWS), the Federal Aviation Administration (FAA), the military weather services, and other aviation oriented groups and individuals. Because of international flights and a need of world-wide weather, foreign weather services also have a vital input into our service. The NWS coordinates weather services, and many NWS products at all echelons are specifically for aviation.

Figure 1-1 is a flow diagram of weather data. This section follows the development and flow of observations, reports, and forecasts through the service to the users.

DATA FLOW

Longline communications providing the flow of data through the system are mostly teletypewriter and facsimile. Teletypewriter circuits collect and distribute weather reports, forecasts, and warnings. Facsimile transmits weather charts.

Each service outlet has a drop on an area teletypewriter circuit which provides complete data within a few hundred miles of the outlet but only sparse data for more remote areas. Reports and forecasts not routinely available on the local area circuit are available on a request/reply circuit.

National Weather Service facsimile distributes graphic weather analyses and prognostic charts. Most service outlets have facsimile.

OBSERVATIONS

Weather observations are measurements and estimates of existing weather. Observations are made at the surface and aloft. When recorded and transmitted, an observation becomes a report. These reports are the basis of all weather analyses and forecasts. Note in figure 1-1 that weather reports flow to all echelons in the aviation weather service.

Surface Observations

Surface aviation observations include weather elements pertinent to flying. A network of airport stations provides routine up-to-date aviation weather reports. Most stations are either NWS or FAA; however, military services and contracted civilians

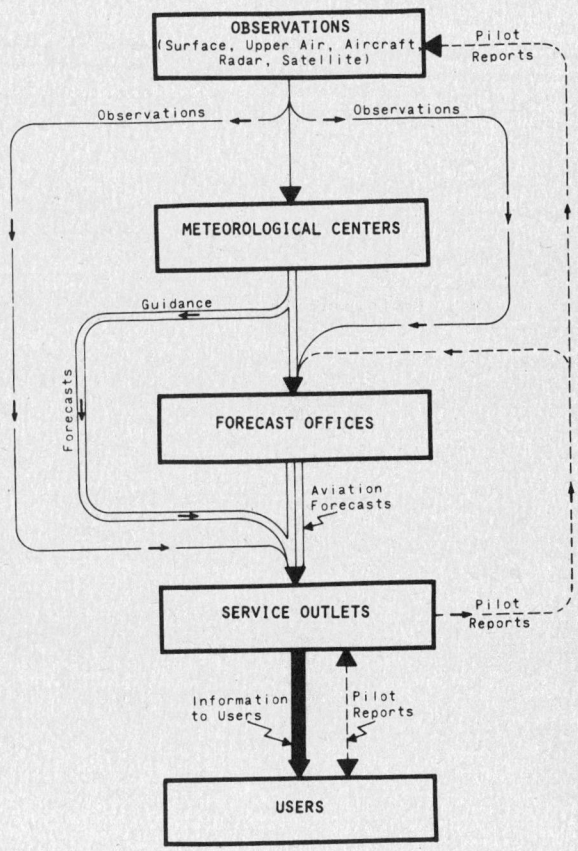

FIGURE 1-1. Data flow in the aviation weather service. All data is available through service outlets. Note the important feedback of pilot reports.

complete the network. All official civilian weather observers must be certified by the National Weather Service.

Radar Observations

Precipitation reflects radar signals, and the reflected signals are displayed as echoes on the radar scope. NWS radar covers nearly all the U.S. east of the Rocky Mountains. Radar coverage over the remainder of the U.S. is largely by Air Route Traffic Control radars. Thus, except for some western mountainous terrain, radar coverage is nearly complete over the contiguous 48 States. Figure 1-2 maps the radar observing network.

The Aviation Weather Service Program

Other Observations

Many other observations have a significant input to the aviation weather service. Upper air observations taken twice daily at specified stations furnish temperature, humidity, pressure, and wind, often to heights above 100,000 feet. Weather satellites scan the Earth providing cloud pictures. These pictures are especially useful in remote areas. Pilots themselves are a vital source of weather observations. In fact, aircraft in flight are the only means of directly observing turbulence, icing, and height of cloud tops.

METEOROLOGICAL CENTERS AND FORECAST OFFICES

Meteorological centers collect and analyze data and prepare forecasts on a national, hemispheric, or global basis. NWS forecast offices prepare forecasts which are generally more detailed.

National Meteorological Center (NMC)

The National Meteorological Center (NMC) of the NWS is the hub of all weather processing. From worldwide weather reports it prepares forecasts and charts of observed and forecast weather. Many of the charts are computer prepared. Others are computer outputs adjusted and annotated by meteorologists. A few are manually prepared by forecasters.

Some NMC products are specifically for aviation. For example, NMC prepares the winds and temperatures aloft forecast. Figure 1-3 is the network of forecast winds and temperatures for the contiguous 48 States.

National Hurricane Center (NHC)

The NWS National Hurricane Center (NHC) develops hurricane forecasting techniques and issues hurricane forecasts for the Atlantic, the Caribbean, the Gulf of Mexico, and adjacent land areas. It is located at Miami, Florida. Hurricane warning centers at San Francisco and Honolulu issue warnings for the eastern and central Pacific.

National Severe Storms Forecast Center (NSSFC)

The NWS National Severe Storms Forecast Center (NSSFC) prepares forecasts of severe convective storms over the contiguous 48 States. It is located at Kansas City, Missouri near the heart of the area most frequented by severe thunderstorms and tornadoes.

National Environmental Satellite Service (NESS)

The National Environmental Satellite Service (NESS) directs the weather satellite program. Through newly developed radiation measuring techniques, it contributes directly to NMC processing. Satellite cloud photographs are available at field facilities by facsimile and at some stations by direct picture reception.

Weather Service Forecast Office (WSFO)

A Weather Service Forecast Office (WSFO) issues forecasts, advisories and warnings for its area. Figure 1-4 (Alaska, fig. 1-10) shows locations of WSFOs, their areas of responsibility, and the airports for which each office prepares terminal forecasts. Selected WSFOs issue area forecasts. Figure 1-5 (Alaska, fig. 1-10) shows locations of these offices and their forecast areas.

Weather Service Office (WSO)

A Weather Service Office (WSO) prepares local forecasts and warnings and provides general weather service. It shoulders part of the terminal forecast responsibility. A WSO can adjust the local terminal forecast for a period of two hours or less.

SERVICE OUTLETS

A weather service outlet as used here is any facility, either government or non-government, that provides aviation weather service. This section discusses only FAA and NWS outlets.

Flight Service Stations (FSS)

The FAA Flight Service Station (FSS) provides more aviation weather briefing service than any other government service outlet. It provides preflight and inflight briefings, makes scheduled and unscheduled weather broadcasts, and furnishes weather advisories to known flights in the FSS area.

Selected FSSs also provide transcribed weather dissemination to aid inflight and preflight briefing. By listening to the transcriptions, you can assess any further need for more detailed briefing. There are two types of transcriptions—(1) Transcribed Weather Broadcast (TWEB) and (2) Pilot's Automatic Telephone Weather Answering Service (PATWAS).

The TWEB is a continuous broadcast on low/medium frequencies (200 to 415 kHz) and selected VORs (108.0 to 117.95 mHz). PATWAS is a recorded telephone briefing service. TWEB and PATWAS transcriptions are on a route concept. A few selected stations also prepare transcriptions for a local area—usually within a 50 nautical mile radius of the station.

Order and content of the transcription are as follows:
1. Synopsis
2. Flight Precautions
3. Route Forecasts
4. Outlook (Optional)
5. Winds Aloft Forecast
6. Radar Reports
7. Surface Weather Reports
8. Pilot Reports
9. Notice to Airmen (NOTAMs)

The first five items are forecasts prepared by the NWS and are discussed in detail in section 4. The synopsis and route forecasts are prepared especially for TWEB and PATWAS. Flight precautions, outlook, and winds aloft are adapted respectively from inflight advisories, area forecasts, and the NMC winds aloft forecast. Radar reports and pilot reports are discussed in section 3. Surface reports are the subject of section 2.

Figure 1–6 maps locations of TWEB outlets; and figure 1–7, PATWAS locations. Figure 1–8 shows routes for which forecasts are prepared. The AIRMAN'S INFORMATION MANUAL, Part 3, shows the availability of TWEB at a facility and lists the frequency. Part 2 shows PATWAS telephone numbers in the directory of FSS and Weather Service telephone numbers.

The enroute flight advisory service (Flight Watch) is a weather service on a common frequency of 122.0 mHz from selected FSSs. The Flight Watch specialist maintains a continuous weather watch, provides time-critical assistance to enroute pilots facing hazardous or unknown weather, and may recommend alternate or diversionary routes. Additionally, Flight Watch is a focal point for rapid receipt and dissemination of pilot weather reports. Flight Watch is operational on the West Coast as shown in figure 1–11 and will be expanded throughout the U.S. during the next few years. To avail yourself of this service, call "FLIGHT WATCH" on 122.0 mHz.

Air Route Traffic Control Center (ARTCC)

FAA Air Route Traffic Control Centers (ARTCC) advise air traffic under their control of significant weather. The controller may also advise aircraft of forecast terminal conditions that may cause a change in flight plan.

Terminal Control Facility

The FAA terminal controller becomes familiar with and remains aware of current weather information needed to perform air traffic control duties. He informs arriving and departing aircraft of pertinent local weather conditions. He shares responsibility with the NWS for reporting visibility observations at many facilities. At other facilities he has the full responsibility for observing, reporting, and classifying aviation weather elements.

Weather Service Office

NWS Weather Service Offices provide weather briefings in areas not served by Flight Service Stations and provide local warnings to aviation. They furnish backup assistance to FAA service outlets.

Weather Service Forecast Office

NWS Weather Service Forecast Offices provide some selective pilot briefings and supply backup service to FAA outlets. When getting a briefing from an FSS, you may, if necessary, request a telephone "patch in" to the WSFO forecaster. A few WSFOs make and record PATWAS.

USERS

The ultimate users of the aviation weather service are pilots and dispatchers. Maintenance personnel also may use the service in protecting idle aircraft against storm damage. As a user of the service, you also contribute to it. Send pilot weather reports (PIREPs) to help your fellow pilots, briefers and forecasters. The service can be no better or more complete than the information that goes into it.

In the interest of safety, you should get a complete briefing before each flight. If you have L/MF radio, you can get a preliminary briefing by listening to the TWEB at your home or place of business. If you have no radio and PATWAS is available, dial PATWAS for a briefing. If, after listening to the TWEB or PATWAS, you desire additional information, contact an FSS or WSO for a more complete briefing. The AIRMAN'S INFORMATION MANUAL, and often the local telephone directory, lists numbers to call for aviation weather.

How to Get a Good Weather Briefing

When requesting a briefing, make known you are a pilot. Give clear and concise facts about your flight:
1. Aircraft number or your name
2. Destination, route, and planned altitude
3. Whether flying VFR or IFR
4. Departure time
5. Time enroute or time of arrival
6. Intermediate stops if any

With this background, the briefer can proceed directly with the briefing and concentrate on weather relevant to your flight.

The weather briefing you receive should include:
1. Hazardous weather if any (you may elect to cancel at this point)
2. Weather synopsis (positions of lows, fronts, etc.)
3. Forecast (enroute and destination)
4. Alternate routes (if any)
5. Forecast winds aloft

The FSSs and WSOs are to serve you. You should not hesitate to discuss factors that need elaboration or to ask questions. You have a complete briefing only when you have a clear picture of the weather to expect. It is to your advantage to make a final weather check immediately before departure if at all possible.

Request/Reply Service

The request/reply service mentioned earlier is available at all FSSs, WSOs, and WSFOs. You may request through the service any reports or forecasts not routinely available at your service outlet. Included in the request/reply are route forecasts used in TWEB and PATWAS recorded briefings. You can request a forecast for any numbered route shown in figure 1-8 or any of the longer cross-country routes shown in figure 1-9.

Have an Alternate Plan of Action

When weather is questionable, get a picture of expected weather over a broader area. Preplan a route to take you rapidly away from the weather if it goes sour. When you fly into weather through which you cannot safely continue, you must act quickly. Without preplanning, you may not know the best direction to turn; a wrong turn could lead to disaster. A preplanned diversion beats panic. Better be safe than sorry.

The Aviation Weather Service Program

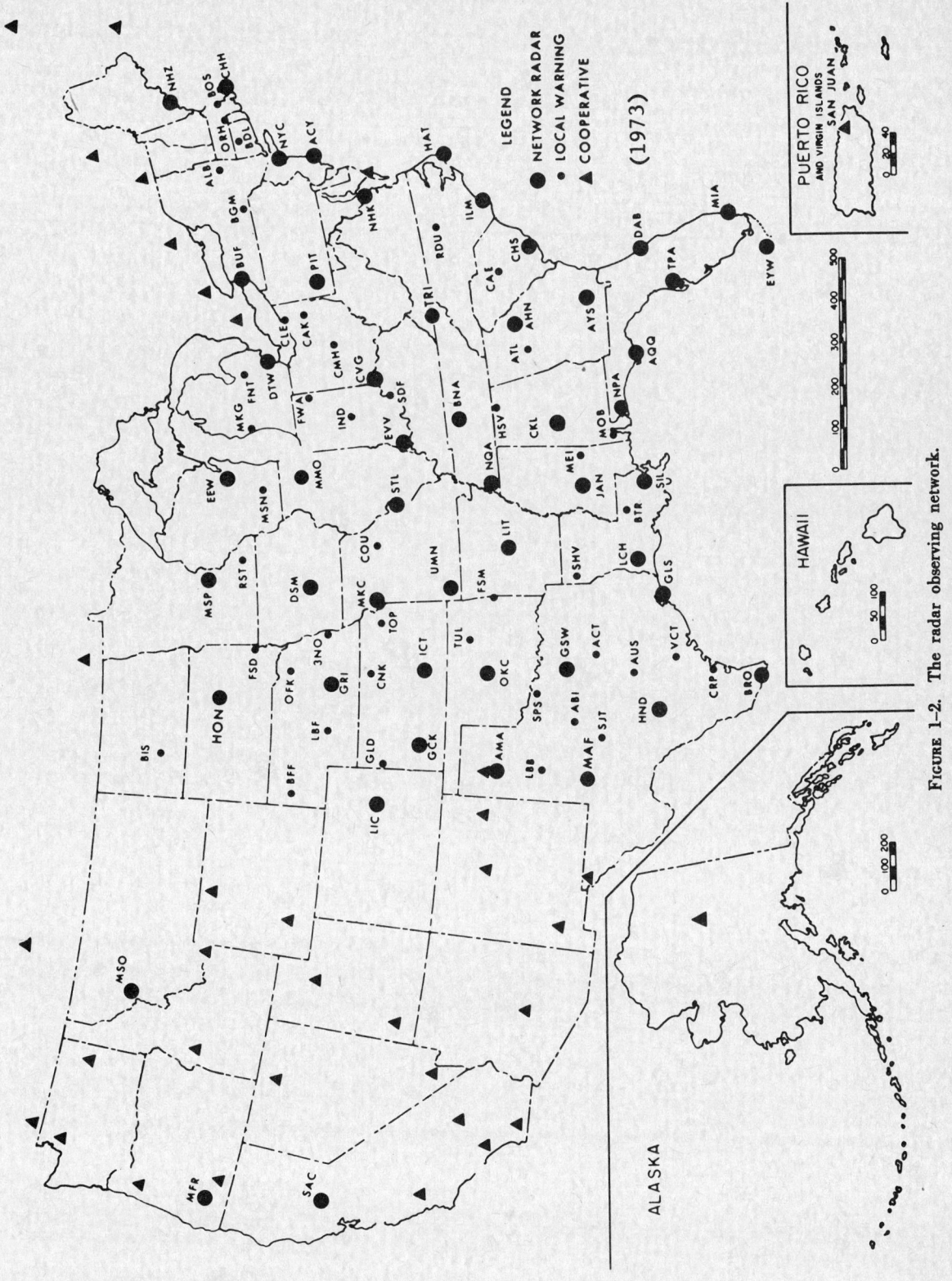

FIGURE 1-2. The radar observing network.

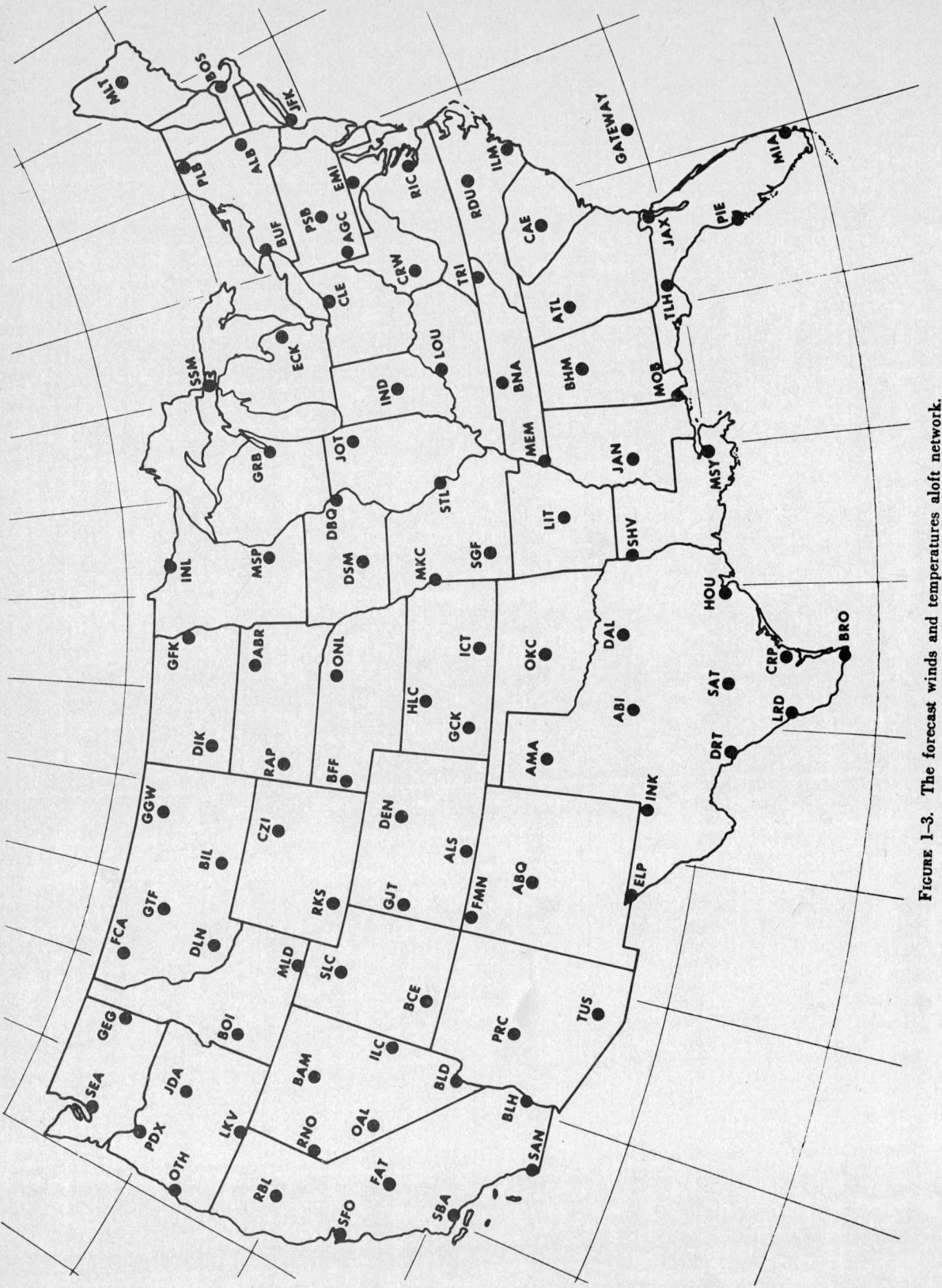

FIGURE 1-3. The forecast winds and temperatures aloft network.

The Aviation Weather Service Program

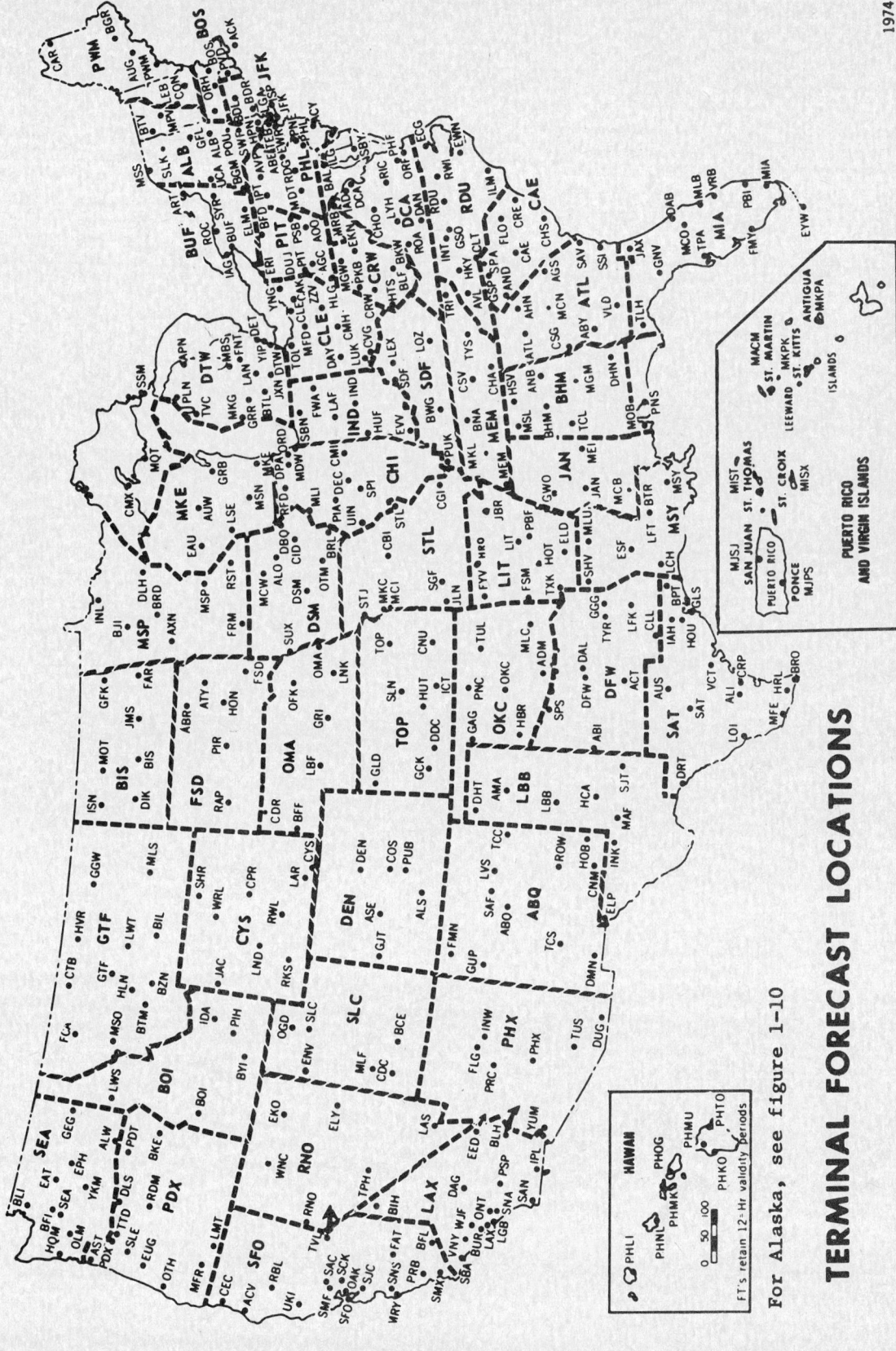

FIGURE 1-4. Locations of WSFOs, their areas of responsibility, and airports for which each prepares terminal forecasts.

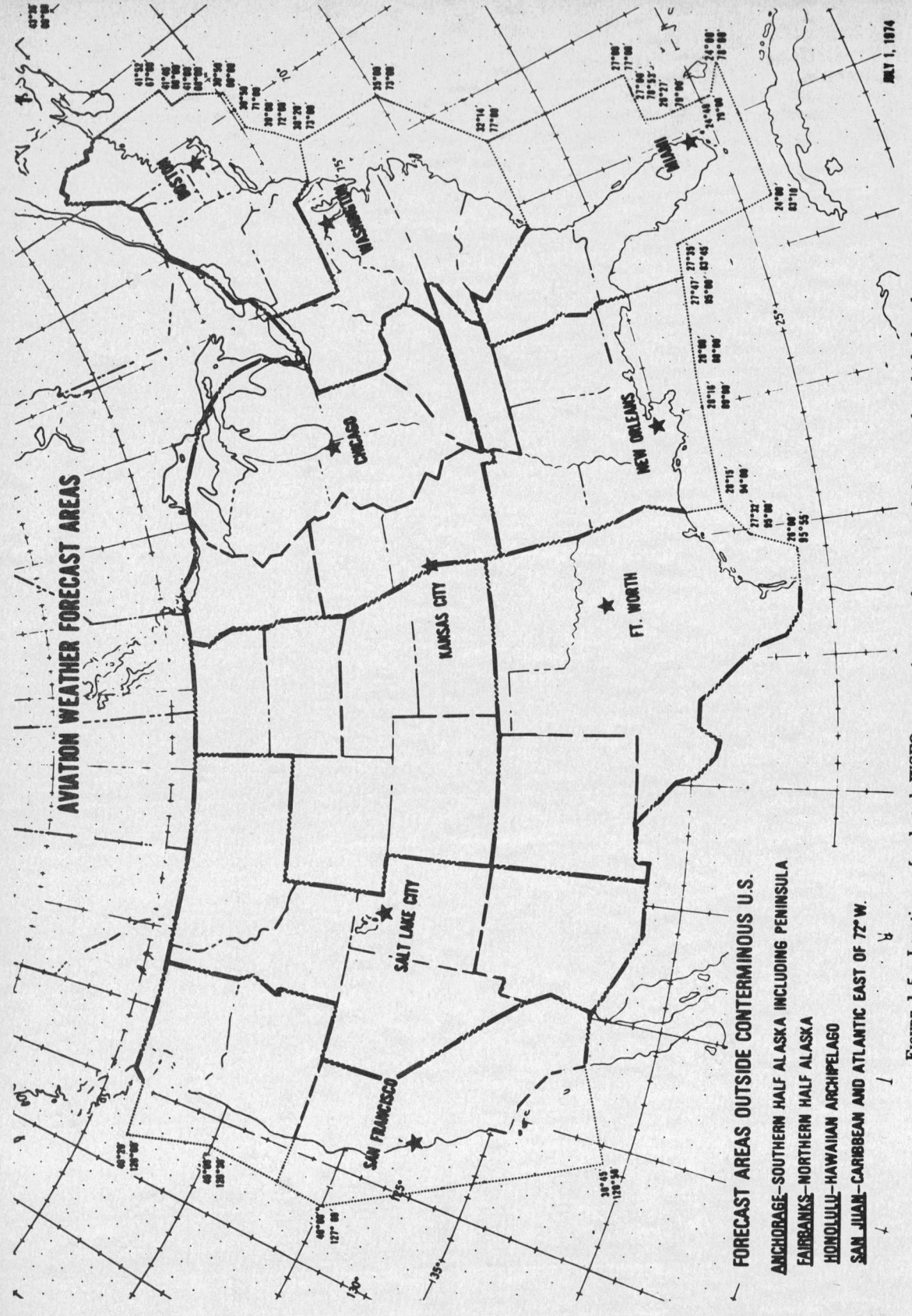

FIGURE 1-5. Locations of selected WSFOs preparing area forecasts and the areas for which they forecast.

The Aviation Weather Service Program

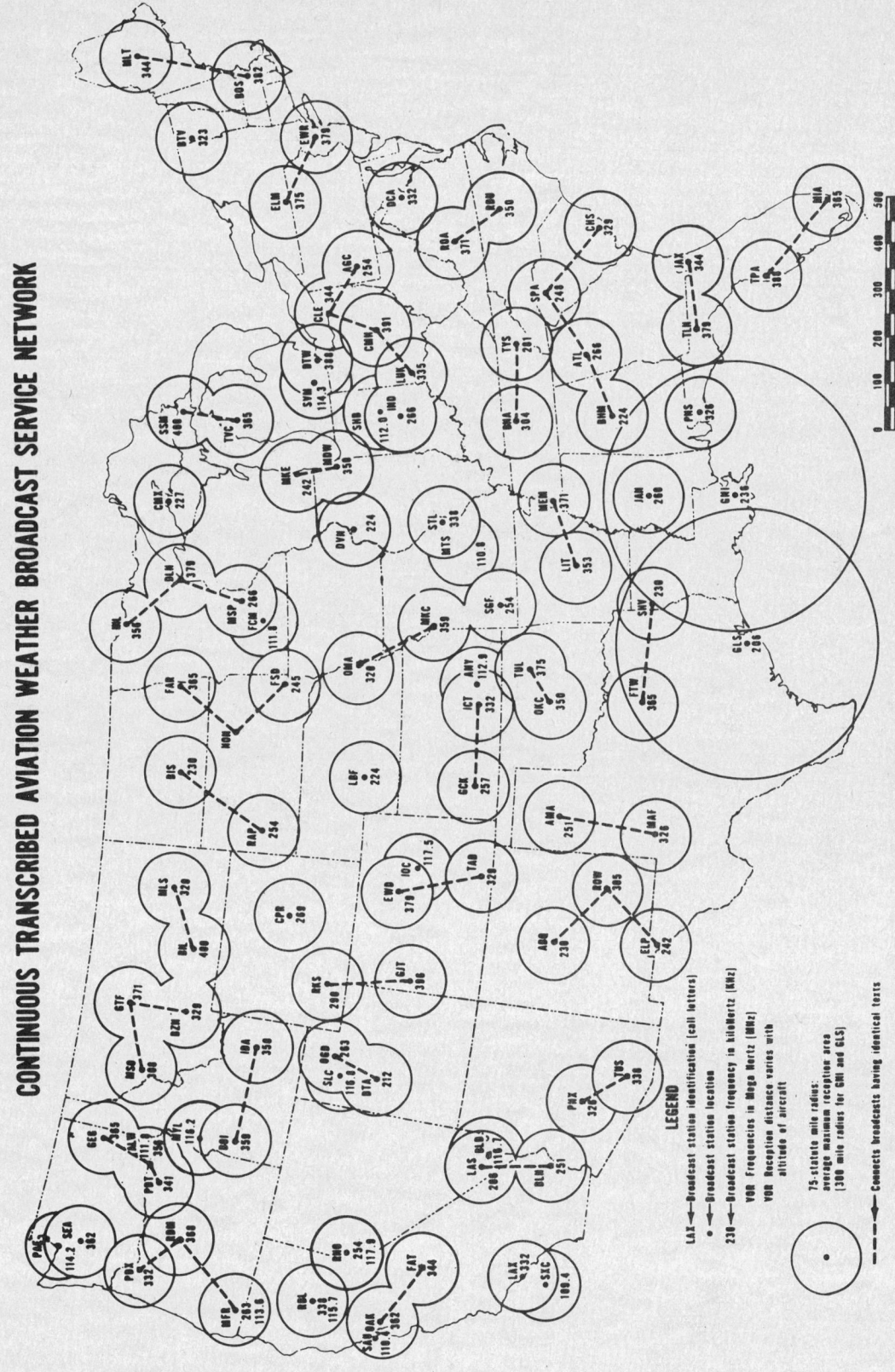

FIGURE 1-6. Locations of selected FSSs providing Transcribed Weather Broadcasts (TWEBs).

143

The Aviation Weather Service Program

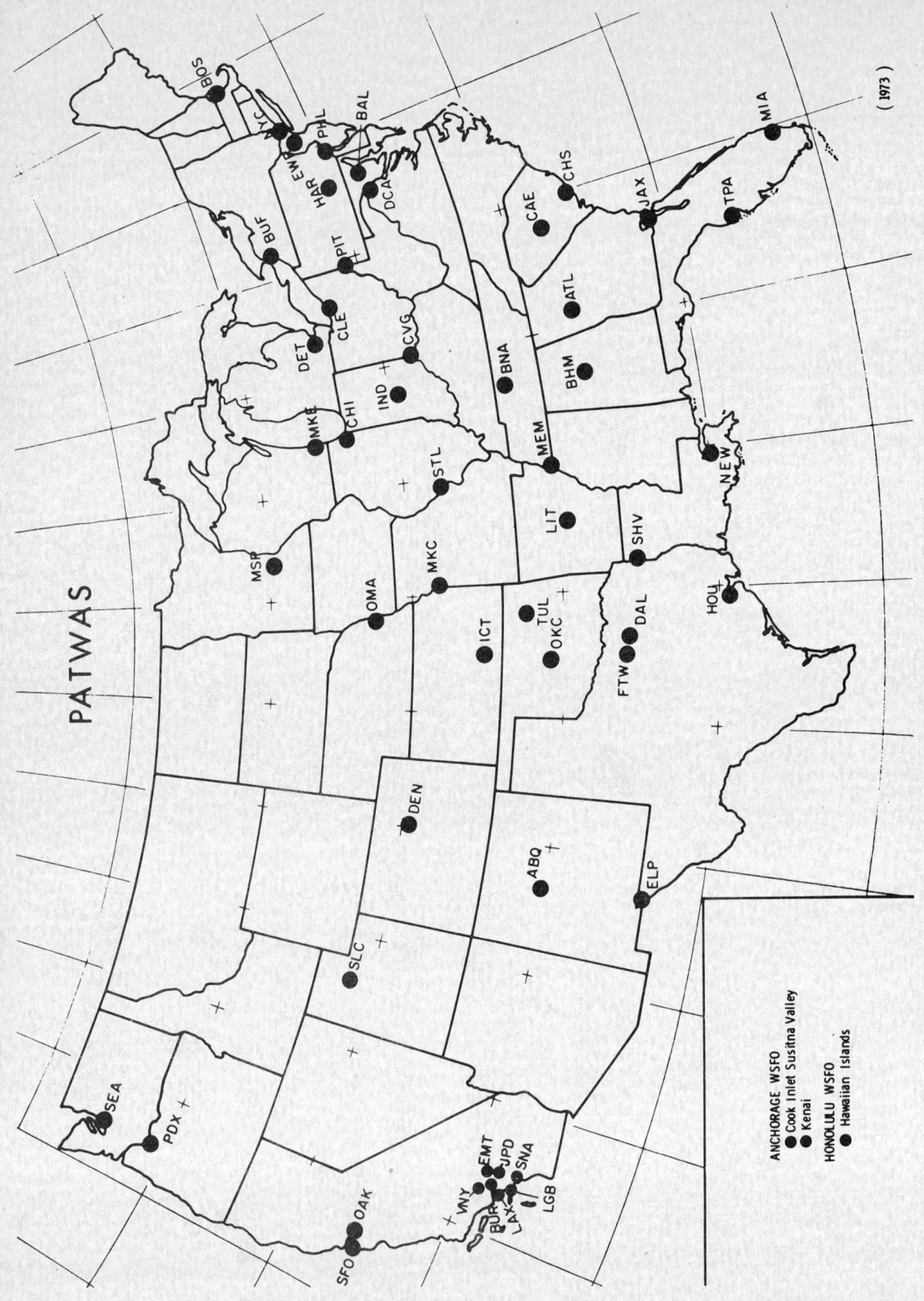

FIGURE 1-7. Locations of Pilot's Automatic Telephone Weather Answering Service (PATWAS).

The Aviation Weather Service Program

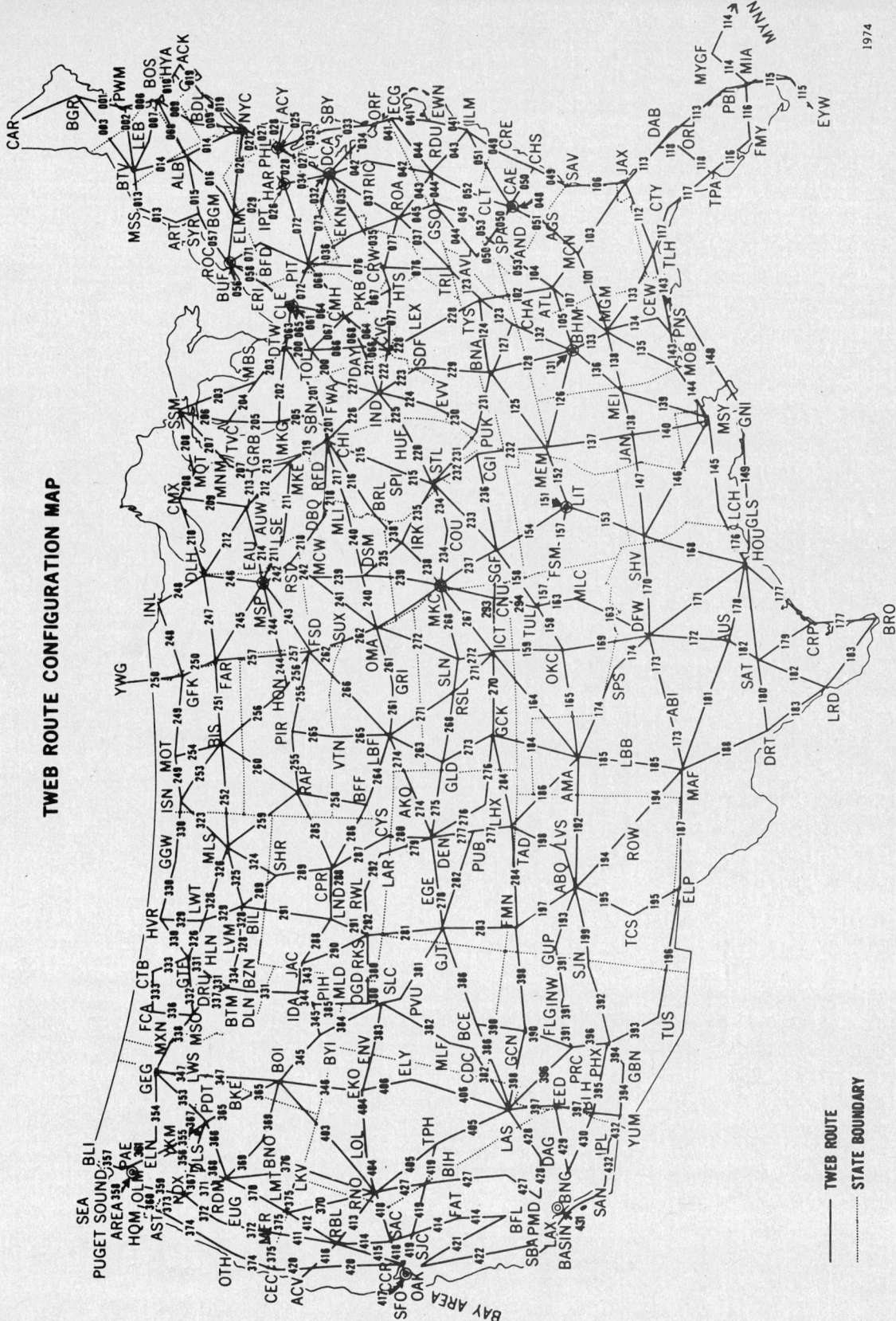

FIGURE 1-8. Numbered routes for which TWEB route forecasts are prepared. Route forecasts may be requested through request/reply service.

CROSS COUNTRY TWEB ROUTES AND RL REQUEST REPLY NUMBERS

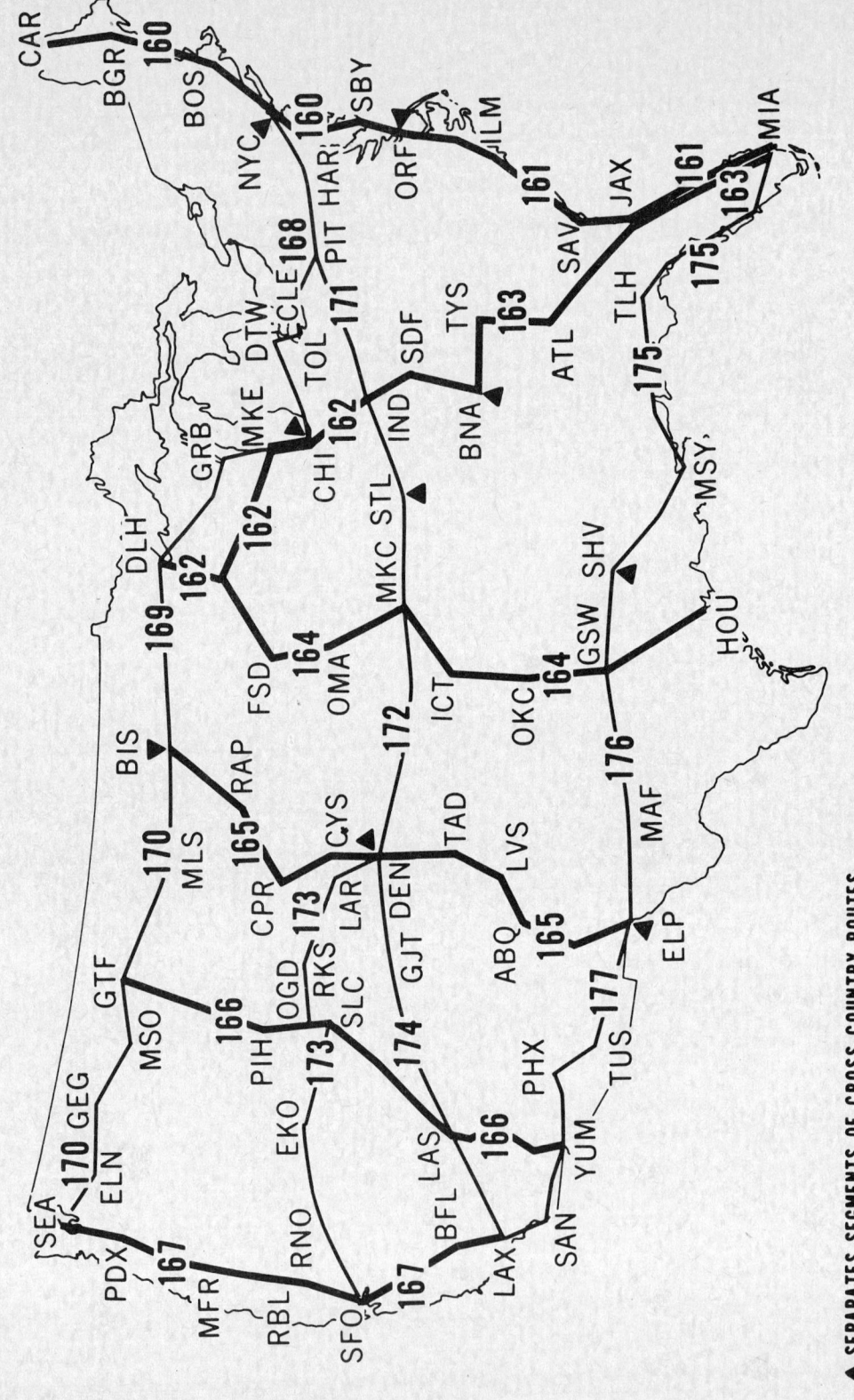

▲ SEPARATES SEGMENTS OF CROSS COUNTRY ROUTES

FIGURE 1-9. Cross country numbered routes for which route forecasts are available through request/reply service.

The Aviation Weather Service Program

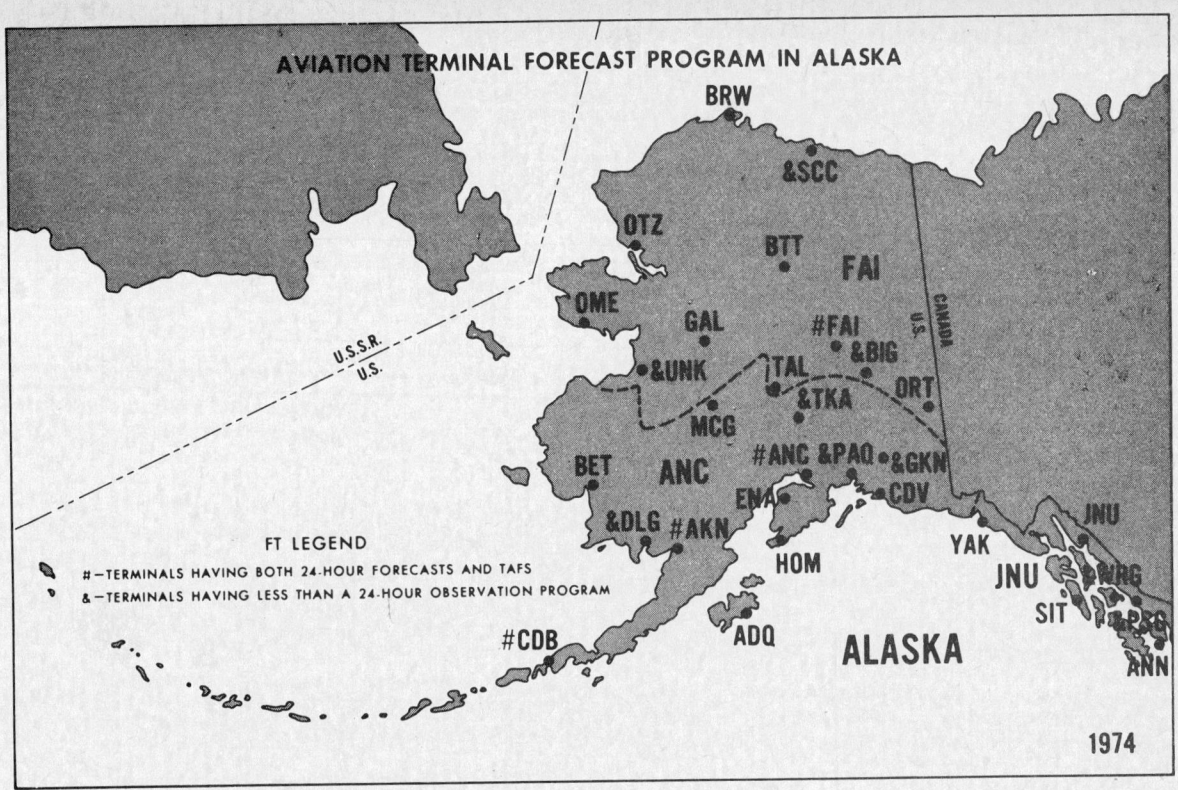

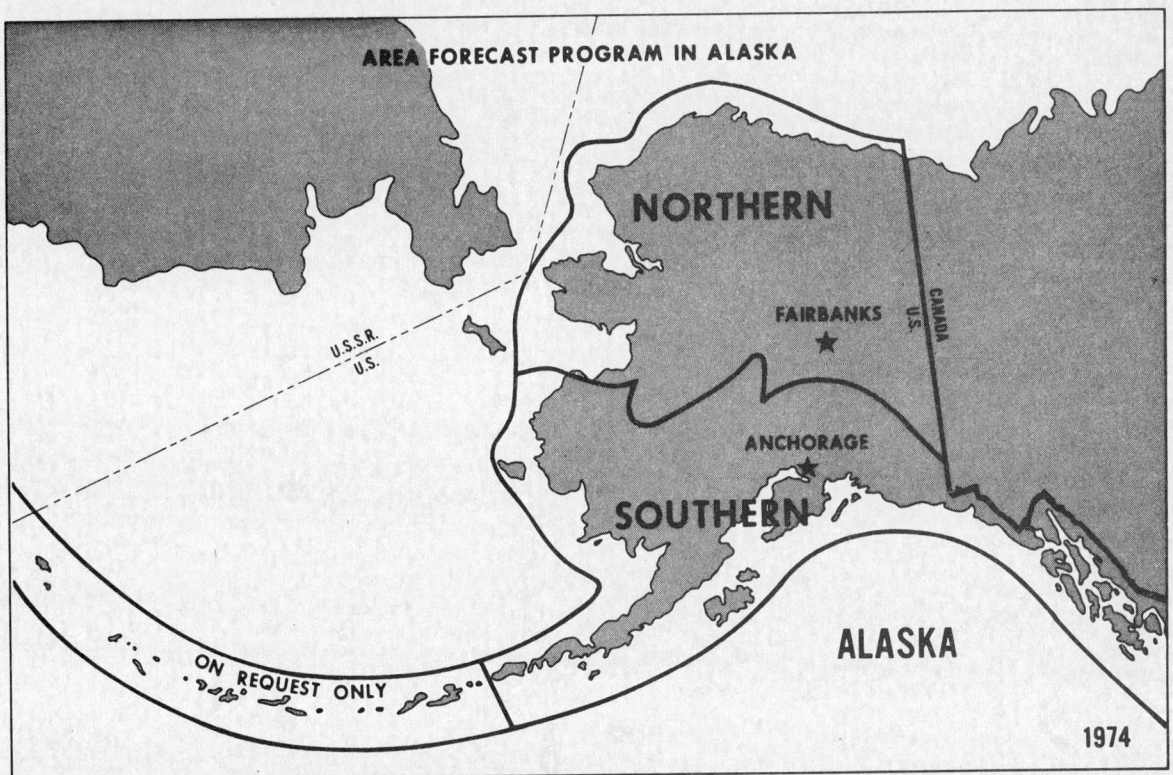

FIGURE 1-10. Alaska WSFOs, locations for which terminal forecasts are prepared (top) and forecast areas (bottom).

The Aviation Weather Service Program

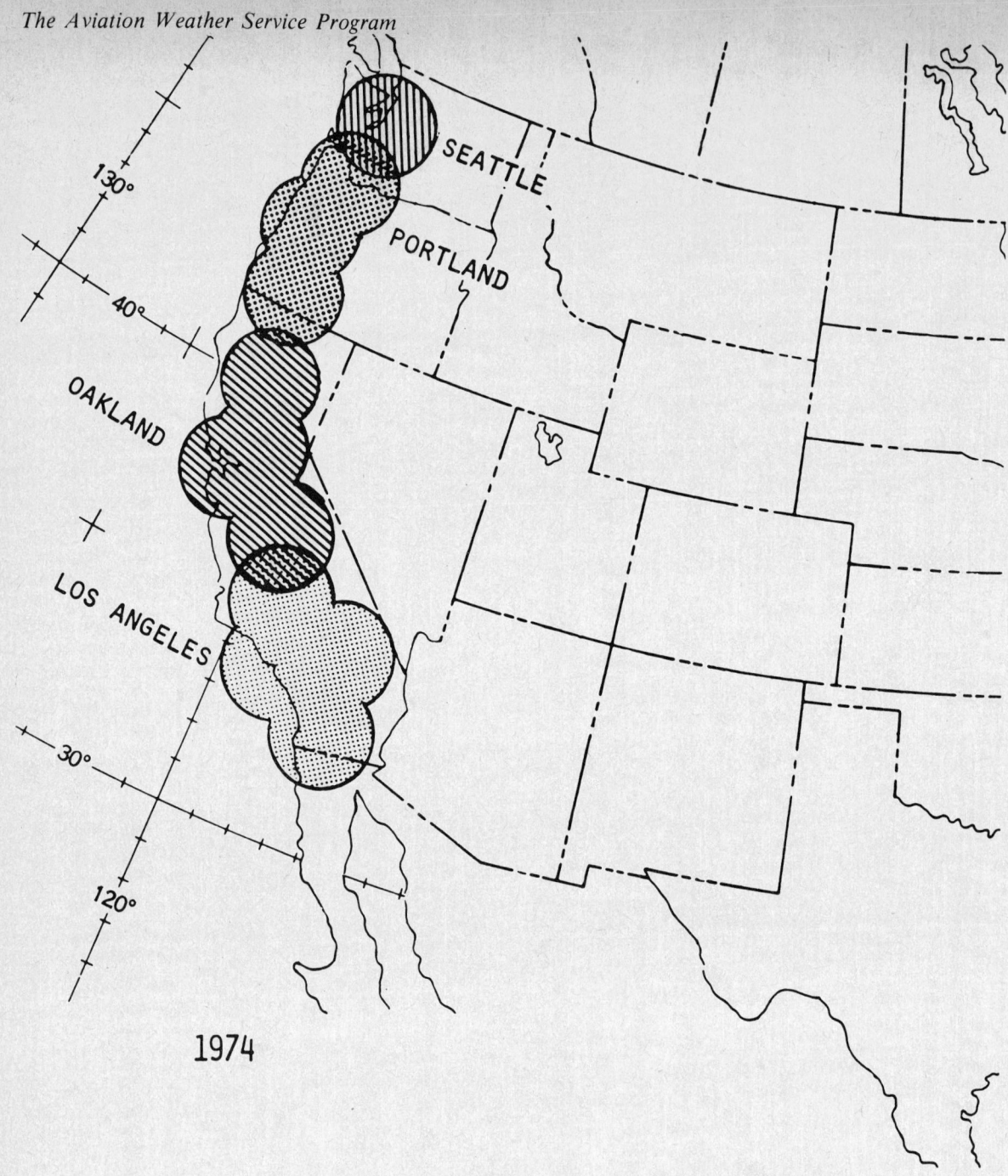

FIGURE 1-11. Four operational Enroute Flight Advisory Service (Flight Watch) facilities and approximate reception areas. All facilities except Seattle have remote transmitter sites. An aircraft at 5,000 feet can receive a transmission to a distance of about 80 miles from any central or remote site.

Chapter 14
SURFACE AVIATION WEATHER REPORTS

When an observation is recorded and transmitted, it is a weather *report*. A surface aviation weather report contains some or all of the following elements:

1. Station Designator
2. Type and Time of Report
3. Sky Condition and Ceiling
4. Visibility
5. Weather and Obstructions to Vision
6. Sea Level Pressure
7. Temperature and Dew Point
8. Wind Direction, Speed, and Character
9. Altimeter Setting
10. Remarks

```
INK   CLR 15 106/77/63/1112G18/000
BOI   150SCT 30 181/62/42/1304/015
LAX   7SCT 250SCT 6KH 129/60/59/2504/991→LAX ↘6/38
MDW SP −X M7OVC 11/2R+F 990/63/61/3205/980/RF2 RB12
JFK SP W5X 1/2F 180/68/64/1804/006/R04RVR22V30 TWR VSBY1/4
```

Those elements not occurring at observation time or not pertinent to the observation are omitted from the report. When an element should be included but is unavailable, the letter "M" is transmitted in lieu of the missing element. Those elements that are included are transmitted in the above sequence.

Following are five reports as transmitted on teletypewriter. These reports are used in discussing the above 10 elements. If you have this reference in a loose leaf binder, you will find it helpful to remove this page and keep it before you as you proceed through the discussion.

STATION DESIGNATOR

The station designator is the three-letter location identifier for the reporting station. These five reports are from Wink, Texas (INK); Boise, Idaho (BOI); Los Angeles, California (LAX); Chicago Midway Airport, Illinois (MDW); and John F. Kennedy Airport, New York City (JFK).

TABLE 2-1. Summary of sky cover designators

Designator	Meaning	Spoken
CLR	CLEAR. (Less than 0.1 sky cover.)	CLEAR
SCT	SCATTERED LAYER ALOFT. (0.1 through 0.5 sky cover.)	SCATTERED
BKN*	BROKEN LAYER ALOFT. (0.6 through 0.9 sky cover.)	BROKEN
OVC*	OVERCAST LAYER ALOFT. (More than 0.9, or 1.0 sky cover.)	OVERCAST
−SCT	THIN SCATTERED. ⎫ At least ½ of the sky cover aloft is	THIN SCATTERED
−BKN	THIN BROKEN. ⎬ transparent at and below the level	THIN BROKEN
−OVC	THIN OVERCAST. ⎭ of the layer aloft.	THIN OVERCAST
X*	SURFACE BASED OBSTRUCTION. (All of sky is hidden by surface based phenomena.)	SKY OBSCURED
−X	SURFACE BASED PARTIAL OBSCURATION. (0.1 or more, but not all, of sky is hidden by surface based phenomena.)	SKY PARTIALLY OBSCURED

* Sky condition represented by this designator may constitute a ceiling layer.

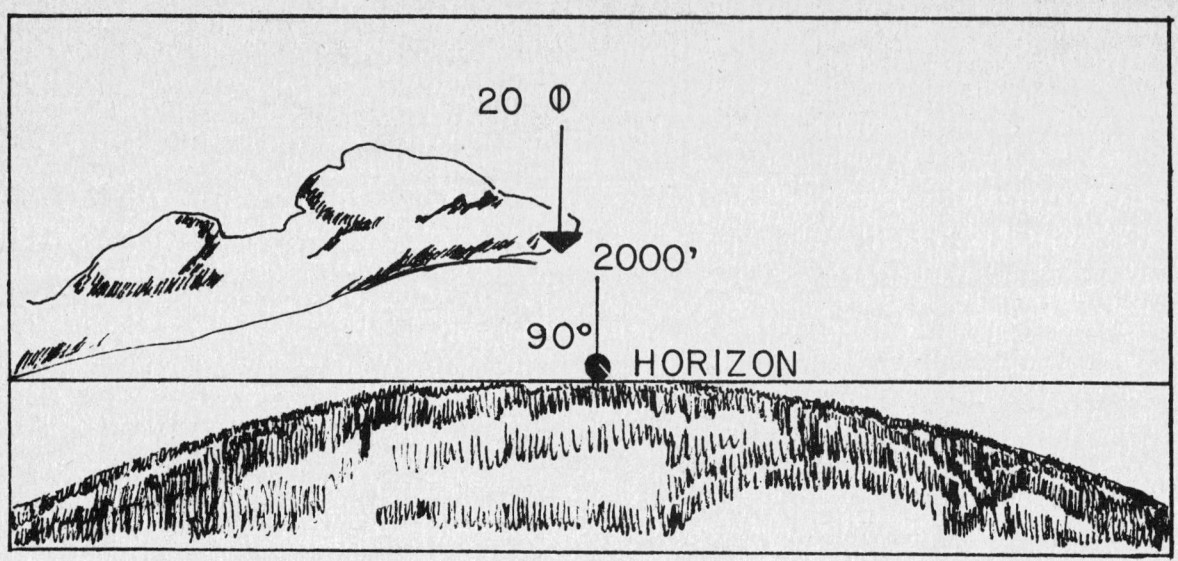

FIGURE 2-1. Scattered sky cover by a single advancing layer. Scattered is 5/10 or less sky cover (5/10 in this example).

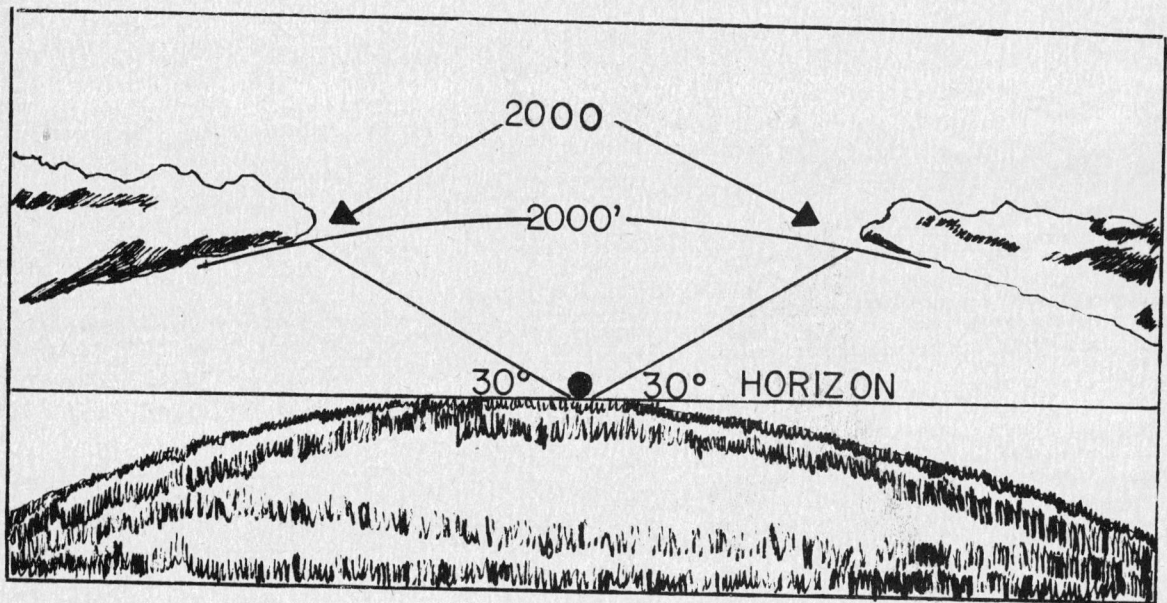

FIGURE 2-2. Scattered sky cover by a single layer surrounding the station (5/10 covered in this example).

TYPE AND TIME OF REPORT

The two basic types of reports are:

1. Record hourly reports of observations taken on the hour and

2. Special reports of observations taken when needed to report significant changes in weather.

Record *hourly* reports are transmitted in sequenced collectives and are identified by sequence headings. The first three reports are of this type (INK, BOI, and LAX). A record *special* is a record hourly that reports a significant change in weather. It is identified by the letters "SP" as shown in the reports from MDW and JFK. The special identifier is the only type-of-report entry that ever appears in an hourly collective. A report transmitted out of sequence must convey the time and type of the observation. These out-of-sequence reports are discussed later.

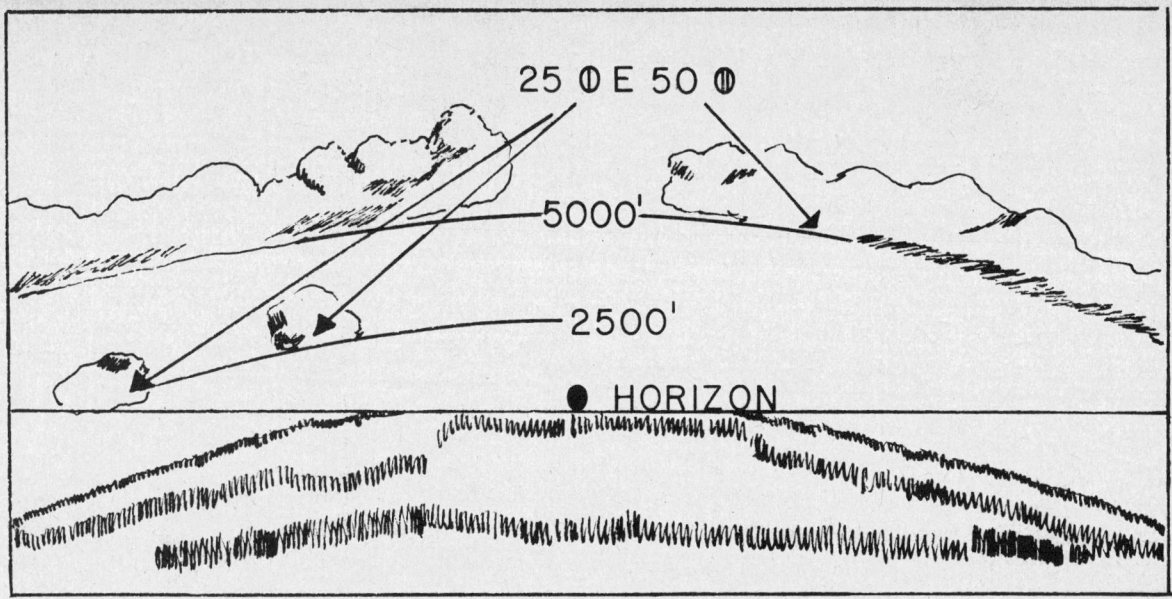

Figure 2-3. Summation of cloud cover in multiple layers.

SKY CONDITION AND CEILING

A clear sky or a layer of clouds or obscuring phenomena *aloft* is reported by one of the first seven *sky cover designators* in table 2-1. A layer is defined as clouds or obscuring phenomena with the base at approximately the same level. Height of the base of a layer precedes the sky cover designator. Height is in hundreds of feet *above ground level*.

Note that INK is reporting sky clear. No height precedes the designator since no sky cover is reported. BOI reports a scattered layer at 15,000 feet above the station. Figures 2-1 and 2-2 illustrate single layers of scattered clouds.

When more than one layer is reported, layers are in ascending order of height. For each layer above a lower layer or layers, the sky cover designator for that layer represents the *total sky* covered by that layer and all lower layers. LAX reports two layers—a scattered layer at 700 feet and a higher layer at 25,000 feet. Total coverage of the two layers does not exceed 5/10 coverage, so the upper layer also is reported as scattered. Figures 2-3 and 2-4 illustrate cloud cover of multiple layers.

"Transparent" sky cover is clouds or obscuring phenomena aloft through which blue sky or higher sky cover is visible. As explained in table 2-1, a scattered, broken, or overcast layer may be reported as "thin". To be classified as thin, a layer must be half or more transparent, and remember that sky cover of a layer includes all sky cover below the layer. For example, if at LAX the sky had been visible through half or more of the total sky cover reported by the higher layer, the report would have been

LAX 7SCT 250—SCT etc.

Any phenomena *based at the surface* and hiding all or part of the sky is reported as SKY OBSCURED* or SKY PARTIALLY OBSCURED* as explained in table 2-1. An obscuration or partial obscuration may be precipitation, fog, dust, blowing snow, etc. No height value precedes the designator for partial obscuration since vertical visibility is not restricted overhead. A height value precedes the designator for an obscuration and denotes vertical visibility into the phenomena.

Ceiling is defined as:

1. Height of the lowest layer of clouds or obscuring phenomena aloft that is reported as broken or overcast and not classified as thin, or

2. Vertical visibility into a surface-based obscuring phenomena that hides all the sky.

Now look at the reports from MDW and JFK. MDW reports a partial obscuration and an overcast at 700 feet. The overcast constitutes a ceiling at 700 feet. Note also that the height of this ceiling layer is preceded by the letter "M". JFK reports a total obscuration, and the height value preceding the sky cover designator represents 500 feet vertical visibility into the obscuring phenomenon. Height of the ceiling value is preceded by the letter "W". The "M" and "W" are "ceiling designators".

---

* Descriptions in capital letters are the usual phraseology in which these reports are broadcast.

Surface Aviation Weather Reports

Table 2-2. Ceiling designators

Coded	Meaning	Spoken
M	MEASURED. Heights determined by ceilometer, ceiling light, cloud detection radar, or by the unobscured portion of a landmark protruding into ceiling layer. (Figure 2-5 illustrates the principle of the ceilometer.)	MEASURED CEILING
E	ESTIMATED. Heights determined from pilot reports, balloons, or other measurements not meeting criteria for measured ceiling.	ESTIMATED CEILING
W	INDEFINITE. Vertical visibility into a surface based obstruction. Regardless of method of determination, vertical visibility is classified as an indefinite ceiling.	INDEFINITE CEILING

A *ceiling designator* always precedes the height of the ceiling layer. Table 2-2 lists and explains ceiling designators. At MDW the ceiling height was measured. JFK had an indefinite ceiling which was vertical visibility into a surface based obscuration.

The sky cover and ceiling as determined from the ground represent as nearly as possible what the pilot should experience in flight. In other words, a pilot flying at or above the reported ceiling layer aloft should see less than half the surface below him. The pilot descending through a surface

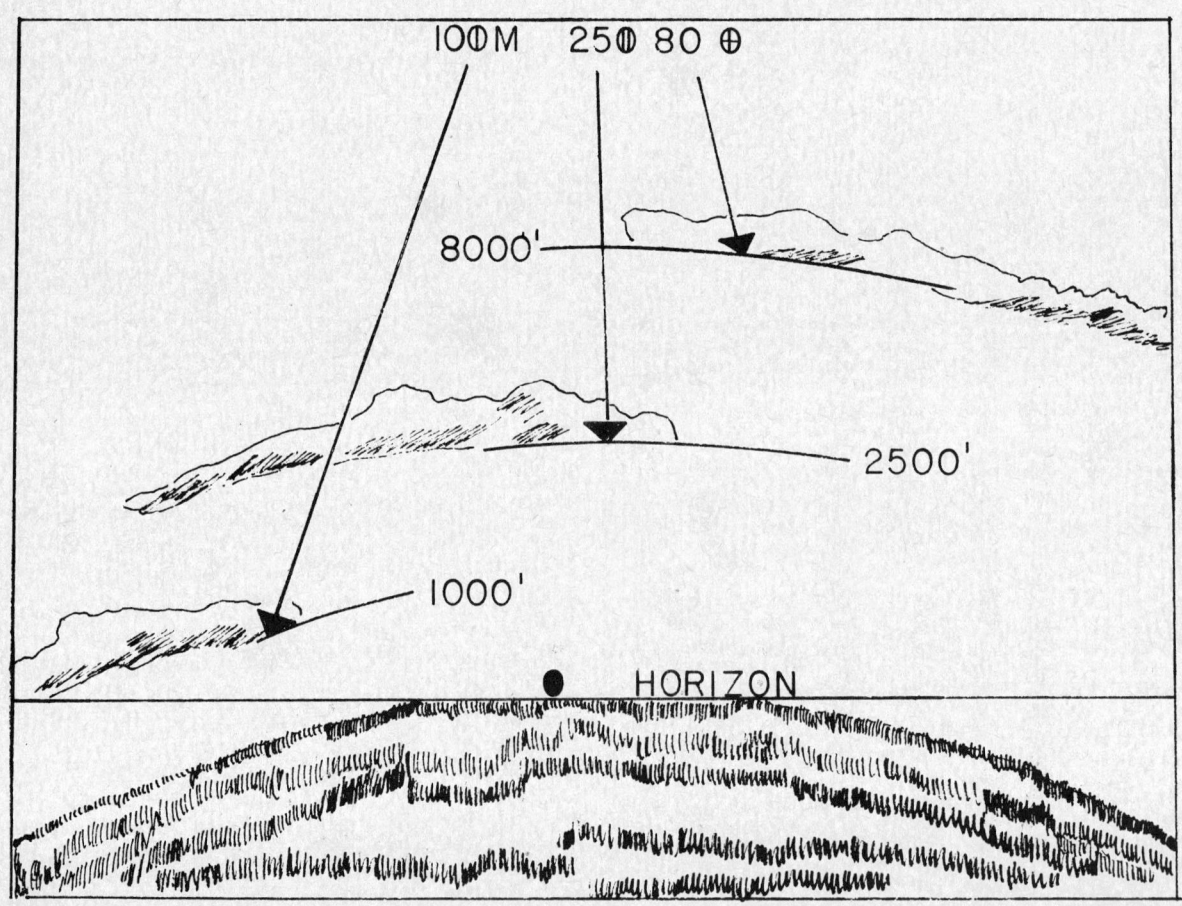

FIGURE 2-4. Summation of cloud cover in multiple layers. Note that at the height of the upper layer, sky cover is reported as overcast even though the upper layer itself covers less than ½ of the sky.

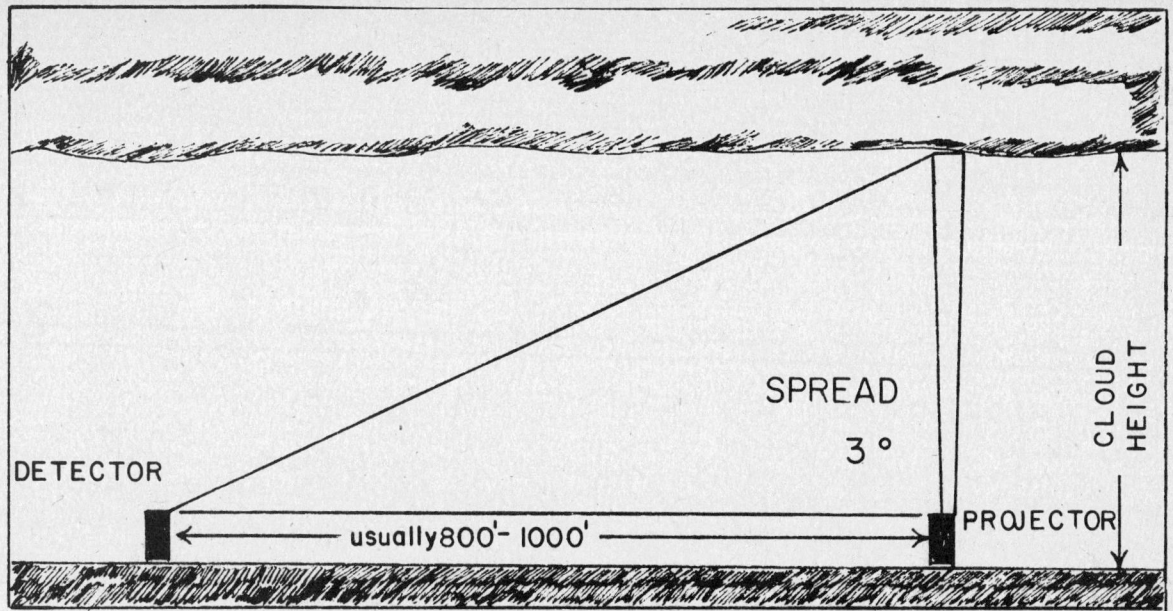

FIGURE 2-5. The rotating beam ceilometer. The projector beams a spot of modulated light on the cloud. The modulated light can be detected day or night. As the projector rotates, the spot moves along the cloud base. When the spot is directly over the detector, it excites a photoelectric cell measuring the angle of the light beam. Height of the cloud is then determined automatically by triangulation. This instrument scans much more rapidly than the older fixed beam ceilometer which is being phased out.

based total obscuration should first see the ground directly below him from the height reported as vertical visibility into the obscuration. However, because of the differing viewing points of the pilot and the observer, these surface reported values do not always exactly agree with what the pilot sees. Figure 2-6 illustrates the effect of an obscured sky on the vision from a descending aircraft.

The letter "V" appended to the ceiling height indicates variable ceiling; the range of variability is shown in remarks. Variable ceiling is reported only when it is critical to terminal operations. As an example,

M12VOVC and in remarks CIG10V13

means MEASURED CEILING ONE THOUSAND TWO HUNDRED VARIABLE OVERCAST, CEILING VARIABLE BETWEEN ONE THOUSAND AND ONE THOUSAND THREE HUNDRED.

Now, let's go back to our five reports and read them through sky and ceiling:

INK CLR	WINK, CLEAR
BOI 150SCT	BOISE, ONE FIVE THOUSAND SCATTERED
LAX 7SCT 250SCT	LOS ANGELES, SEVEN HUNDRED SCATTERED, TWO FIVE THOUSAND SCATTERED
MDW SP −X M7OVC	CHICAGO MIDWAY, SPECIAL, SKY PARTIALLY OBSCURED, MEASURED CEILING SEVEN HUNDRED OVERCAST
JFK SP W5X	NEW YORK KENNEDY, SPECIAL, INDEFINITE CEILING FIVE HUNDRED SKY OBSCURED

VISIBILITY

Prevailing visibility at the observation site immediately follows sky and ceiling in the report. Prevailing visibility is the greatest distance objects can be seen and identified through at least 180° of the horizon. It is reported in statute miles and fractions.

Prevailing visibilities in the five reports are:

INK	VISIBILITY ONE FIVE
BOI	VISIBILITY THREE ZERO
LAX	VISIBILITY SIX
MDW	VISIBILITY ONE AND ONE-HALF
JFK	VISIBILITY ONE-HALF

When visibility is critical at an airport with a weather observing station and a control tower, both take visibility observations. Of the two observa-

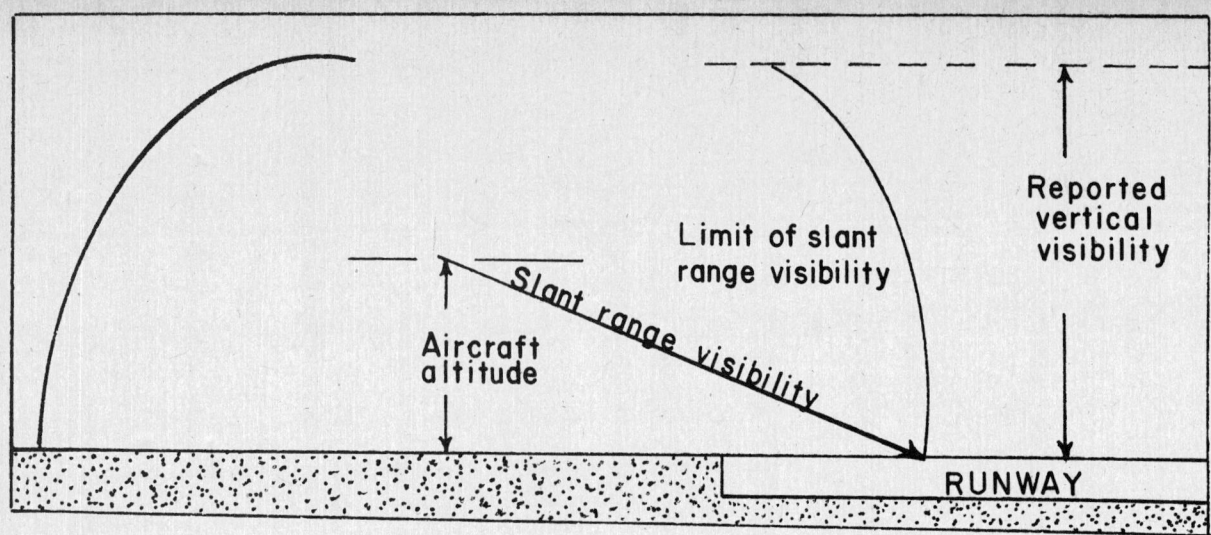
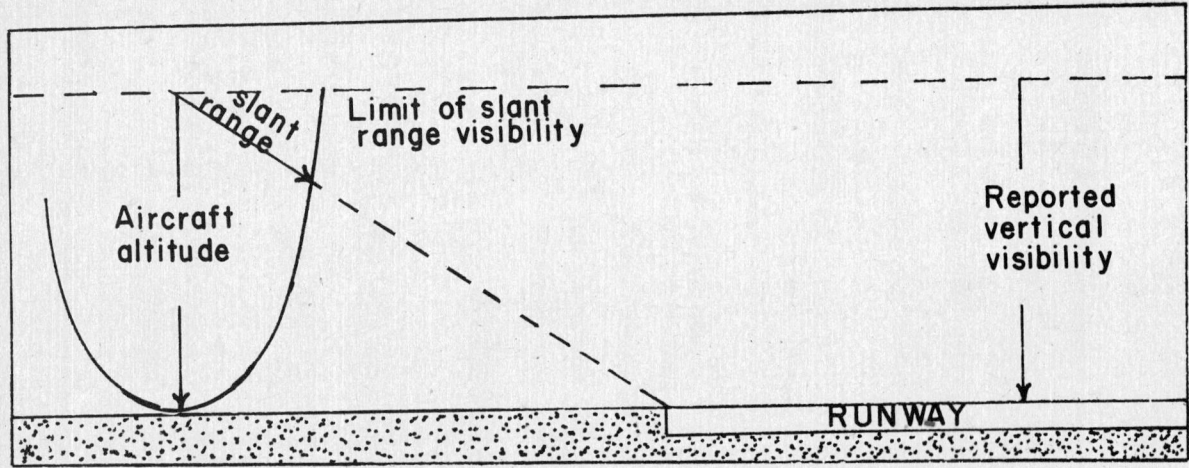

FIGURE 2-6. Vertical visibility is the altitude above the ground from which a pilot should first see the ground directly below him (top). His real concern is slant range visibility which most often is less than vertical visibility. He usually must descend to a lower altitude (bottom) before he sees a representative surface and can fly by visual reference to the ground.

tions, the one most representative is usually reported as prevailing visibility. If the other is operationally significant, it is reported in remarks. Note that the report from JFK has a remark,

TWR VSBY1/4

meaning TOWER VISIBILITY ONE-QUARTER.

The letter "V" suffixed to prevailing visibility denotes variable visibility; the range of visibility is shown in remarks. Variable visibility is reported only when critical to aircraft operations. As an example,

3/4 and in remarks VSBY1/2V1

means VISIBILITY THREE QUARTERS VARIABLE . . . VISIBILITY VARIABLE BETWEEN ONE-HALF AND ONE.

Visibility in some directions may differ significantly from prevailing visibility. These significant differences are reported in remarks. For example, prevailing visibility is reported as 1½ miles with a remark,

VSBY NE2 1/2SW3/4

which means visibility to the northeast is 2½ miles; and to the southwest, it is ¾ of a mile.

WEATHER AND OBSTRUCTIONS TO VISION

Weather and obstructions to vision when occurring at the station at observation time are reported immediately following visibility. If observed at a distance from the station, they are reported in remarks.

The term *weather* as used for this element refers only to those items listed in table 2-3 rather than to the more general meaning of all atmospheric phenomena. Weather includes all forms of precipitation plus thunderstorm, tornado, funnel cloud, and waterspout.

Precipitation is reported in one of four intensities. The intensity symbol follows the weather symbol with meanings as follows:

Very light	$--$
Light	$-$
Moderate	(no sign)
Heavy	$+$

No intensity is reported for hail (A) or ice crystals (IC).

A thunderstorm is reported as "T" and a severe thunderstorm, as "T+". A *severe thunderstorm* is one in which surface wind is 50 knots or greater and/or hail is 3/4 inch or more in diameter.

Obstructions to vision include the phenomena listed in table 2-4. No intensities are reported for obstructions to vision.

Now referring back to our initial five reports, INK and BOI report no weather or obstructions to vision, and no entries appear in the reports. LAX reports two obstructions to vision, smoke and haze. MDW reports heavy rain as weather and fog as an obstruction to vision. JFK reports fog; is this weather or obstruction to vision?

Two types of remarks concern obscuring phenomena surface and aloft. These remarks we study here.

When obscuring phenomena is surface based and partially obscures the sky, a remark reports tenths of sky hidden. For example,

K6

means 6/10 of the sky is hidden by smoke. Now look at the report from MDW; how much of the sky is hidden and by what obscuring phenomena? Note the remark

RF2

which means 2/10 of the sky is hidden by rain and fog.

A layer of obscuring phenomena aloft is reported in the sky and ceiling portion the same as a layer of cloud cover. A remark identifies the layer as obscuring phenomena. For example,

20—BKN and a remark K20—BKN

means a broken layer of smoke based on 2,000 feet above the surface and not concealing the sky (thin).

SEA LEVEL PRESSURE

Sea level pressure is separated from the preceding elements by a space. It is transmitted in

TABLE 2-3. Weather symbols and meanings

Coded	Spoken
Tornado	TORNADO
Funnel Cloud	FUNNEL CLOUD
Waterspout	WATERSPOUT
T	THUNDERSTORM
T+	SEVERE THUNDERSTORM
R	RAIN
RW	RAIN SHOWER
L	DRIZZLE
ZR	FREEZING RAIN
ZL	FREEZING DRIZZLE
A	HAIL
IP	ICE PELLETS
IPW	ICE PELLET SHOWERS
S	SNOW
SW	SNOW SHOWERS
SP	SNOW PELLETS
SG	SNOW GRAINS
IC	ICE CRYSTALS

TABLE 2-4. Obstructions to vision—symbols and meanings

Coded	Spoken
BD	BLOWING DUST
BN	BLOWING SAND
BS	BLOWING SNOW
BY	BLOWING SPRAY
D	DUST
F	FOG
GF	GROUND FOG
H	HAZE
IF	ICE FOG
K	SMOKE

record hourly reports only. It is in three digits to the nearest tenth millibar with the decimal point omitted. Sea level pressure usually is greater than 960.0 millibars and less than 1050.0 millibars. The first 9 or 10 is omitted. To decode, prefix a 9 or 10 whichever brings it closer to 1000.0 millibars. Again going back to our five reports, sea level pressures are:

INK	1010.6 millibars
BOI	1018.1
LAX	1012.9
MDW	999.0
JFK	1018.0

TEMPERATURE AND DEW POINT

Temperature and dew point are in whole degrees Fahrenheit. They are separated from sea level

Surface Aviation Weather Reports

pressure by a slash (/). If sea level pressure is not transmitted, temperature is separated from preceding elements by a space. Temperature and dew point are separated also by a slash. A minus sign precedes a temperature or dew point when below 0°F. From our five reports, we have:

INK ... 77/63	WINK ...	TEMPERATURE SEVEN SEVEN, DEW POINT SIX THREE
BOI ... 62/42	BOISE ...	TEMPERATURE SIX TWO, DEW POINT FOUR TWO
LAX ... 60/59	LOS ANGELES ...	TEMPERATURE SIX ZERO, DEW POINT FIVE NINER
MDW ... 63/61	CHICAGO MIDWAY ...	TEMPERATURE SIX THREE, DEW POINT SIX ONE
JFK ... 68/64	NEW YORK KENNEDY ...	TEMPERATURE SIX EIGHT DEW POINT SIX FOUR

WIND

Wind follows dew point and is separated from it by a slash. Average one minute direction and speed are in four digits. The first two digits are direction *from* which the wind is blowing. It is in tens of degrees referenced to true North*, i.e., 01 is 10°; 21 is 210°; 36 is 360° or North. The second two digits are speed in knots. A calm wind is reported as 0000.

If windspeed is 100 knots or greater, 50 is added to the direction code and the hundreds digit of speed is omitted. Example,

5908

means

090° (09+50=59) at 108 knots.

A *gust* is a variation in windspeed of at least 10 knots between peaks and lulls. A *squall* is a sudden increase in speed of at least 15 knots to a sustained speed of 20 knots or more lasting for at least one minute. Gusts or squalls are reported by the letter "G" or "Q" respectively following the average one-minute speed and followed by the peak speed in knots. For example,

1522Q37

means

wind 150° at 22 knots with peak speed in squalls to 37 knots.

Winds decoded from our five reports are

INK	WIND ONE ONE ZERO DEGREES AT ONE TWO PEAK GUSTS ONE EIGHT
BOI	WIND ONE THREE ZERO DEGREES AT FOUR
LAX	WIND TWO FIVE ZERO DEGREES AT FOUR
MDW	WIND THREE TWO ZERO DEGREES AT FIVE
JFK	WIND ONE EIGHT ZERO DEGREES AT FOUR

When any part of the wind report is *estimated* (direction, speed, peak speed in gusts or squalls),

the letter "E" precedes the wind group. Example, E1522G28

is decoded WIND ONE FIVE ZERO DEGREES ESTIMATED TWO TWO PEAK GUSTS ESTIMATED TWO EIGHT.

A few stations do not transmit sea level pressure, temperature, and dew point; and these elements usually are not included in a special. When the elements are not transmitted, the wind group is separated from the preceding element by a space; i.e.,

CSM SP W5X 2F 1705/990

is a record special from Clinton-Sherman Oklahoma (CSM) *not* transmitting sea level pressure, temperature, or dew point.

ALTIMETER SETTING

Altimeter setting follows the wind group and is separated from it by a slash. Normal range of altimeter settings is from 28.00 inches to 31.00 inches of mercury. The last three digits are transmitted with the decimal point omitted. To decode, prefix to the coded value either a 2 or a 3 whichever brings it closer to 30.00 inches. Examples,

996 means ALTIMETER TWO NINER NINER SIX, (29.96 inches)

013 means ALTIMETER THREE ZERO ONE THREE (30.13 inches)

An estimated altimeter is read from an instrument not compared to a standard instrument as recently as required (see AVIATION WEATHER, Chapter 3). It is reported by prefixing an "E" to the coded value. Example,

E035 means ALTIMETER ESTIMATED THREE ZERO THREE FIVE

REMARKS

Remarks, if any, follow altimeter setting separated from it by a slash. Certain remarks should be reported routinely; others the observer may include when considered significant to aviation. Often, some of the most important information in an observation may be the remarks portion discussed in succeeding paragraphs.

---

* Wind direction for the local station is *broadcast* in degrees magnetic.

Surface Aviation Weather Reports

Runway Visibility and Runway Visual Range

The first remark, when transmitted, should be runway visibility or runway visual range. Figure 2–7 illustrates the difference. The terms are defined as follows:

Runway visibility—the visibility from a particular location along an identified runway, usually determined by transmissometer instrument. It is in miles and fractions. Figure 2–8 diagrams the principle of the transmissometer.

Runway visual range—the maximum horizontal distance down a specified instrument runway at which a pilot can see and identify standard high intensity runway lights. It is always determined using a transmissometer and is reported in hundreds of feet.

The report consists of a runway designator and the contraction "VV" or "VR" followed by the appropriate visibility or visual range. Both the VV and the VR report are for a 10-minute period preceding observation time. The remark usually reports the 10-minute extremes separated by the letter "V". However, if the visual range or visibility has not changed significantly during the 10 minutes, a single value is sent indicating that the value has remained constant.

The following examples show several reports and their decoding:

R36VV11/2 — RUNWAY THREE SIX, VISIBILITY ONE AND ONE-HALF. (Visibility remained constant during the 10-minute period.)

R05LVV1V2 — RUNWAY FIVE LEFT, VISIBILITY VARIABLE BETWEEN ONE AND TWO.

R18VR20V30 — RUNWAY ONE EIGHT, VISUAL RANGE VARIABLE BETWEEN TWO THOUSAND FEET AND THREE THOUSAND FEET.

R26RVR24 — RUNWAY TWO SIX RIGHT, VISUAL RANGE TWO THOUSAND FOUR HUNDRED FEET. (Visual range remained constant during the 10-minute period.)

Runway visual range in excess of 6,000 feet is written 60+. VR less than the minimum value that can be observed by the instrument is encoded

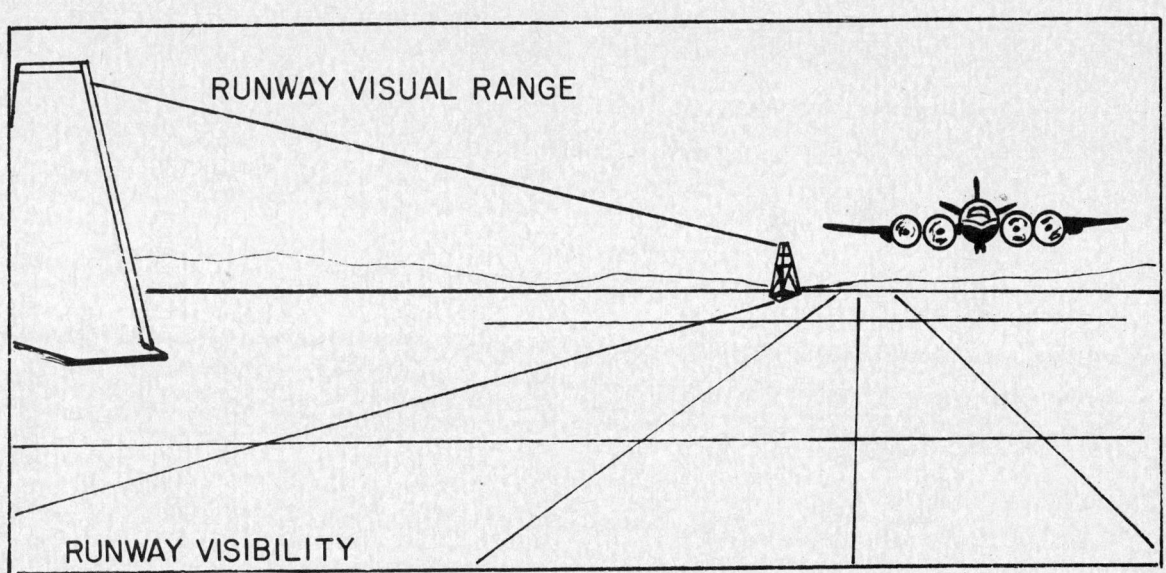

FIGURE 2–7. Difference between *runway visibility* and *runway visual range*. Runway visibility is the distance down the runway the pilot can see unlighted objects or unfocused lights of moderate intensity. Runway visual range is the distance he can see high intensity runway lights. Visual range usually is greater than visibility because the high intensity lights penetrate farther into the obscuring phenomena.

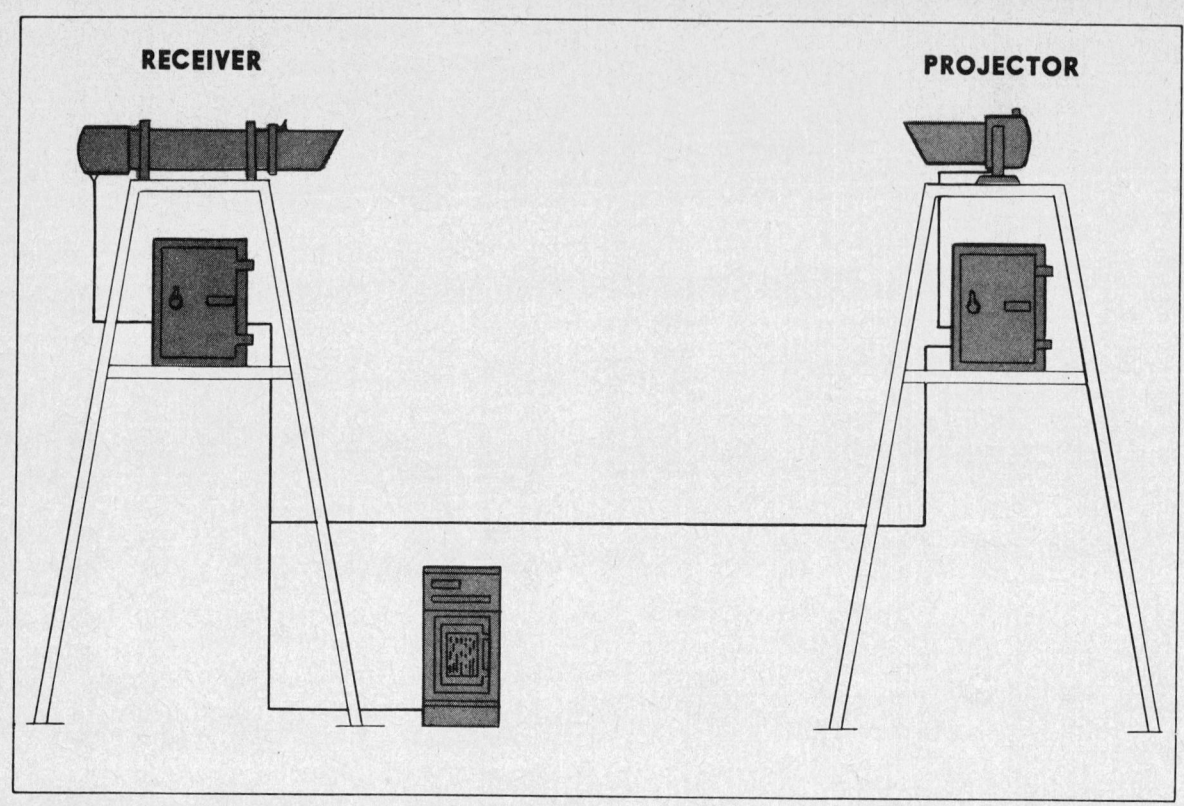

FIGURE 2-8. The transmissometer. The projector beams light toward the receiver. Obscuring phenomena in the path of the beam absorbs some of the light. A photoelectric cell in the receiver measures the amount of light penetrating through the obscuring phenomena. The amount received is converted into visibility.

as the minimum suffixed by a minus sign. For example:

R36LVR10—V25

is decoded RUNWAY THREE SIX LEFT, VISUAL RANGE VARIABLE FROM LESS THAN ONE THOUSAND FEET TO TWO THOUSAND FIVE HUNDRED FEET.

Heights of Bases and Tops of Sky Cover Layers

Bases and tops of clouds or obscuring phenomena may be reported. These remarks originate from pilots. Heights are above MSL.

Coded Elements	Coded Remarks
/UA BKN 50	Top broken layer 5,000 feet (MSL)
/UA OVC 30/60 OVC	Top lower overcast 3,000 feet, base of higher overcast 6,000 feet.

Clarification of Coded Data

Following, by category, are coded remarks clarifying or expanding on coded elements:

SKY AND CEILING

Coded Elements	Coded Remarks	Coded Elements	Coded Remarks
FEW CU	Few cumulus clouds	30SCT V BKN	Scattered layer at 3000 feet variable to broken
HIR CLDS VSB	Higher clouds visible		
BRKHIC	Breaks in higher overcast	SC BANK NW	Stratocumulus cloud bank northwest
BINOVC	Breaks in overcast		
BRKS N	Breaks north	TCU W*	Towering cumulus clouds west
BKN V OVC	Broken layer variable to overcast	CB N MOVG E*	Cumulonimbus north moving east
CIG 14V19	Ceiling variable between 1400 feet and 1900 feet	CBMAM OVHD-W*	Cumulonimbus mamma overhead to west

Surface Aviation Weather Reports

SKY AND CEILING—Continued

Coded Elements	Coded Remarks	Coded Elements	Coded Remarks
ACCAS ALQDS	Altocumulus castellanus all quadrants	CONTRAILS N 420 MSL	Condensation trails **north** at 42,000 feet MSL
ACSL SW-NW	Standing lenticular altocumulus southwest to northwest	CLDS TPG MTNS SW	Clouds topping **mountains** southwest
ROTOR CLDS NW	Rotor clouds northwest	RDGS OBSCD W-N	Ridges obscured west **through** north
VIRGA E-SE	Virga (precipitation not reaching the ground) east through southeast	CUFRA W APCHG STN	Cumulus fractus clouds **west** approaching station
		LWR CLDS NE	Lower clouds northeast

OBSCURING PHENOMENA

D5	Dust obscuring 5/10 of the sky	K20SCT	Scattered layer of smoke aloft based at 2000 feet **above** the surface
S7	Snow obscuring 7/10 of the sky		
BS3	Blowing snow obscuring 3/10 of the sky	THIN FOG NW	Thin fog northwest (from reporting station)
FK4	Fog and smoke obscuring 4/10 of the sky		

VISIBILITY (STATUTE MILES)

VSBY S1W1/4	Visibility south 1, west 1/4	TWR VSBY3/4	Tower visibility 3/4
VSBY1V3	Visibility variable between 1 and 3	SFC VSBY1/2	Surface visibility 1/2

WEATHER AND OBSTRUCTIONS TO VISION

T W FQT LTGCG	Thunderstorm west, frequent lightning cloud to ground	OCNL RW	Occasional moderate rain shower
RB30	Rain began 30 minutes after the hour	WET SNW	Wet snow
SB15E40	Snow began 15, ended 40 minutes after the hour	SNOINCR 5	Snow increase 5 inches during past hour
UNCONFIRMED TORNADO 15W OKC MOVG NE 2000	Unconfirmed tornado 15 (nautical miles) west of Oklahoma City, moving northeast, sighted at 2000Z	R− OCNLY R+	Light rain occasionally heavy rain
		RWU	Rain showers of unknown intensity
T OVHD MOVG E	Thunderstorm overhead, moving east	F DSIPTG	Fog dissipating
		K DRFTG OVR FLD	Smoke drifting over field
OCNL DSNT LTG NW	Occasional distant lightning northwest	KOCTY	Smoke over city
		SHLW GFDEP 4	Shallow ground fog 4 feet deep
HLSTO 2	Hailstones 2 inches in diameter	DUST DEVILS NW	Dust devils northwest
INTMT R−	Intermittent light rain	PATCH GF S	Patch ground fog south

WIND

WSHFT30	Wind shifted at 30 minutes past the hour	PK WND 3348/22	Peak wind within the past hour from 330° at 48 knots occurred 22 minutes past the hour
WND 27V33	Wind variable between 270° and 330°		

PRESSURE

PRESRR	Pressure rising rapidly	PRJMP 8/1012/18	Pressure jump (sudden increase) .08 inches began 1012 GMT, ended 1018 GMT
PRESFR	Pressure falling rapidly		
LOWEST PRES 631 1745	Lowest pressure (sea level) 963.1 millibars at 1745 GMT		

Surface Aviation Weather Reports

Freezing Level Data

Upper air (rawinsonde) observation stations append in remarks *freezing level data*. The coded remark is appended to the first record report transmitted after the information becomes available. Code for the remark is as follows:

RADAT UU (D) ($h_p h_p h_p$) ($h_p h_p h_p$) (/n)

(a) RADAT—a contraction identifying the remark as "freezing level data".

(b) UU—relative humidity at the freezing level in percent. When more than one level is sent, "UU" is highest relative humidity observed at any of the levels transmitted.

(c) (D)—a coded letter "L", "M", or "H" to indicate that relative humidity is for the "lowest", "middle", or "highest" level coded. This letter is omitted when only one level is sent.

(d) ($h_p h_p h_p$)—a height in hundreds of feet above MSL at which the upper air sounding crossed the 0° C isotherm. No more than three levels are coded. If the sounding crosses the 0° C isotherm more than three times, the levels coded are the lowest and the top two levels.

(e) (/n)—indicator to show the number of crossings of the 0° C isotherm, other than those coded. The indicator is omitted when all levels are coded.

Examples:

RADAT 87045 — Relative humidity 87%, only crossing of 0° C isotherm was 4,500 feet MSL.

RADAT 87L024105 — Relative humidity 87% at the lowest (L) crossing. Two crossings occurred at 2,400 and 10,500 feet MSL.

RADAT 84M019045051/1 — Relative humidity 84% at the middle (M) crossing of the three coded crossings. Coded crossings were at 1,900, 4,500, and 5,100 feet. The 84% humidity was at 4,500 feet MSL. "/1" indicates one additional crossing and it was between 1,900 and 4,500 feet.

RADAT MISG — The sounding terminated below the first crossing of the 0° C isotherm—temperatures were all above freezing.

RADAT ZERO — The entire sounding was below 0° C.

Icing Data

When the rawinsonde observer determines definitely that icing was occurring on his instruments, he enters the data in the following code:

RAICG HHMSL (SNW)

(a) RAICG—indicates icing data follows.

(b) HH—height in hundreds of feet at which icing occurred. "MSL" is always appended to the height.

(c) (SNW)—used to indicate that snow is causing a reduced balloon ascension rate. (Omitted otherwise.)

Examples:

RAICG 12MSL—Icing at 1,200 feet MSL.

RAICG 24MSL SNW—Icing at 2,400 feet MSL in snow.

Other Information

A group or groups of numerically coded data may appear in remarks. These data are primarily of concern to the meteorologist and are not discussed here.

A printed arrow marks the end of weather information and signifies that the rest of the report is notice(s) to airmen (NOTAM). The NOTAM code is explained in the AIRMAN'S INFORMATION MANUAL.

REPORT IDENTIFIERS

A heading begins the record hourly collective on the local circuit identifying the type of message, the circuit number, and the date and time of observations making up the collective reports. For example,

SA21 271900

means surface aviation reports (SA); 21 is the circuit number; 27 is the day of the month; and the observations were made at 1900 GMT.

A slightly different heading begins each relay. It identifies the location of reporting stations either (1) by States or (2) in relation to the local circuit. It indicates the time the relay began. Example,

SA NEAR EAST201904

means surface aviation reports; area covered by the relay is just east of the local circuit area; day of the month is the 20th (20); time of observations is 1900 GMT (19); and the relay began 4 minutes past the hour (04).

Relay designators other than States are INTERMEDIATE EAST, FAR EAST, NEAR NORTH, etc. The relay collectives are assembled by a centralized computer and are unique to each circuit.

Individual reports must each convey the time and type of report. These reports include specials, corrected reports, and supplemental reports. Following are examples:

Example 1

INK 1100.....

indicates a relayed report from Wink, Texas for 1100 GMT (all times transmitted in teletypewriter reports are GMT). Since the time is on the hour, it signifies a record hourly so that further identification is unnecessary.

Example 2

INK COR 1100.....

signifies a correction to the 1100 GMT record hourly report as originally transmitted. The correction may transmit the complete corrected report, or it may contain only the corrected element or elements.

Example 3

INK SP 2315.....

indicates a special report of an observation taken at 2315 GMT to report a significant change in weather.

Example 4

INK COR SP 2315.....

indicates a correction to the special report in example 3.

Example 5

BVO SW 1130.....

indicates a Supplemental Aviation Weather Reporting Station (SAWRS) report by the contraction "SW". SAWRS reports are unscheduled and are made by non-Government observers at airports not served by a regularly reporting weather station. Observations are taken during commercial aircraft operations. Type and time are transmitted. This report was from Bartlesville, Oklahoma at 1130 GMT.

READING THE SURFACE AVIATION WEATHER REPORT

Now that we have studied the individual elements and their decoding, let's read completely each of the five reports. Capitalized phrases are those elements which *normally* are broadcast by the station at or near the airport where the observation was made:

INK CLR 15 106/77/63/1112G18/000

WINK, WINK, CLEAR, VISIBILITY ONE FIVE, pressure 1010.6 millibars, TEMPERATURE SEVEN SEVEN, dew point six three, WIND ONE ONE ZERO DEGREES AT ONE TWO PEAK GUSTS ONE EIGHT, ALTIMETER THREE ZERO ZERO ZERO.

BOI 150SCT 30 181/62/42/1304/015

BOISE, BOISE, ONE FIVE THOUSAND SCATTERED, VISIBILITY THREE ZERO, pressure 1018.1 millibars, TEMPERATURE SIX TWO, dew point four two, WIND ONE THREE ZERO DEGREES AT FOUR, ALTIMETER THREE ZERO ONE FIVE.

LAX 7SCT 250SCT 6HK 129/60/59/2504/991→LAX↘6/38

LOS ANGELES, LOS ANGELES, SEVEN HUNDRED SCATTERED TWO FIVE THOUSAND SCATTERED, VISIBILITY SIX, HAZE, SMOKE, pressure 1012.9 millibars, TEMPERATURE SIX ZERO, DEW POINT FIVE NINER, WIND TWO FIVE ZERO DEGREES AT FOUR, ALTIMETER TWO NINER NINER ONE.

Note that nothing past the arrow was read. The arrow indicates that NOTAM information follows and is not part of the *weather* report.

MDW SP −X M7OVC 11/2R+F 990/63/61/3205/950/RF2 RB12

CHICAGO, CHICAGO MIDWAY, SPECIAL, SKY PARTIALLY OBSCURED, MEASURED CEILING SEVEN HUNDRED OVERCAST, VISIBILITY ONE AND ONE-HALF, HEAVY RAIN, FOG, pressure 999.0 millibars, TEMPERATURE SIX THREE, DEW POINT SIX ONE, WIND THREE TWO ZERO DEGREES AT FIVE, ALTIMETER TWO NINER FIVE ZERO, TWO TENTHS SKY OBSCURED BY RAIN AND FOG, rain began 12 minutes past the hour.

JFK SP W5X 1/2F 180/68/64/1804/006/R04RVR22V30 TWR VSBY1/4

NEW YORK, NEW YORK KENNEDY, SPECIAL, INDEFINITE CEILING FIVE HUNDRED SKY OBSCURED, VISIBILITY ONE-HALF, FOG, pressure 1018.0 millibars, TEMPERATURE SIX EIGHT, DEW POINT SIX FOUR, WIND ONE EIGHT ZERO DEGREES AT FOUR, ALTIMETER THREE ZERO ZERO SIX, RUNWAY FOUR RIGHT VISUAL RANGE VARIABLE BETWEEN TWO THOUSAND TWO HUNDRED FEET AND THREE THOUSAND FEET, TOWER VISIBILITY ONE QUARTER.

Chapter 15
PILOT AND RADAR REPORTS

The preceding section explained the decoding of surface aviation weather reports. However, these spot reports only sample the total weather picture. Pilot and radar reports help fill the gaps between stations.

PILOT WEATHER REPORTS (PIREPS)

No observation is more timely than the one you make from your cockpit. In fact, aircraft in flight are the only means of directly observing cloud tops, icing, and turbulence. Your fellow pilots welcome your PIREP as well as do the briefer and forecaster. Help yourself and the aviation weather service by sending pilot reports!

A PIREP usually is transmitted by teletypewriter in a prescribed format. The letters "UA" identify the message as a pilot report. Next in order are location; time; phenomena encountered; altitude; and, if the report is turbulence or icing, the type of aircraft. All altitude references are MSL unless noted, distances are in nautical miles, and time is in GMT.

A PIREP is transmitted over teletypewriter as a single message, in a group of PIREPs collated by States (UBUS 1 bulletins), or as a remark appended to a surface aviation weather report. Description of the phenomena may be sent as received or coded in contractions and symbols.

Let's read a PIREP message transmitted by Washington, D.C. (DCA):

DCA UA 20 S DCA 1620 MDT RIME ICE 50 BE18

> "A pilot 20 nautical miles south of Washington at 1620 GMT encountered moderate rime icing at 5,000 feet MSL. The aircraft was a BE-18."

Most contractions in PIREP messages are self-explanatory. Icing and turbulence reports state intensities using standard terminology when possible. Intensity tables for turbulence and icing are in section 16. If a pilot's description of an icing or turbulence encounter cannot readily be translated into standard terminology, the pilot's description is transmitted verbatim.

The following excerpts may assist you in reading transmitted pilot weather reports:

UA RDU . . . . DURGD OAOI 150 OI 80 . . . .

means ". . . during descent on and off instruments at 15,000 feet; on instruments at 8,000 feet . . . ."

UA MGW . . . . 10–2515/–2 . . . .

is decoded ". . . . wind at 10,000 feet, 250° at 15; temperature –2° C. . . ."

UA MRB . . . INTMTLY BL MDT TURBC R 60 TURBC INCRS WWD . . .

states ". . . . intermittently between layers (contraction BL); moderate turbulence, moderate rain at 6,000 feet; turbulence increases westward . . ."

UA ABQ 1845 TIJERAS PASS CLOSED DUE TO FOG AND LOW CLDS. UNABLE VFR RTNG ABQ.

is self-explanatory. Information of this type is helpful to others planning VFR flight in the area.

UA CLE OVR TOL 2200 MDT CAT 350–390 B707

means " . . . over Toledo at 2200 GMT a pilot reports moderate clear air turbulence from flight level 35,000 to 39,000; aircraft is a Boeing 707." The report was transmitted by Cleveland.

To lessen the chance of misinterpretation by others, you are urged to report icing and turbulence in standard terminology (intensity tables for turbulence and icing, section 16). If your report cannot be translated into standard terminology, it is transmitted verbatim. This PIREP stated,

. . . . PRETTY ROUGH AT 6,500, SMOOTH AT 8,500 PA24

Would a report of "light", "moderate", or "severe" turbulence at 6,500 have meant more to you?

Pilot reports of cloud bases and tops are usually in symbols and are often appended to surface aviation weather reports. Height of cloud base precedes the sky cover symbol, and top follows the symbol. For example,

38 BKN 70

DSM M8OVC 3R–F 132/45/44/3213/992/UA 1735 12NW DSM OVC 65/80 OVC 140
is decoded ".... pilot reports at 1735 GMT 12 (nautical miles) northwest of Des Moines, top of the lower overcast 6,500 MSL; base of a second layer (overcast) at 8,000 and top, 14,000 feet MSL."

Pilot reports of a non-meteorological nature sometimes help air traffic controllers. This "plain language" report stated:

... 3N PNS LRG FLOCK OF GOOSEY LOOKING BIRDS HDG GNLY NORTH MAY BE SEAGULLS FORMATION LOUSY COURSE ERRATIC ....

While in humorous vein, this PIREP alerted pilots and controllers to a bird hazard.

Your PIREP always helps someone else and becomes a part of the aviation weather service. Please report anything you observe that may be of concern to other pilots.

means base of a broken layer at 3,800 feet and top 7,000 feet (all MSL).

The following example appended to an aviation weather report,

RADAR WEATHER REPORTS (RAREPS)

Thunderstorms and general areas of precipitation can be observed by radar. Radar weather reports are routinely transmitted by teletypewriter and some are included in scheduled weather broadcasts by Flight Service Stations.

Most radar stations report each hour with intervening special reports as required. They report location of precipitation along with type, intensity and trend. Table 3–1 explains symbols denoting intensity and trend. Table 3–2 summarizes the order and content of a radar weather report.

To assist you in interpreting RAREPs, five examples are decoded into plain language:

LIT 1133 AREA 4TRW+/+ 22/100 88/170 196/180 220/115 CELLS 2425 MT 310 AT 162/110
  Little Rock, Arkansas radar weather observation at 1133 GMT
  An area of echoes, four-tenths coverage, containing thunderstorms and heavy rainshowers, increasing in intensity
  Area is defined by points (referenced LIT radar) at 22°, 100 NM (nautical miles); 88°, 170 NM; 196°, 180 NM; and 220°, 115 NM. (These points plotted on a map and connected with a line outline the area of echoes.)
  Individual cells are moving from 240° at 25 knots
  Maximum tops (MT) are 31,000 feet located at 162° and 110 NM from LIT

JAN 1935 SPL LN 10TRWX/NC 86/40 164/60 199/115 12W CELLS 2430 MT 440 AT 159/65 D10
  Jackson, Mississippi, a 1935 special radar report
  Line of echoes, ten-tenths coverage, thunderstorm, intense rainshowers, no change in intensity
  Center of the line extends from 86°, 40 NM; 164°, 60 NM; and 199°, 115 NM. The line is 12 NM wide (12W). (To display graphically, plot the center points on a map and connect the points with a line; since the thunderstorm line is 12 miles wide, it extends 6 miles either side of your plotted line.)
  Thunderstorm cells are moving from 240° at 30 knots
  Maximum tops are 44,000 feet, centered at 159°, 65 NM from Jackson. Diameter of this cell is 10 NM (D10)

MAF 1130 AREA 2S 27/80 90/125 196/50 268/100 2410 MT 100 UNIFORM
  Midland, Texas radar weather report at 1130 GMT
  An area, two-tenths coverage, of snow
  Area is bounded by points 27°, 80 NM; 90°, 125 NM; 196°, 50 NM; and 268°, 100 NM
  Movement is from 240° at 10 knots
  Maximum tops are 10,000 feet; tops are uniform (smooth)

Pilot and Radar Reports

HDO 1132 AREA 2TRW++ 6R−/NC 67/130 308/45 105W CELLS 2240 MT 380 at 66/54

Hondo, Texas radar weather report at 1132 GMT

An area of echoes containing two-tenths coverage of thunderstorms, very heavy rainshowers, and six-tenths coverage light rain. No intensity change. (This report suggests thunderstorms embedded in a general area of light rain.)

Although the pattern is an "area", only two points are given followed by "105W". This means the area lies 52½ miles either side of the line defined by the two points—67°, 130 NM and 308°, 45 NM

Thunderstorm cells are moving from 220° at 40 knots

Maximum tops are 38,000 feet at 66°, 54 NM

TABLE 3–1. Precipitation intensity and intensity trend

Intensity		Intensity Trend	
Symbol	Intensity	Symbol	Trend
− (none) + ++ X XX U	Light Moderate Heavy Very heavy Intense Extreme Unknown	+ − NC NEW	Increasing Decreasing No change New echo

TABLE 3–2. Order and content of a radar weather report

OKC 1934 LN 8TRW++/+ 86/40 164/60/ 199/115 15W 2425
MT 570 AT 159/65 2 INCH HAIL RPRTD THIS ECHO

OKC 1934	LN	8	TRW++/+	86/40 164/60 199/115	15W	2425
a.	b.	c.	d.	e.	f.	g.

MT 570 AT 159/65	2 INCH HAIL RPRTD THIS ECHO
h.	i.

a. Location identifier and time of radar observation (GMT)
b. Echo pattern[1] (line in this example)
c. Coverage in tenths (8/10 of this example)
d. Type, intensity, and trend of weather[2] (thunderstorm (T), very heavy rainshowers (RW++), increasing in intensity (/+))
e. Azimuth (reference true N) and range in nautical miles (NM) of points defining the echo pattern
f. Dimension of echo pattern[3] (15 NM wide)
g. Pattern movement (line moving from 240° at 25 knots); may also show movement of individual storms or "cells"
h. Maximum tops and location (57,000 feet)
i. Remarks; self-explanatory in plain language contractions.

[1] Echo pattern may be a line (LN), fine line (FINE LN), area (AREA), spiral band area (SPIRAL BAND), or single cell (CELL).

[2] Teletypewriter weather symbols are used. See Table 3–1 for intensity and intensity trend symbols.

[3] Dimension of an echo pattern is given when azimuth and range define only the center or center line of the pattern.

When a radar report is transmitted but contains no encoded weather observation, a contraction is sent which indicates operational status of the radar. Table 3–3 explains the contractions.

OKC 1135 PPINE

Oklahoma City, Oklahoma radar at 1135 GMT detects no echoes

Radar weather reports also contain groups of digits, i.e., 00220 00221, etc., which are entered on a line following the RAREP. This manually digitized radar information is omitted from the foregoing examples since it is used primarily by meteorologists and hydrologists for estimating amount of rainfall.

A radar weather report may contain remarks in addition to the coded observation. Certain types of severe storms produce distinctive patterns on the radar scope. For example, a hook-shaped echo may be associated with a tornado; and a spiral band with a hurricane. If hail, strong winds, tornado activity, or other adverse weather is known to be associated with identified echoes on the radar scope, the location and type of phenomena are included as a remark. Examples of remarks are: "HAIL REPORTED THIS ECHO"; "TORNADO ON GROUND AT 338/15"; and "HOOK ECHO 243/18".

When using hourly and special radar weather reports in preflight planning, note the location and coverage of echoes, the type of weather reported, the intensity trend, and especially the direction of movement. A word of caution—remember that radar detects only thunderstorms and general areas of precipitation; it is *not* designed to detect enroute ceiling and visibility. An area may be blanketed with fog or low stratus, but unless precipitation is also present, the radar scope will be clear of echoes. Use radar reports along with PIREPs and aviation weather reports and forecasts.

RAREPs help you to plan ahead to avoid thunderstorm areas. Once airborne, however, you must depend on visual sighting or airborne radar to evade individual storms.

TABLE 3–3. Contractions reporting operational status of radar

Contraction	Operational status
PPINE	Equipment normal and operating in PPI (Plan Position Indicator) mode; no echoes observed.
PPIOM	Radar inoperative or out of service for preventative maintenance.
PPINA	Observations omitted or not available for reasons other than PPINE or PPIOM.
ROBEPS	Radar operating below performance standards.
ARNO	"A" scope or azimuth/range indicator inoperative.
RHINO	Radar cannot be operated in RHI (Range-height indicator) mode. Height data not available.

Chapter 16
AVIATION WEATHER FORECASTS

Good flight planning considers forecast weather. This section explains the following aviation forecasts:

1. Terminal Forecasts
   (a) Domestic (FT)
   (b) International (ICAO TAF)
2. Area Forecast (FA)
3. TWEB Route Forecast and Synopsis
4. SIGMET and AIRMET (WS, WA, and WAC)
5. Winds and Temperatures Aloft Forecast (FD)
6. Special Flight Forecast

Also discussed are the following general forecasts which may aid in flight planning:

1. Hurricane Advisory (WH)
2. Convective Outlook (AC)
3. Severe Weather Watch Bulletin (WW)

U.S. terminal and area forecasts group ceiling and visibility into the following categories:

LIFR (Low IFR) —Ceiling less than 500 feet and/or visibility less than 1 mile.

IFR —Ceiling 500 to less than 1,000 feet and/or visibility 1 to less than 3 miles.

MVFR (Marginal VFR) —Ceiling 1,000 to 3,000 feet and/or visibility 3 to 5 miles inclusive.

VFR —Ceiling greater than 3,000 feet and visibility greater than 5 miles; includes sky clear.

These categorical groupings are used for the outlook portions of the forecasts extending beyond 18 hours. They enable the forecaster to more realistically describe conditions in the outlook period intended primarily for advanced operational planning.

The cause of LIFR, IFR, or MVFR is also given by either ceiling or visibility restrictions or both. The contraction "CIG" and/or weather and obstruction to vision symbols are used. If winds or gusts of 25 knots or greater are forecast for the outlook period, the word "WIND" is also included for all categories including VFR. Examples:

LIFR CIG —Low IFR due to low ceiling.

IFR F —IFR due to visibility restricted by fog.

MVFR CIG H K—Marginal VFR due both to ceiling and to visibility restricted by haze and smoke.

IFR CIG R WIND—IFR due both to low ceiling and to visibility restricted by rain; wind expected to be 25 knots or greater.

You should memorize the categories and their defining ceiling and visibility limits. Knowing them is mandatory to readily interpreting an outlook into operational planning.

Forecasts are regularly scheduled in collectives; but occasionally an unscheduled forecast must be transmitted out of collective. A forecast transmitted out of a collective must be identified by one of the following contractions with meaning as noted:

RTD—Routine delayed weather bulletin
COR—Correction bulletin
AMD—Amendment bulletin

TERMINAL FORECASTS

A terminal forecast is for a specific airdrome. In the U.S., it is for an area within a 5-mile radius of the center of the runway complex. Terminal forecasts are in both the domestic U.S. code (FT)

Aviation Weather Forecasts

and the ICAO (TAF) code. Scheduled terminal forecasts are valid for 24 hours.

U.S. Terminal Forecast Code (FT)

Terminal forecasts in the U.S. code (FT) are issued three times daily by WSFOs. Figure 1–4, section 1, shows the FT network. Issue and valid times are according to time zones of the issuing WSFO (see table in the next column).

WSFO Location (time zone)	Issue time	Valid period
Eastern/Central	0940Z	10Z–10Z
	1440Z	15Z–15Z
	2140Z	22Z–22Z
Mountain/Pacific	0940Z	10Z–10Z
	1540Z	16Z–16Z
	2240Z	23Z–23Z

Format of the FT is essentially the same as that of the SA report. Following is an FT:

STL 251010 C5X 1/2S–BS 3325G35 OCNL C0X 0S+BS. 16Z C30BKN 3BS BRF SW–. 22Z 30SCT 3315. 00Z CLR. 04Z VFR WIND . .

To aid in the discussion, we have divided the forecast into the following elements lettered "a" through "i"

STL	251010	C5X	1/2	S–BS	3325G35	OCNL C0X 0S+BS.
a.	b.	c.	d.	e.	f.	g.

16Z C30BKN 3BS BRF SW–. 22Z 30SCT 3315. 00Z CLR.
h.

04Z VFR WIND . .
i.

a. *Station identifier.* "STL" identifies St. Louis, Missouri. The forecast is for St. Louis.

b. *Date-time group.* "251010" is date and valid times. The forecast is valid beginning on the 25th day of the month at 1000Z valid until 1000Z the following day.

c. *Sky and ceiling.* "C5X" means ceiling 500 feet, sky obscured. The letter "C" always identifies a *forecast* ceiling layer.

d. *Visibility.* "1/2" means visibility ½ mile. Visibility is in statute miles and fractions. Absence of a visibility entry specifically implies visibility more than 6 miles.

e. *Weather and obstructions to vision.* "S–BS" means light snow and blowing snow. These elements are in symbols identical to those used in SA reports and entered only when expected.

f. *Wind.* "3325G35" means wind from 330° at 25 knots gusting to 35 knots—the same as in SAs. Omission of a wind entry specifically implies wind less than 10 knots.

g. *Remarks.* "OCNL C0X 0S+BS" means occasional ceiling zero, sky obscured, visibility zero, heavy snow and blowing snow. Remarks may be added to more completely describe expected weather.

h. *Expected changes.* When changes are expected, preceding conditions are followed by a period and the time and conditions of the expected change. "16Z C30BKN 3BS 3320 BRF SW–. 22Z 30SCT 3315. 00Z CLR." means by 1600Z, ceiling 3,000 broken, visibility 3, blowing snow, wind 330° at 20 knots, brief light snow showers. By 2200Z, 3,000 scattered, visibility more than 6 (implied), wind 330° at 15 knots. By 0000Z sky clear, visibility more than 6, wind less than 10 knots (implied).

i. *6-hour categorical outlook.* The last 6 hours of the forecast is a categorical outlook as explained on page 35. "04Z VFR WIND . ." means that from 0400Z until 1000Z—the end of the forecast period—weather will be ceiling more than 3,000 and visibility greater than 5 (VFR); wind will be 25 knots or stronger. The double period (. .) signifies the end of the forecast for the specific terminal.

Scheduled FT Collectives and Relays. Scheduled FTs are collected on area teletypewriter circuits. Selected FTs are then relayed to the area circuit from surrounding areas. The coverage of FTs becomes more sparse as the relays become more remote from the circuit area.

The heading of an FT collective identifies the message as an FT with a 6-digit date-time group giving the transmission time. For example, "FT130940" means a collective transmitted on the 13th at 0940Z.

Relay headings on some circuits are by States. On other circuits a relay heading identifies location relative to the area circuit. For example, "FT NEAR WEST 130941" means the terminals are just west of the circuit area; the relay was transmitted on the 13th at 0941Z. Other relay areas

Aviation Weather Forecasts

are NEAR EAST, NEAR NORTH, INTERMEDIATE SOUTH, FAR WEST, etc. These FT relay areas are the same as covered by SA relays whether by State or by areas relative to the local circuit. For example, if the SA report for station XYZ is in the FAR EAST relay, the FT for XYZ also will be found in the FAR EAST relay of FTs. The relays are assembled by computer and are unique to each circuit; they do not coincide with adjacent circuit collectives and relays.

Out of Sequence FTs. A delayed, corrected, or amended FT is identified in the message rather than in the heading. Following are a delayed FT for Binghamton, New York, a corrected FT for Memphis, Tennessee, and an amended FT for Lufkin, Texas:

BGM FT RTD 131615 1620Z 100SCT 250SCT 1810. 18Z 50SCT 100SCT 1913 CHC C30BKN 3TRW AFT 20Z. 03Z 100SCT C250BKN. 09Z VFR . .

MEM FT COR 132222 2230Z 40SCT 300SCT CHC TRW. 02Z CLR. 16Z VFR . .

LFK FT AMD 1 131410 1425Z C8OVC 4F OVC V BKN. 15Z 20SCT 250−BKN. 19Z 40SCT 120SCT CHC C30BKN 3TRW. 04Z MVFR CIG F . .

Note in each forecast a time group following the valid period; this is the issue time. Note also that the amended forecast for LFK has the entry "AMD 1". Amended FTs for each terminal are numbered sequentially starting after each scheduled forecast.

ICAO Terminal Forecast (TAF)

Terminal forecasts for international flights (TAF) are in an alphanumeric code. They are scheduled four times daily at 0000Z, 0600Z, 1200Z and 1800Z.

Format. The TAF is a series of groups made up of digits and letters. An individual group is identified by its position in the sequence, by its alphanumeric coding, by its length, or by a numerical indicator. Listed below are a few contractions used in the TAF. Some of the contractions are followed by time entries indicated by "tt" or "tttt" or by probability, "pp":

GRADU tttt—A gradual change occurring during a period in excess of one-half hour. "tttt" are the beginning and ending times of the expected change to the nearest hour; i.e., "GRADU 1213" means the transition will occur between 1200Z and 1300Z.

RAPID tt —A rapid change occurring in one-half hour or less. "tt" is the time to the nearest hour of the change; i.e., "RAPID 23" means the change will occur about 2300Z.

TEMPO tttt—Temporary changes from prevailing conditions lasting less than one hour. "tttt" are the earliest and latest times during which the temporary changes are expected; i.e., "TEMPO 0107" means the temporary changes may occur between 0100Z and 0700Z.

INTER tttt —Changes from prevailing conditions are expected to occur frequently and briefly. "tttt" are the earliest and latest times the brief changes are expected; i.e., "INTER 1518" means that the brief changes may occur between 1500Z and 1800Z.

PROBpp —Probability of conditions occurring. "pp" is the probability in percent; i.e., "PROB20" means a 20% probability of the conditions occurring.

FRONT tttt—Frontal passage. "tttt" is the time in hours and minutes of expected frontal passage; i.e., "FRONT 1645" means an expected frontal passage at 1645Z.

CAVOK —No clouds below 5,000 feet and visibility 6 miles or greater. No precipitation or thunderstorms.

WX NIL —No significant weather or obstructions to vision.

SKC —Sky clear.

Following is a St. Louis forecast in TAF code. It is the same as the preceding FT example except that it begins 2 hours later.

KSTL 1212 33025/35 0800 71SN 9//005 INTER 1215 0000 39BLSN 9//000 GRADU 1516 33020 4800 38BLSN 7SC030 TEMPO 1620 85SNSH GRADU 2122 33015 9999 WX NIL 3SC030 RAPID 00 VRB05 9999 SKC

Aviation Weather Forecasts

The forecast is broken down into the elements lettered "a" to "k" to aid in the discussion. Not included in the example but explained at the end are three optional forecast groups for "l" icing, "m" turbulence, and "n" temperature.

KSTL	1212	33025/35	0800	71SN	9//005
a.	b.	c.	d.	e.	f.

<u>INTER 1215 0000 39BLSN 9//000</u>
g.

<u>GRADU 1516 33020 4800 38BLSN 7SC030</u>
h.

<u>TEMPO 1620 85SNSH</u>
i.

<u>GRADU 2122 33015 9999 WX NIL 3SC030</u>
j.

<u>RAPID 00 VRB05 9999 SKC</u>
k.

a. *Station identifier.* The TAF code uses ICAO 4-letter station identifiers. In the contiguous 48 States the 3-letter identifier is prefixed with a "K"; i.e., the 3-letter identifier for Seattle is SEA while the ICAO identifier is KSEA. Elsewhere, the first two letters of the ICAO identifier tell what region the station is in. "MB" means Panama/Canal Zone (MBHO is Howard AFB); "MI" means Virgin Islands (MISX is St. Croix); "MJ" is Puerto Rico (MJSJ is San Juan); "PA" is Alaska (PACD is Cold Bay); "PH" is Hawaii (PHTO is Hilo).

b. *Valid time.* Valid time of the forecast follows station identifier. "1212" means a 24-hour forecast valid from 1200Z until 1200Z the following day.

c. *Wind.* Wind is forecast usually by a 5-digit group giving degrees in 3 digits and speed in 2 digits. When wind is expected to be 100 knots or more, the group is 6 digits with speed given in 3 digits. When speed is gusty or variable, peak speed is separated from average speed with a slash. For example, in the KSTL TAF, "33025/35" means wind 330°, average speed 25 knots, peak speed 35 knots. A group "160115/130" means wind 160°, 115 knots, peak speed 130 knots. "00000" means calm; "VRB" followed by speed indicates direction variable; i.e., "VRB10" means wind direction variable at 10 knots.

d. *Visibility.* Visibility is in meters. Table 4–2 is a table for converting meters to miles and fractions. "0800" means 800 meters converted from the table to ½ mile.

e. *Significant weather.* Significant weather is decoded using table 4–1. Groups in the table are numbered sequentially. Each number is followed by an acronym suggestive of the weather; you can soon learn to read most of the acronyms without reference to the table. Examples: "17TS", thunderstorm; "18SQ", squall; "31SA", sandstorm; "60RA", rain; "85SNSH", snow shower. "XX" between the number and acronym means "heavy". Examples: "33XXSA", heavy sandstorm; "67XXFZRA", heavy freezing rain. In the KSTL forecast, "71SN" means light snow. The TAF encodes only the single most significant type of weather; the U.S. domestic FT permits encoding of multiple weather types.

f. *Clouds.* A cloud group is a 6-character group. The first digit is coverage in octas (eighths) as shown in the top of table 4–3. The two letters identify cloud type as shown in the bottom of the table. The last three digits are cloud height in hundreds of feet. In the KSTL TAF, "9//005" means sky obscured (9), clouds not observed (//), vertical visibility 500 feet (005). The TAF may include as many cloud groups as necessary to describe expected sky condition.

g. and i. *Variation from prevailing conditions.* Variations from prevailing conditions are identified by the contractions INTER and TEMPO as defined earlier. In the KSTL TAF, "INTER 1215 0000 39BLSN 9//000" means intermittently from 1200Z to 1500Z (1215) visibility zero meters (0000) or zero miles, blowing snow (39BLSN), sky obscured, clouds not observed, vertical visibility zero (9//000). "TEMPO 1620 85SNSH" means between 1600Z and 2000Z, temporary, or brief, snow showers. Omission of other groups imply no significant change in wind, visibility, or cloud cover.

h, j, and k. *An expected change in prevailing conditions.* An expected change in prevailing conditions is indicated by the contraction GRADU, RAPID, or FRONT as defined earlier. In the KSTL TAF, "GRADU 1516 33020 4800 38BLSN 7SC030" means a gradual change between 1500Z and 1600Z to wind 330° at 20 knots, visibility 4,800 meters or 3 miles (table 4–2), blowing snow, ⅞ stratocumulus (table 4–3) at 3,000 feet. "GRADU 2122 33015 9999 WX NIL 3SC030" means a gradual change between 2100Z and 2200Z to wind 330° at 15 knots, visibility 10 kilometers

Aviation Weather Forecasts

or more (more than 6 miles), no significant weather, ⅜ stratocumulus at 3,000 feet. "RAPID 00 VRB05 9999 SKC" means a rapid change about 0000Z to wind direction variable at 5 knots, visibility more than 6 miles, sky clear.

l. Icing. An icing group may be included. It is a 6-digit group. The first digit is 6 identifying it as an icing group. The second digit is the type of ice accretion from table 4–4, top. The next three digits are height of the base of the icing layer in hundreds of feet. The last digit is the thickness of the layer in *thousands* of feet. For example, let's decode the group "680304". "6" indicates an icing forecast; "8" indicates severe icing in cloud (table 4–4); "030" says the base of the icing is at 3,000 feet; and "4" specifies a layer 4,000 feet thick.

m. Turbulence. A turbulence group also may be included. It also is a 6-digit group coded the same as the icing group except a "5" identifies the group as a turbulence forecast, and type of turbulence is from table 4–4, bottom. Decode the group "590359". "5" identifies a turbulence forecast; "9" specifies frequent severe turbulence in cloud; (table 4–4) "035" says the base of the turbulent layer is 3,500 feet; "9" specifies that the turbulence layer is 9,000 feet thick.

When either an icing layer or a turbulent layer is expected to be more than 9,000 feet thick, multiple groups are used; the top specified in one group is coincident with the base in the following group. Let's assume a cloud base at 5,000 feet and the forecaster expects frequent turbulence in thunderstorms from the surface to 45,000 feet; the most hazardous turbulence is at mid-levels. This could be encoded 530005 550509 591409 592309 553209 554104. While you most likely will never see such a complex coding with this many groups, the flexible TAF code permits it.

n. Temperature. A temperature code is seldom included in a terminal forecast. However, it may be included if critical to aviation. It may be used to alert the pilot to high density altitude or possible frost when on the ground. The temperature group is identified by the digit "0". The next two digits are time to the nearest hour GMT at which the temperature will occur. The last two digits are temperature in degrees Celsius. A minus temperature is preceded by the letter "M". Examples: "02137" means temperature at 2100Z is expected to be 37° C, about 99° F; "012M02" means temperature at 1200Z is expected to be minus 2° C. A forecast may include more than one temperature group.

TABLE 4–1. TAF weather codes

Code	Decode	Code	Decode
04FU	Smoke. Visibility reduced by smoke, e.g., veldt or forest fires, industrial smoke or volcanic ashes	17TS	Thunderstorms. Thunderstorm, but no precipitation at the time of observation
06HZ	Dust haze. Widespread dust in suspension in the air, not raised by wind at or near the station at the time of observation	18SQ	Squall. Squalls at or within sight of the station during the preceding hour or at the time of observation
08PO	Dust devils. Well-developed dust whirl(s) or sand whirl(s) seen at or near the station during the preceding hour or at the time of observation, but no duststorm or sandstorm	19FC	Funnel cloud. Funnel cloud(s) (tornado cloud or waterspout) at or within sight of the station during the preceding hour or at the time of observation
11MIFG	Shallow fog. Patches of shallow fog or ice fog at the station, whether on land or sea, not deeper than about 2 metres on land or 10 metres at sea	30SA	Duststorm or sandstorm. Slight or moderate duststorm or sandstorm—has decreased during the preceding hour
12MIFG	Shallow fog. More or less continuous shallow fog or ice fog at the station, whether on land or sea, not deeper than about 2 metres on land or 10 metres at sea	31SA	Duststorm or sandstorm. Slight or moderate duststorm or sandstorm—no appreciable change during the preceding hour

TABLE 4-1. TAF weather codes—Continued

Code	Decode
32SA	Duststorm or sandstorm. Slight or moderate duststorm or sandstorm—has begun or has increased during the preceding hour
33XXSA	Heavy duststorm or sandstorm. Severe duststorm or sandstorm—has decreased during the preceding hour
34XXSA	Heavy duststorm or sandstorm. Severe duststorm or sandstorm—no appreciable change during the preceding hour
35XXSA	Heavy duststorm or sandstorm. Severe duststorm or sandstorm—has begun or has increased during the preceding hour
36DRSN	Low drifting snow. Slight or moderate drifting snow—generally low (below eye level)
37DRSN	Low drifting snow. Heavy drifting snow generally low (below eye level)
38BLSN	Blowing snow. Slight or moderate blowing snow—generally high (above eye level)
39BLSN	Blowing snow. Heavy blowing snow—generally high (above eye level)
40BCFG	Fog patches. Fog or ice fog at a distance at the time of observation, but not at the station during the preceding hour, the fog or ice fog extending to a level above that of the observer
41BCFG	Fog patches. Fog or ice fog in patches
42FG	Fog. Fog or ice fog, sky visible—has become thinner during the preceding hour
43FG	Fog. Fog or ice fog, sky invisible—has become thinner during the preceding hour
44FG	Fog. Fog or ice fog, sky visible—no appreciable change during the preceding hour
45FG	Fog. Fog or ice fog, sky invisible—no appreciable change during the preceding hour
46FG	Fog. Fog or ice fog, sky visible—has begun or has become thicker during the preceding hour
47FG	Fog. Fog or ice fog, sky invisible—has begun or has become thicker during the preceding hour
48FZFG	Freezing fog. Fog, depositing rime, sky visible
49FZFG	Freezing fog. Fog, depositing rime, sky invisible
50DZ	Drizzle. Drizzle, not freezing, intermittent—slight at time of observation
51DZ	Drizzle. Drizzle, not freezing, continuous—slight at time of observation
52DZ	Drizzle. Drizzle, not freezing, intermittent—moderate at time of observation
53DZ	Drizzle. Drizzle, not freezing, continuous—moderate at time of observation
54XXDZ	Heavy drizzle. Drizzle, not freezing, intermittent—heavy (dense) at time of observation
55XXDZ	Heavy drizzle. Drizzle, not freezing, continuous—heavy (dense) at time of observation
56FZDZ	Freezing drizzle. Drizzle, freezing, slight
57XXFZDZ	Heavy freezing drizzle. Drizzle, freezing, moderate or heavy (dense)
58RA	Rain. Drizzle and rain, slight
59RA	Rain. Drizzle and rain, moderate or heavy
60RA	Rain. Rain, not freezing, intermittent—slight at time of observation
61RA	Rain. Rain, not freezing, continuous—slight at time of observation
62RA	Rain. Rain, not freezing, intermittent—moderate at time of observation
63RA	Rain. Rain, not freezing, continuous—moderate at time of observation
64XXRA	Heavy rain. Rain, not freezing, intermittent—heavy at time of observation
65XXRA	Heavy rain. Rain, not freezing, continuous—heavy at time of observation
66FZRA	Freezing rain. Rain, freezing, slight

TABLE 4-1. *TAF weather codes*—Continued

Code	Decode
67XXFZRA	Heavy freezing rain. Rain, freezing, moderate or heavy
68RASN	Rain and snow. Rain or drizzle and snow, slight
69XXRASN	Heavy rain and snow. Rain or drizzle and snow, moderate or heavy
70SN	Snow. Intermittent fall of snowflakes—slight at time of observation
71SN	Snow. Continuous fall of snowflakes—slight at time of observation
72SN	Snow. Intermittent fall of snowflakes—moderate at time of observation
73SN	Snow. Continuous fall of snowflakes—moderate at time of observation
74XXSN	Heavy snow. Intermittent fall of snowflakes—heavy at time of observation
75XXSN	Heavy snow. Continuous fall of snowflakes—heavy at time of observation
77SN	Snow. Snow grains (with or without fog)
79PE	Ice pellets. Ice pellets, type (a)
80RASH	Showers. Rain shower(s), slight
81XXSH	Heavy showers. Rain shower(s), moderate or heavy
82XXSH	Heavy showers. Rain shower(s), violent
83RASN	Shower(s) of rain and snow mixed, slight
84XXRASN	Heavy showers of rain and snow. Shower(s) of rain and snow mixed, moderate or heavy
85SNSH	Snow showers. Snow shower(s), slight
86XXSN	Heavy snow showers. Snow shower(s), moderate or heavy
87GR	Soft hail. Shower(s) of snow pellets or ice pellets, type (b), with or without rain or rain and snow mixed—slight
88GR	Soft hail. Shower(s) of snow pellets or ice pellets, type (b), with or without rain or rain and snow mixed—moderate or heavy
89GR	Hail. Shower(s) of hail (hail, ice pellets, type (b), snow pellets), with or without rain and snow mixed, not associated with thunder—slight
90XXGR	Heavy hail. Shower(s) of hail (hail, ice pellets, type (b), snow pellets), with or without rain or rain and snow mixed, not associated with thunder—moderate or heavy
91RA	Rain. Slight rain at time of observation—thunderstorm during the preceding hour but not at time of observation
92XXRA	Heavy rain. Moderate or heavy rain at time of observation—thunderstorm during the preceding hour but not at time of observation
93GR	Hail. Slight snow, or rain and snow mixed or hail (hail, ice pellets, type (b), snow pellets) at time of observation—thunderstorm during the preceding hour but not at time of observation
94XXGR	Heavy hail. Moderate or heavy snow, or rain and snow mixed or hail (hail, ice pellets, type (b), snow pellets) at time of observation—thunderstorm during the preceding hour but not at time of observation
95TS	Thunderstorm. Thunderstorm, slight or moderate, without hail (hail, ice pellets, type (b), snow pellets) but with rain and/or snow at time of observation
96TSGR	Thunderstorm with hail. Thunderstorm slight or moderate with hail (hail, ice pellets, type (b), snow pellets) at time of observation
97XXTS	Heavy thunderstorm. Thunderstorm, heavy, without hail (hail, ice pellets, type (b), snow pellets) but with rain and/or snow at time of observation
98TSSA	Thunderstorm with duststorm or sandstorm. Thunderstorm combined with duststorm or sandstorm at time of observation
99XXTSGR	Heavy thunderstorm with hail. Thunderstorm, heavy, with hail (hail, ice pellets, type (b), snow pellets) at time of observation

TABLE 4-2. Visibility conversion—TAF code to miles

Meters	Miles
0000	0
0100	1/16
0200	1/8
0300	3/16
0400	1/4
0500	5/16
0600	3/8
0800	1/2
1000	5/8
1200	3/4
1400	7/8
1600	1
1800	1 1/8
2000	1 1/4
2200	1 3/8
2400	1 1/2
2600	1 5/8
2800	1 3/4
3000	1 7/8
3200	2
3600	2 1/4
4000	2 1/2
4800	3
6000	4
8000	5
9000	6
9999	more than 6

TABLE 4-3. TAF cloud code

Cloud amount		Cloud type	
0	0 (Clear)	CI	Cirrus
1	1 octa or less but not zero	CC	Cirrocumulus
		CS	Cirrostratus
2	2 octas	AC	Altocumulus
3	3 octas	AS	Altostratus
4	4 octas	NS	Nimbostratus
5	5 octas	SC	Stratocumulus
6	6 octas	ST	Stratus
7	7 octas or more but not 8 octas	CU	Cumulus
		CB	Cumulonimbus
8	8 octas (Overcast)	//	Cloud not visible due to darkness or obscuring phenomena
9	Sky obscured, or cloud amount not estimated		

TABLE 4-4. TAF icing and turbulence

Figure Code	Amount of ice accretion (TAF group 6)
0	No icing
1	Light icing
2	Light icing in cloud
3	Light icing in precipitation
4	Moderate icing
5	Moderate icing in cloud
6	Moderate icing in precipitation
7	Severe icing
8	Severe icing in cloud
9	Severe icing in precipitation

Figure Code	Turbulence (TAF group 5)
0	None
1	Light turbulence
2	Moderate turbulence in clear air, infrequent
3	Moderate turbulence in clear air, frequent
4	Moderate turbulence in cloud, infrequent
5	Moderate turbulence in cloud, frequent
6	Severe turbulence in clear air, infrequent
7	Severe turbulence in clear air, frequent
8	Severe turbulence in cloud, infrequent
9	Severe turbulence in cloud, frequent

AREA FORECAST (FA)

An area forecast (FA) is a forecast of general weather conditions over an area the size of several States. It is used to determine forecast enroute weather and to interpolate conditions at an airport for which no FT is issued. Figure 1-5, section 1, maps FA areas.

Aviation Weather Forecasts

Example of an FA:

MIA FA 200040.
01Z FRI—19Z FRI.
OTLK 19Z FRI—07Z SAT.
FLA E OF 85 DEGS GA AND CSTL WTRS . . .
HGTS ASL UNLESS NOTED . . .
SYNOPSIS . . . STNRY HI PRES RDG NCAR CST EWD OVR ATLC. E TO SE FLO CONTG OVR FLA AND GA . . .
SIGCLDS AND WX . . .
NRN AND CNTRL GA . . .
40 SCT VRBL BKN LYRD TO 140. AFT 07Z OCNL CIGS BLO 10 VSBYS BLO 5HK. CONDS IMPVG BY 15Z TO CIGS ABV 15 VSBY 5HK. OTLK. VFR . . .
E CST SECS CNTRL SRN FLA AND ADJ CSTL WTRS . . .
GENLY 25 SCT VRBL BKN TOPS 100–120. SCT SHWRS OVR WTRS DRFTG WWD OCNLY MOVG ONSHR CSTL AREAS WITH CONDS LCLY CIG 25 BKN 2RW TOPS 180. OTLK. VFR . . .
SRN GA AND RMNDR FLA AND ADJ WTRS . . .
NO SIGCLD AND WX. OTLK. VFR . . .
ICG . . . LCL MDT IN TCU/RW. FRZG LVL 110 N GA TO 140 S FLA.

FAs are scheduled every 12 hours. They cover an 18-hour period with an additional 12-hour outlook. All times are GMT in whole hours (two digits), i.e., 13Z. Wind speed is in knots; and wind direction, in degrees true. All distances except visibility are in nautical miles; visibility is in statute miles.

Each FA has the sections:
1. Heading
2. Forecast Area
3. Height Statement
4. Synopsis
5. Significant Clouds and Weather Plus Outlook
6. Icing and Freezing Level

Heading

The heading identifies an area forecast, the originating WSFO, the date and time of issue, and the valid periods of the forecast and outlook. For example,

FA MKC 131240
13Z THU–07Z FRI
OTLK 07Z FRI–19Z FRI

states that the FA was issued by Kansas City (MKC) on the 13th day of the month at 1240Z. The forecast is valid from 1300Z Thursday until 0700Z Friday with a categorical outlook from 0700Z Friday until 1900Z Friday.

Forecast Area

The area is in contractions identifying States; portions of States; and, where applicable, adjacent waters. For example,

TENN ARK LA MISS ALA FLA W OF 85 DEGS CSTL WTRS

means Tennessee, Arkansas, Louisiana, Mississippi, Alabama, Florida west of 85° Longitude and coastal waters adjacent to the area.

Height Statement

Each FA contains the statement,

HGTS ASL UNLESS NOTED

to alert the user that heights for the most part are above sea level. For example, "3 THSD BKN TOPS 100 HIR TRRN OBSCD" means broken clouds 3,000 feet tops 10,000 feet—all heights ASL; terrain above 3,000 feet will be obscured. Tops of clouds and bases and tops of icing are always ASL.

Heights *above ground level* may be denoted in either of two ways. (1) Ceiling by definition is above ground. Therefore, the contraction "CIG" indicates above ground. For example "CIGS GENLY BLO 1 THSD" means that ceilings are expected to be generally below 1,000 feet. (2) The contraction "AGL" means above ground level. "SCT 2 THSD AGL" means scattered clouds, bases 2,000 feet above ground level.

Synopsis

The synopsis briefly summarizes locations and movements of fronts, pressure systems, and circulation patterns. It also may give moisture and stability conditions.

Significant Clouds And Weather

The significant clouds and weather section, identified by the contraction,

SIGCLD AND WX

forecasts, in broad terms, cloudiness and weather significant to flight operations. Table 4-5 defines the contractions and compares them to the designators used in the FT.

Obstructions to vision are included when forecast visibility is 6 miles or less. Expected precipitation and thunderstorms are always included. Table 4-6 gives expected coverage indicated by the terms "isolated," "few," "scattered," and "numerous."

The SIGCLD and WX section usually is several paragraphs. The breakdown may be by States, by well known geographical areas, or in reference to location and movement of a pressure system or front. Figure 4-1 is a map to assist in identifying geographical areas.

A categorical outlook, identified by "OTLK", is included for each area breakdown. Examples "OTLK. VFR BCMG MVFR CIG F AFT 09Z" means that weather is expected to be VFR becoming marginal VFR due to low ceiling and to visibility restricted by fog after 0900Z.

TABLE 4-5. Contractions in FA

Contraction	FT Designator	Definition
CLR	CLR	Sky clear
SCT	SCT	Scattered
BKN	BKN	Broken
OVC	OVC	Overcast
OBSC	X	Obscured, obscure, or obscuring
PTLY OBSC	−X	Partly obscured
THN	−	Thin
VRBL	V	Variable
CIG	C	Ceiling
INDEF	W	Indefinite

TABLE 4-6. Areal coverage of showers and thunderstorms

Adjective	Coverage
Isolated	Extremely small number
Few	15% or less of area or line
Scattered	16% to 45% of area or line
Numerous	More than 45% of area of line

Icing

The contraction,

ICG

identifies the icing section which gives location, type, and extent of expected icing. It always includes the freezing level in hundreds of feet ASL. It may contain qualifying terms such as "ICG LKLY", icing likely; "MDT MXD ICGIC ABV FRZLVL", moderate mixed icing in clouds above the freezing level.

Amended Area Forecasts

Amendments to the FA are issued as needed. Only that portion of the FA being revised is transmitted as an amendment. Area forecasts are also amended and updated by inflight advisories.

TWEB ROUTE FORECASTS AND SYNOPSIS

The TWEB Route Forecast is similar to the Area Forecast (FA) except more specific information is contained in a route format. Forecast sky cover (height and amount of cloud bases), cloud tops, visibility (including vertical visibility), weather and obstructions to vision are described for a corridor 25 miles either side of the route. Cloud bases and tops are always ASL unless noted. Ceilings are always above ground.

The Synopsis is a brief statement of frontal and pressure systems affecting the route during the forecast validity period.

The TWEB Route Forecasts are prepared by the WSFOs for more than 300 selected short-leg and cross-country routes over the contiguous U.S., figure 1-8, section 1. WSFOs prepare synopses for the routes in their areas. These forecasts go into the Transcribed Weather Broadcasts (TWEB) and the Pilot's Automatic Telephone Weather Answering Service (PATWAS) transcriptions described in section 1. Individual route forecasts and synopses are also available by request/reply teletypewriter through any FSS or WSO.

The TWEB Route Forecasts and Synopses are issued by the WSFOs three times per day according to time zone. The early morning and midday forecasts are valid for 12 hours; the evening forecast, for 18 hours:

WSFO Location (time zone)	Issue time	Valid period
Eastern/Central	1040Z	11Z–23Z
	1740Z	18Z–06Z
	2240Z	23Z–17Z
Mountain/Pacific	1140Z	12Z–00Z
	1840Z	19Z–07Z
	2340Z	00Z–18Z

This schedule provides 24-hour coverage with most frequent updating during the hours of greatest general aviation activity.

Aviation Weather Forecasts

Example of a TWEB Synopsis:

BIS SYNS 252317. LO PRESS TROF MVG ACRS NDAK TDA AND TNGT. HI PRESS MVG SEWD FM CANADA INTO NWRN NDAK BY TNGT AND OVR MST OF NDAK BY WED MRNG.

BIS—Bismarck, N.D. WSFO issuing Synopsis and Route Forecasts

SYNS—Synopsis for the area covered by the Route Forecasts

25—25th day of the month

2317—Valid 23Z on the 25th to 17Z on the 26th (18 hours)

(Rest of Message)—LOW PRESSURE TROUGH MOVING ACROSS N. DAKOTA TODAY AND TONIGHT. HIGH PRESSURE MOVING SOUTHEASTWARD FROM CANADA INTO NORTHWESTERN N. DAKOTA BY TONIGHT AND MOST OF N. DAKOTA BY WEDNESDAY MORNING.

Example of a TWEB Route Forecast:

249 TWEB 252317 GFK MOT ISN. GFK VCNTY CIGS AOA 5 THSD TILL 12Z OTRW OVR RTE CIGS 1 TO 3 THSD VSBY 3 TO 5 MI IN LGT SNW WITH CONDS BRFLY LWR IN HVYR SNW SHWRS

249—Route number

TWEB—TWEB Route Forecast

25—25th day of month

2317—Valid 23Z on the 15th to 17Z on the 16th (18 hours)

GFK MOT ISN—Route: Grand Forks to Minot to Williston, N.D.

(Rest of Message)—GRAND FORKS VICINITY CEILINGS AT OR ABOVE 5000 FEET UNTIL 1200Z OTHERWISE OVER ROUTE CEILINGS 1 TO 3 THOUSAND FEET VISIBILITY 3 TO 5 MILES IN LIGHT SNOW WITH CONDITIONS BRIEFLY LOWER IN HEAVIER SNOW SHOWERS.

When visibility is not stated it is implied to be 7 miles or greater.

Because of their varied accessibility and route format, these forecasts perhaps are the most important and useful weather information available to the pilot today for flight operations and planning. You should become familiar with them and use them regularly.

INFLIGHT ADVISORIES (WS, WA, WAC)

Inflight advisories are unscheduled forecasts to advise enroute aircraft of development of potentially hazardous weather. They are also excellent for preflight planning and briefing. All heights are ASL unless noted; i.e., ceiling heights are always AGL. The advisories are of three types—SIGMET (WS), AIRMET (WA), and CONTINUOUS AIRMET (WAC).

SIGMET (WS)

A SIGMET advises of weather potentially hazardous to all categories of aircraft, specifically:

1. Tornadoes
2. Lines of thunderstorms (squall lines)
3. Embedded thunderstorms
4. Hail of ¾" or greater in diameter
5. Severe or extreme turbulence
6. Severe icing
7. Widespread sandstorms/duststorms lowering visibilities to below 3 miles

AIRMET (WA)

An AIRMET is for weather that may be hazardous to single engine and light aircraft and in some cases to other aircraft as well, specifically:

1. Moderate icing
2. Moderate turbulence
3. Sustained winds of 30 knots or greater at or within 2,000 feet of the surface
4. Onset of extensive areas of visibility below 3 miles and/or ceilings less than 1,000 feet, including mountain ridges and passes

AIRMET Continued (WAC)

AIRMET Continued is issued for:

1. Continued moderate turbulence over mountainous terrain
2. Continued ceilings below 1,000 feet and/or visibility less than 3 miles over an extensive area.

Aviation Weather Forecasts

Valid Period

A WAC remains valid until cancelled. Valid period of a WS or WA is specifically stated.

Format

Format of an advisory consists of a heading and text.

Heading. The heading identifies the (1) issuing WFSO, (2) type of advisory, and (3) valid period. Examples:

BOS WS 202210
202210–210300Z

A SIGMET issued by Boston WSFO on the 20th at 2210Z valid from 2210Z on the 20th until 0300Z on the 21st.

MKC WA 052055
052100–060100Z

An AIRMET issued by Kansas City WSFO on the 5th at 2055Z valid from 2100Z on the 5th until 0100Z on the 6th.

SLC WAC 060800
060800–UFN

An AIRMET Continued issued by Salt Lake City WSFO on the 6th at 0800Z valid from 0800Z on the 6th until further notice (UFN).

Text. The text of the advisory contains (1) a message identifier, (2) a flight precautions statement, and (3) further details if necessary.

Message identifier. A WSFO identifies each hazardous area by a phonetic identifier (ALFA, BRAVO, CHARLIE, etc.). Advisories for each hazardous area are numbered sequentially (ALFA 1, ALFA 2, ALFA 3, etc.; or BRAVO 1, BRAVO 2, BRAVO 3, etc.). A new advisory of the same alphabetic series by the same WSFO automatically cancels preceding advisories of the same series; i.e., ALFA 2 cancels ALFA 1, BRAVO 3 cancels BRAVO 2, etc. A new issuance by one WSFO does not cancel an advisory by another WSFO unless specifically stated; i.e., ALFA 2 by Kansas City *does not* cancel ALFA 1 by Fort Worth.

Flight precautions statement (FLT PRCTN). A flight precautions statement in each advisory states location and kind of hazard, and it also gives the onset time if the hazard is not already occurring. This flight precaution is used in the flight precaution statement of the TWEB and PATWAS.

Further details. Further details describe the hazard when necessary. If the hazard is expected to continue beyond the valid period shown in the heading of a WS or WA, the fact is stated. For example, "CONDS CONTG BYD 02Z".

Following are examples of each of the three types of inflight advisories:

BOS WS 202210
202210–210300Z

SIGMET ALFA 1. FLT PRCTN. SVR ICG NH WRN ME BLO 160. OVRNG CONDS RESULTING IN SVR ICGICIP SFC–160. CONDS CONTG BYD 03Z.

MKC WA 052055
052100–060100Z

AIRMET FOXTROT 6. FLT PRCTN. SE COLO SRN AND WRN KAN CIG GENLY BLO 10 VSBY BLO 3 IN FOG AND SNW. LCL FRZG PCPN SRN KAN. MDT MXD ICGICIP. CONDS CONTG BYD 01Z.

SLC WAC 060800
060800–UFN

AIRMET ECHO 1 FLT PRCTN. FQT MDT TURB LCLY STG UDDF OVR AND NR MTNS CNTRL AND SRN UTAH WILL CONT UFN. CONT AIRMET UNTIL CNCL NOTICE IS RCVD

WINDS AND TEMPERATURES ALOFT FORECAST (FD)

Winds and temperatures aloft are forecast for specific locations in the contiguous U.S. as shown in figure 1-3, section 1. FD forecasts are also prepared for a network of locations in Alaska.

Below is a sample FD message containing a heading and six FD locations. The heading always includes time during which the FD may be used (1800–0300Z in the example) and a notation "TEMPS NEG ABV 2400". Since temperatures above 24,000 are always negative, the minus sign is omitted.

FD WBC 151745
BASED ON 151200Z DATA
VALID 1600Z FOR USE 1800–0300Z. TEMPS NEG ABV 24000

Aviation Weather Forecasts

FT	3000	6000	9000	12000	18000	24000	30000	34000	39000
ALS			2420	2635−08	2535−18	2444−30	245945	246755	246862
AMA		2714	2725+00	2625−04	2531−15	2542−27	265842	256352	256762
DEN			2321−04	2532−08	2434−19	2441−31	235347	236056	236262
HLC		1707−01	2113−03	2219−07	2330−17	2435−30	244145	244854	245561
MKC	0507	2006+03	2215−01	2322−06	2338−17	2348−29	236143	237252	238160
STL	2113	2325+07	2332+02	2339−04	2356−16	2373−27	239440	730649	731960

Forecast Levels

The line labelled "FT" shows the 9 standard FD levels. Through 12,000 feet the levels are true altitude; 18,000 feet and above are pressure altitude. The FD locations are transmitted in alphabetical order.

Note that some lower level groups are omitted. No winds are forecast within 1,500 feet of station elevation. No temperatures are forecast for the 3,000-foot level or for a level within 2,500 feet of station elevation.

Decoding

A 4-digit group shows wind direction (reference true North) and windspeed. Look at the St. Louis (STL) forecast for 3,000 feet. The group 2113 means wind from 210° at 13 knots. The first two digits give direction in tens of degrees; the second two, speed in knots.

A 6-digit group includes forecast temperature. In the STL forecast, the coded group for 9,000 feet is 2332+02; wind is 230° at 32 knots temperature +2° C.

Encoded windspeed 100 to 199 knots have 50 added to the direction code and 100 subtracted from the speed. The STL forecast for 39,000 feet is "731960". Wind is 230° at 119 knots, temperature −60.

How do you recognize when coded direction has been increased by 50? Coded direction (in tens of degrees) ranges from 01 (010°) to 36 (360°). Thus, a coded direction of more than "36" indicates winds 100 knots or more; the coded direction will range from 51 through 86.

If windspeed is forecast at 200 knots or greater, the wind group is coded as 199 knots; i.e., "7799" is decoded 270° at 199 knots or greater. When forecast speed is less than 5 knots, the coded group is "9900" and read, "LIGHT AND VARIABLE."

Examples of decoding FD winds and temperatures:

Coded	Decoded
9900+00	Wind light and variable, temperature 0° C
2707	270° at 7 knots
850552	350° (85−50=35) at 105 knots (05+100=105), temperature −52° C.

SPECIAL FLIGHT FORECAST

When planning a *special category* flight and scheduled forecasts are insufficient to meet your needs, you may request a special flight forecast through any FSS or WSO. The contact forwards the request to a WSFO and receives the printed forecast via teletypewriter. Special category flights are hospital or rescue flights; experimental, photographic, or test flight; record attempts; and mass flights such as air tours, air races, and fly-aways from special events.

Make your request far enough in advance to allow ample time for preparing and transmitting the forecast. Advance notice of 6 hours is desirable. In making a request, give the:

1. Aircraft mission
2. Number and type of aircraft
3. Point of departure
4. Route of flight (including intermediate stops, destination, alternates)
5. Estimated time of departure
6. Time enroute
7. Flight restrictions (such as VFR, below certain altitudes, etc.)
8. Time forecast is needed

The forecast is written in plain language contractions as in the examples:

SPL FLT FCST ABQ-PHOTO MISSION-ABQ 121500Z. THIN CI CLDS AVGG LESS THAN TWO TENTHS CVR. VSBY MORE THAN 30. WNDS AND TEMPS ALF 10-2320+03. ABQ WSFO 121300Z.

SPL FLT OTLK MKC-RST 062100Z-062400Z. CIG 2 THSD OVC OR BTR. WNDS ALF 3-2320. MKC WSFO 052300Z.

HURRICANE ADVISORY (WH)

When a hurricane threatens, an abbreviated hurricane advisory (WH) is issued to alert aviation interests. The advisory gives location of the storm center, its expected movement and maximum winds in and near the storm center. It does not contain details of associated weather; specific ceilings, visibilities, weather, and hazards are in area and terminal forecasts and inflight advisories.

An example of an abbreviated aviation hurricane advisory:

WH MIA 181010
HURCN IONE AT 1105Z CNTRD 29.4N 75.2W OR 400 NMI E OF JACKSONVILLE FLA EXPCTD TO MOV N ABT 12 KT. MAX WNDS 110 KT OVR SML AREA NEAR CNTR AND HURCN WNDS WITHIN 55-75 MIS.

CONVECTIVE OUTLOOK (AC)

A convective outlook (AC) describes prospects of both severe and general thunderstorms during the following 24 hours. Use the outlook primarily for planning flights later in the day. Outlooks are transmitted by the National Severe Storms Forecast Center (NSSFC) about 0900Z and 1500Z.

A notation, "ABV SELS LIMITS", means activity probably will meet the criteria for a severe weather watch. Expected conditions requiring a severe weather watch are:

1. Severe thunderstorms—one or both of the following:

   (a) Damaging surface wind with gusts of 50 knots or more

   (b) Hail 3/4 inch or more in diameter

2. Tornado activity

Following is a convective outlook:

AC MKC 020840

MKC AC 020840
VALID 021200-031200Z

SQLN CRNTLY IN ECNTRL TEX PNHDL EXTNDS NWD INTO NRN TEX PNHDL AND SW KANS AS LN OF OVRRNG TSTMS. THIS LN MOVG EWD 30 KT WILL GRDLY INTSFY DURG THE FRNN WITH SVR TSTM ACTVTY WELL ABV SELS LIMITS EXPCD BGN BY LATE FRNN CNTRL OKLA TO N CNTRL TEX MOVG EWD DURG AFTN THRU ERN OKLA SE KANS AND NE TEX INTO MOST OF ARK SW AND SRN MO DURG AFTN AND EVE BFR DMNSHG. INTSFY LATE EVE NE ARK SE MO W KY AND W TENN.

TSTMS DURG PRD EXPCD TO RT OF LN DRT INK HOB GCK STJ BRL DAY HTS LOZ BNA LFK SAT DRT. FEW TSHWRS CRNTLY IN NEW ENG WILL DMNSH DURG FRNN HWVR ISOLD TSHWRS ALSO EXPCD DURG THE PRD TO RT OF LN DOV IPT MSS.

FORECASTER (NAME)

SEVERE WEATHER WATCH BULLETIN (WW)

A severe weather watch bulletin (WW) defines areas of possible severe thunderstorms or tornado activity. The bulletins are issued by the National Severe Storms Forecast Center at Kansas City, Mo. WWs are unscheduled and are issued as required. On the next page is a severe weather watch bulletin.

A severe thunderstorm watch describes expected areal coverage of thunderstorms using the density adjectives listed in table 4-6. A tornado watch simply states that the threat of tornadoes exists in the designated watch area. Forecasters do not attempt to indicate areal coverage of these extremely localized storms.

Status reports are issued as needed to show progress of storms and to delineate areas no longer under the threat of severe storm activity. Cancellation bulletins are issued when it becomes evident that no severe weather will develop or that storms have subsided and are no longer severe. The bulletins are self-explanatory.

When tornadoes or severe thunderstorms have developed, local WSOs and WSFOs issue local warnings.

BULLETIN
TORNADO WATCH NUMBER 451
ISSUED 455 PM CDT SEPT 29 1972

A . . . THE NATIONAL WEATHER SERVICE HAS ISSUED A TORNADO WATCH FOR . . .

NORTHERN MISSISSIPPI

SOUTHEASTERN ARKANSAS

NORTHEAST LOUISIANA
THE THREAT OF TORNADOES AND SEVERE THUNDERSTORMS WITH LARGE HAIL AND DAMAGING WINDS WILL EXIST IN THESE AREAS FROM 5 PM CDT UNTIL 9 PM CDT THIS FRIDAY EVENING. THE GREATEST THREAT OF TORNADOES AND SEVERE THUNDERSTORMS IS IN AN AREA ALONG AND 70 MILES . . . 60 NAUTICAL . . . EITHER SIDE OF A LINE FROM 30 MILES . . . 25 NAUTICAL . . . NORTH OF COLUMBUS MISSISSIPPI TO EL DORADO ARKANSAS

PERSONS IN OR CLOSE TO THE TORNADO WATCH AREA ARE ADVISED TO BE ON THE WATCH FOR LOCAL WEATHER DEVELOPMENTS AND FOR LATER STATEMENTS AND WARNINGS.

B . . . OTHER WATCH INFORMATION . . .

THIS TORNADO WATCH REPLACES TORNADO WATCH NUMBER 449 ISSUED AT 130 PM CDT . . . WATCH NUMBER 449 WILL NOT BE EFFECTIVE AFTER 5 PM CDT.

C . . . TORNADOES AND A FEW SVR TSTMS WITH HAIL SFC AND ALF TO 1 IN. EXTRM TURBC AND SFC WND GUSTS TO 65 K. SCTD CBS WITH MAX TOPS TO 550. MEAN WIND VECTOR 24535.

D . . . INSTBLTY LN NOW FM WRN TENN ACRS CNTRL ARK INTO NE TEX MOVG SEWD ABT 25 K.

E . . . OTHER TSTMS. CONTD RMNDR AC.

F . . . FORECASTER (NAME)

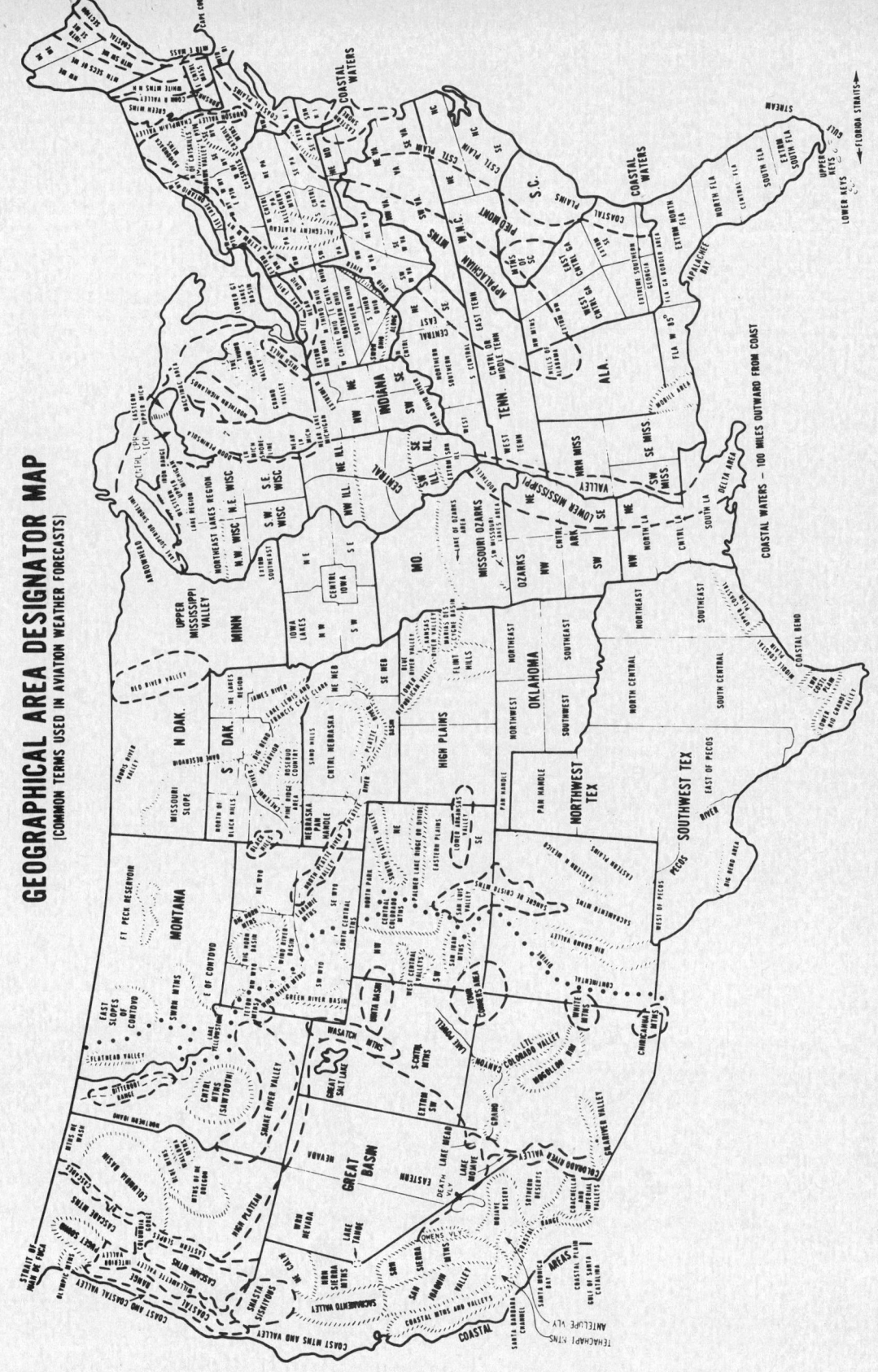

FIGURE 4-1. Geographical areas and terrain features. Forecasts often best locate weather by reference to terrain.

Chapter 17
SURFACE ANALYSIS

A surface analysis is commonly referred to as a surface weather map. In the contiguous 48 States a map covering these States and adjacent areas is transmitted every three hours. Other areas with facsimile receive surface weather maps appropriate to their areas at regularly scheduled intervals. Figure 5-1 is a section of a surface weather map and figure 5-2 illustrates symbols depicting fronts and pressure centers. The following explains contents of the chart.

VALID TIME

Valid time of the map corresponds to the time of the plotted observations. A date-time group in Greenwich Mean Time tells the user when conditions portrayed on the map were occurring.

ISOBARS

Isobars are solid lines depicting the pressure pattern. They are usually spaced at 4 millibar intervals. When pressure gradient is weak, dashed isobars are sometimes inserted at 2 millibar intervals to more clearly define the pressure pattern. Each isobar is labelled by a two-digit number. For example, 32 signifies 1032.0 mb; 00 is 1000.0 mb; and 92 is 992.0 mb.

PRESSURE SYSTEMS

The letter "L" denotes a low pressure center and an "H" marks a high pressure center. The pressure at each center is indicated by a two-digit underlined number which is interpreted the same as isobar labels.

FRONTS

The analysis shows frontal positions by the symbols in figure 5-2. The "pips" on the frontal symbols indicate the type of front and point the direction of movement. Pips on either side of the symbol of a stationary front suggest little or no movement. Briefing offices sometimes color the symbols to facilitate use of the map.

A three-digit number entered along a frontal symbol classifies the front as to type, table 5-1; intensity, table 5-2; and character, table 5-3. For example, the front at the lower left of figure 5-1 is labelled "463" meaning a cold front at the surface ("4" in table 5-1); moderate, increasing ("6" in table 5-2); and frontal activity increasing ("3" in table 5-3). Two short lines across a front indicate change in classification. Note in figure 5-1 the two lines crossing the cold front where its classification changes from "463" to "452".

TABLE 5-1. Type of front

Code Figure	Description
0	Quasi-stationary at surface
1	Quasi-stationary above surface
2	Warm front at surface
3	Warm front above surface
4	Cold front at surface
5	Cold front above surface
6	Occlusion
7	Instability line
8	Intertropical front
9	Convergence line

TABLE 5-2. Intensity of front

Code Figure	Description
0	No specification
1	Weak, decreasing
2	Weak, little or no change
3	Weak, increasing
4	Moderate, decreasing
5	Moderate, little or no change
6	Moderate, increasing
7	Strong, decreasing
8	Strong, little or no change
9	Strong, increasing

Surface Analysis

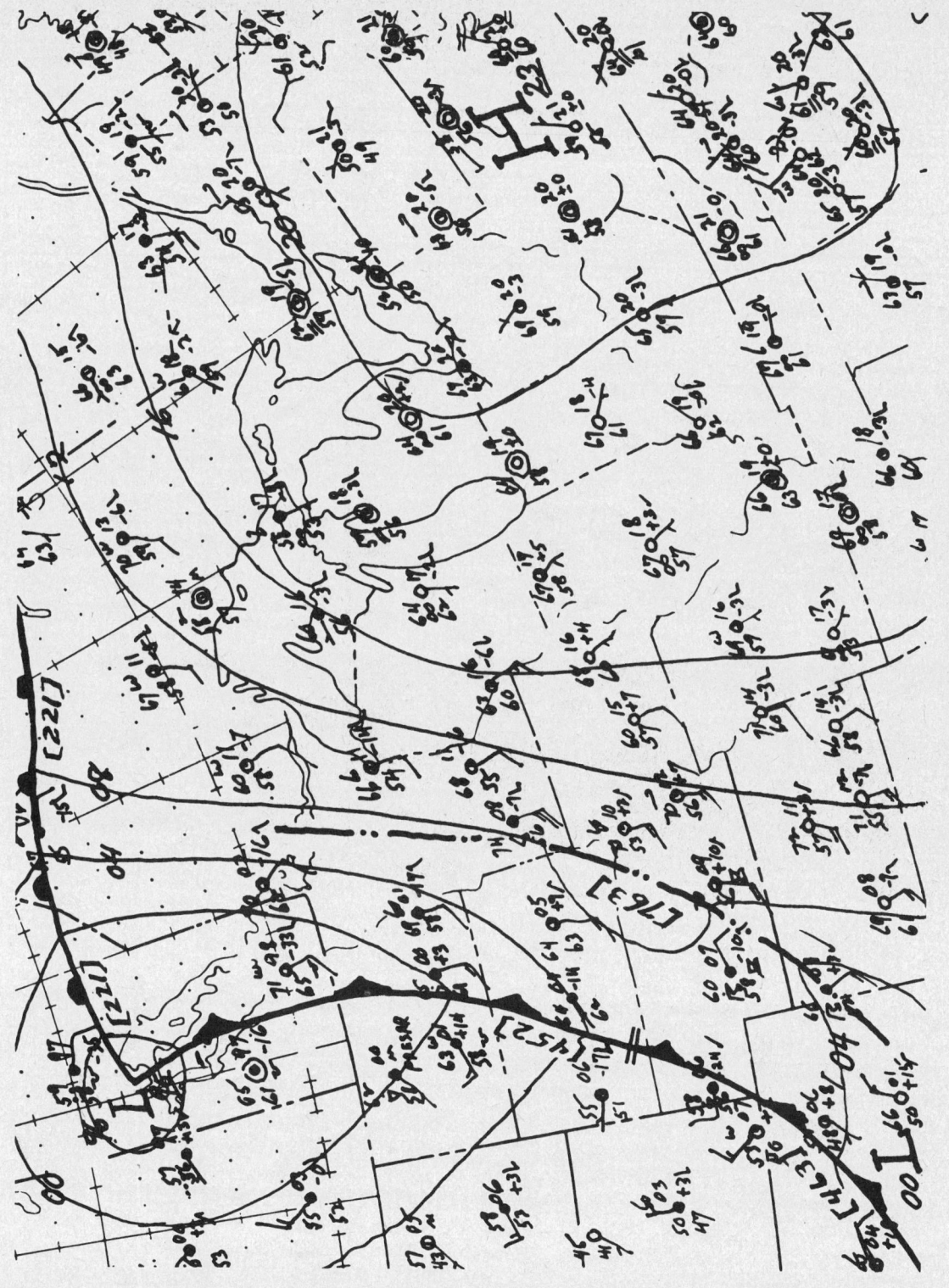

FIGURE 5-1. Section of a Surface Weather Analysis.

Surface Analysis

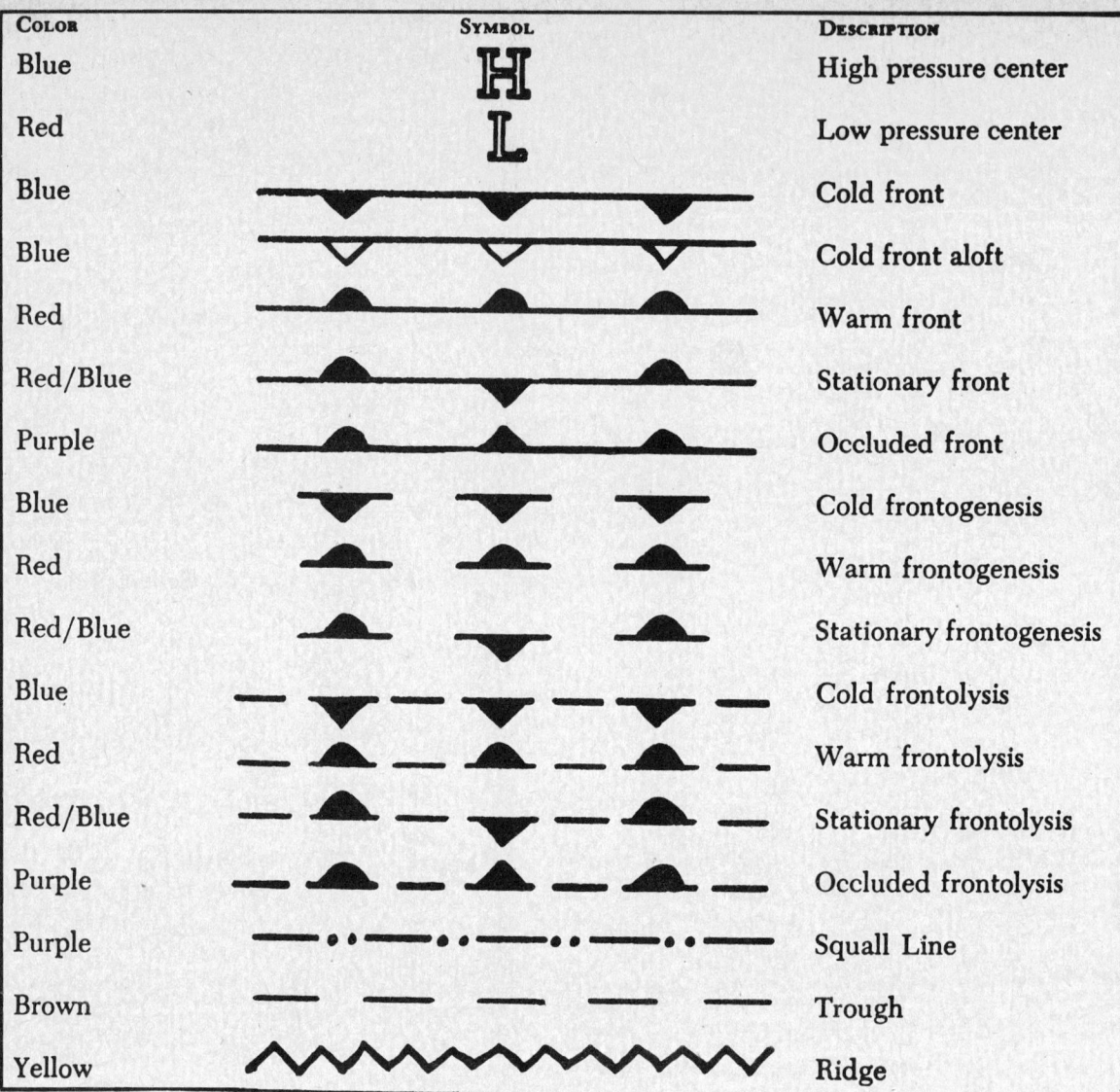

FIGURE 5-2. List of symbols on surface analyses. Colors are those suggested for on-station use. NOTE: A trough line usually is further identified by the coded group "830XX". Do not attempt to decode this group using the tables showing frontal classification. A trough line is not a front.

TABLE 5-3. Character of front

Code Figure	Description
0	No specification
1	Frontal area activity decreasing
2	Frontal area activity, little change
3	Frontal area activity increasing
4	Intertropical
5	Forming or existence expected
6	Quasi-stationary
7	With waves
8	Diffuse
9	Position doubtful

OTHER INFORMATION

Figure 5-3 shows an abbreviated station model which explains how to read temperature, dew point, and wind from the surface map. A complete station model plot contains detailed weather information more conveniently available from other facsimile charts and from teletypewriter data.

USING THE CHART

The surface analysis provides you a ready means of locating pressure systems and fronts and also gives you an overview of winds, temperatures, and dew points *as of map time*. When using the map, keep in mind that weather moves and conditions change. For example, a front located over northern Kansas may be nearing Oklahoma by the time you see the map. Using the surface map in conjunction with other charts such as weather depiction, radar summary, upper air, and prognostics (forecast charts) gives a more complete weather picture.

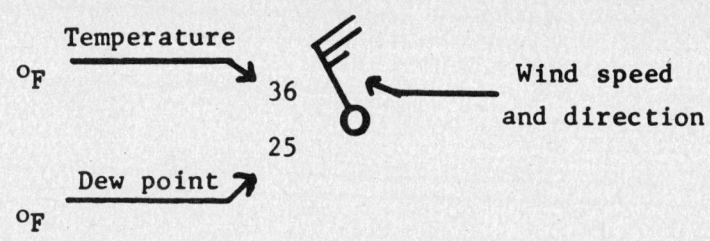

FIGURE 5-3. Abbreviated station model and explanation.

Chapter 18
WEATHER DEPICTION CHART

The weather depiction chart is prepared from surface aviation (SA) reports to give a quick picture of conditions as of valid time of the chart. Figure 6-1 is a weather depiction chart.

PLOTTED DATA

Shown for each plotted station as appropriate are:

1. Total sky cover
2. Height of cloud or ceiling
3. Weather and obstructions to vision, and
4. Visibility.

Total Sky Cover

Total sky cover is shown by the station circle shaded as in table 6-1.

TABLE 6-1. Total sky cover

Symbol	Total sky cover
○	Sky clear
ⓘ	Less than 1/10 (Few)
◐	1/10 to 5/10 inclusive (Scattered)
◕	6/10 to 9/10 inclusive (Broken)
⦿	10/10 with breaks (BINOVC)
●	10/10 (Overcast)
⊗	Sky obscured or partially obscured

Cloud Height or Ceiling

Cloud height is entered under the station circle in hundreds of feet—the same as coded in an SA report. If total sky cover is few or scattered, the height is the base of the lowest layer. If total sky cover is broken or greater, the height is the ceiling. Broken or greater sky cover without a height entry indicates *thin* sky cover. Partially or totally obscured sky is shown by the same sky cover symbol. Partial obscuration is denoted by absence of a height entry; total obscuration has a height entry denoting the ceiling (vertical visibility into the obscuration).

Weather and Obstructions to Vision

Weather and obstructions to vision are entered just to the left of the station circle using the same letter designators as used in SA reports. Precipitation intensity is not entered. When several types of weather and/or obstructions are reported at a station, only the most significant one or two types are entered. When an SA reports clouds topping ridges, a symbol unique to the weather depiction chart is entered to the left of the station circle:

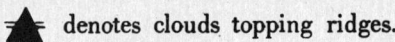 denotes clouds topping ridges.

Visibility

When visibility is less than 7 miles, it is entered to the left of weather and obstructions to vision. It is in miles and fractions.

Table 6-2 shows examples of plotted data.

TABLE 6-2. Examples of plotting on the Weather Depiction Chart

Plotted	Interpreted
ⓘ 8	Few clouds, base 800 feet, visibility more than 6
RW ◕ 12	Broken sky cover, ceiling 1,200 feet, rain shower
SH ⦿	Thin overcast with breaks, visibility 5 in haze
▲ ◐ 30	Scattered at 3,000 feet, clouds topping ridges
2F ○	Sky clear, visibility 2, ground fog or fog
1/2 BS ⊗	Sky partially obscured, visibility 1/2, blowing snow
1/4 S ⊗ 5	Sky obscured, ceiling 500, visibility 1/4, snow
1 TR ● 12	Overcast, ceiling 1,200 feet, thunderstorm, rain, visibility 1

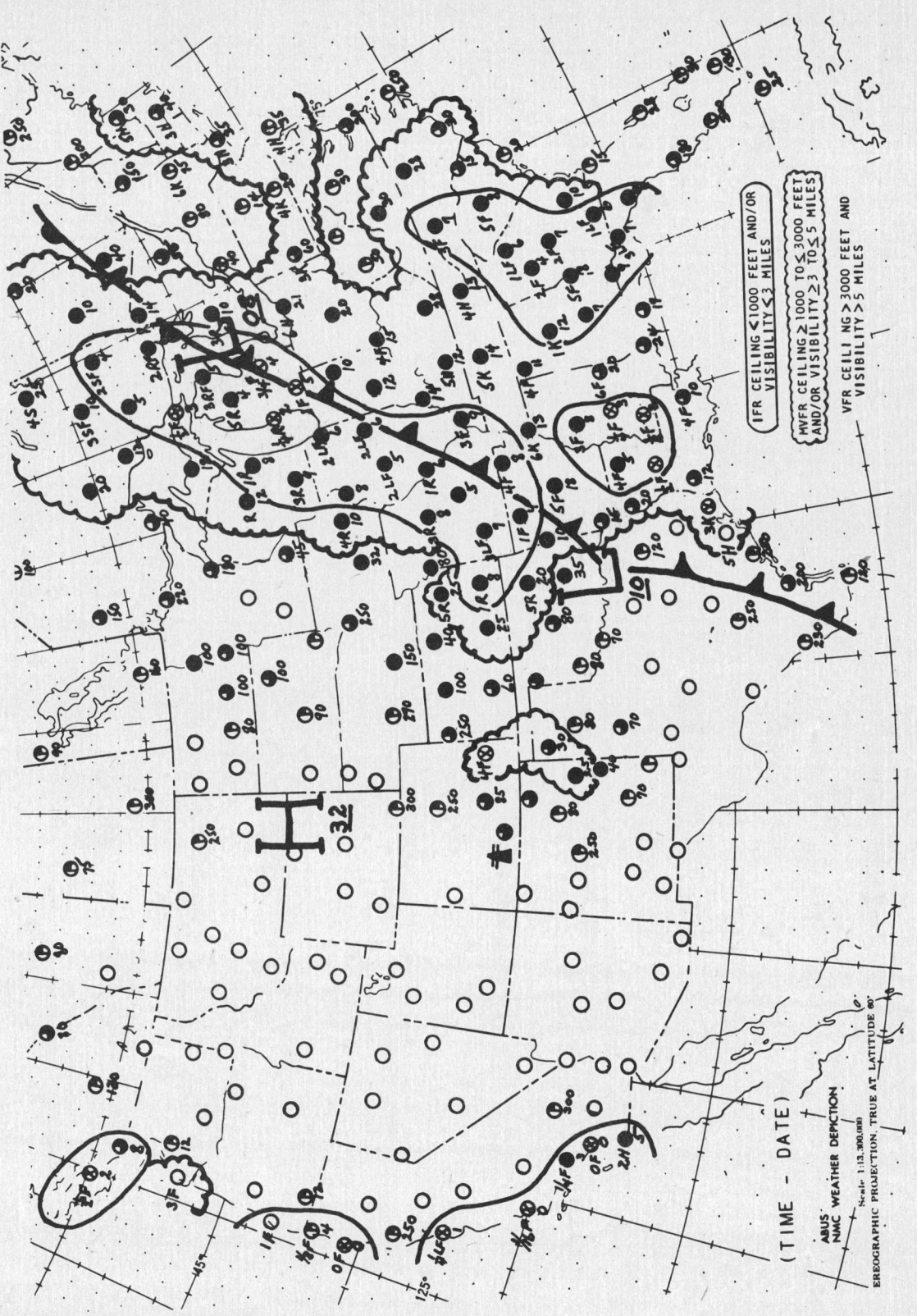

FIGURE 6-1. A Weather Depiction Chart.

Weather Depiction Chart

ANALYSIS

The chart shows observed ceiling and visibility by categories as follows:

1. IFR—Ceiling less than 1,000 feet and/or visibility less than 3 miles, outlined by a *smooth* line.
2. MVFR (Marginal VFR)—Ceiling 1,000 feet to 3,000 feet inclusive and/or visibility 3 to 5 miles inclusive, outlined by a *scalloped* line.
3. VFR—Ceiling greater than 3,000 feet or unlimited and visibility greater than 5 miles, *not* outlined.

In addition, the chart shows major fronts and high and low pressure centers from the surface analysis for the preceding hour. These features are depicted the same as on the surface chart.

USING THE CHART

The weather depiction chart is a choice place to begin your weather briefing and flight planning. From it, you can determine general weather conditions more readily than from any other source. It gives you a "bird's eye" view at map time of areas of favorable and adverse weather and pictures frontal and pressure systems associated with the weather.

The chart may not completely represent enroute conditions because of variations in terrain and weather between stations. Furthermore, weather changes; by the time the chart is available, plotted data around the stations have been superseded by SA reports. After you initially size up the general picture, your final flight planning must consider forecasts, progs, and the latest pilot, radar, and surface weather reports.

Chapter 19
RADAR SUMMARY CHART

A radar summary chart graphically displays a collection of radar reports. Valid time is time of the radar observations. Figure 7-1 is a radar summary chart. It shows precipitation echoes indicating their location, coverage, movement, and tops along with other pertinent weather information associated with the echoes. This section explains chart annotations, symbols, and use.

ECHO PATTERN AND COVERAGE

The *echo pattern* is the arrangement of echoes. A pattern may be (1) a line of echoes, (2) an area of echoes, or (3) an isolated cell. A cell is a solid convective mass normally 20 nautical miles or less in diameter. *Echo coverage* is the areal coverage of echoes or cells within an area or line. Table 7-1 shows depiction and symbols used to denote echo pattern and coverage.

TABLE 7-1. Echo coverage symbols on the Radar Summary Chart

Symbol	Meaning	Called
	A line of echoes	Line
	An area of echoes	Area
⊕	Over 9/10 coverage	Solid
	6/10 to 9/10 coverage	Broken
	1/10 to 5/10 coverage	Scattered
⊙	Less than 1/10 coverage	Widely scattered
O	Isolated cell	Cell
✱	Strong cell detected by two or more radars	Cell

WEATHER ASSOCIATED WITH ECHOES

Weather radar primarily detects particles of precipitation size within a cloud or falling from a cloud. The echo from an aggregate of particles does not specifically identify the type of precipitation. However, the radar observer usually can determine precipitation type from other sources. Table 7-2 lists symbols identifying type of precipitation associated with echoes.

TABLE 7-2. Weather symbols

Symbol	Meaning
R	Rain
RW	Rain showers
A	Hail
S	Snow
IP	Ice Pellets
SW	Snow showers
L	Drizzle
T	Thunderstorm
ZR, ZL	Freezing precipitation

INTENSITY AND TREND OF PRECIPITATION

Type of precipitation is further annotated to show *intensity* and *intensity trend*. Intensity follows the precipitation symbol, and a solidus (/) separates intensity from intensity trend. Table 7-3 lists symbols for intensity; table 7-4, for intensity trend.

TABLE 7-3. Echo intensity

Symbol	Echo intensity	Estimated precipitation
—	Weak	Light
(none)	Moderate	Moderate
+	Strong	Heavy
++	Very strong	Very heavy
X	Intense	Intense
XX	Extreme	Extreme
U	Unknown	Unknown

Radar Summary Chart

TABLE 7-4. Intensity trend

Symbol	Meaning
+	Increasing
−	Decreasing
NC	No change
NEW	New

Examples of precipitation type, intensity, and trend are:

R−/+	Light rain, increasing in intensity.
TRW+/−	Thunderstorm, heavy rain shower, decreasing in intensity.
RW/NC	Moderate rain shower, no change in intensity.
TRW−/NEW	Thunderstorm, light rain shower, newly developed.
TRWXX/NC	Thunderstorm, rain shower extreme intensity, no change.
S	Snow. (No intensity or characteristic is shown for frozen precipitation.)

HEIGHTS OF ECHO BASES AND TOPS

Heights in hundreds of feet MSL are entered above and/or below a line to denote echo tops and bases respectively. Examples are:

450	Average tops 45,000 feet
220/80	Bases 8,000 feet; tops 22,000 feet
330	Top of an individual cell, 33,000 feet
\650/	Maximum tops, 65,000 feet
A350	Tops 35,000 feet reported by aircraft

Absence of a figure below the line indicates that the echo base was not reported. Radar detects tops more readily than bases since precipitation usually reaches the ground. Also, curvature of the earth prohibits the detection of bases of distant precipitation.

Vertical extent of echoes is measured by a range-height indicator. Primary weather radar has this feature, but Air Traffic Control (ATC) radar does not. Information from ATC radar shows tops only when reported by aircraft. Most radar weather reports across the intermountain regions of the western United States are from ATC radar.

MOVEMENT OF ECHOES

Movement of echoes is also shown. Movement of individual storms within a line or area often differs from the movement of the overall storm pattern. Movement of individual echoes is shown by a direction arrow and a number representing speed in knots. Movement of a line or area is shown by an arrow with flags, a full flag for 10 knots and a half flag for 5 knots.

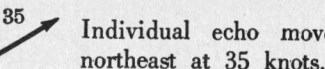

Individual echo movement to the northeast at 35 knots.

Line or area movement to the east at 20 knots.

ADDITIONAL INFORMATION

"Boxes" enclosed by a dashed line indicate a severe weather watch in effect. Refer to the latest severe weather watch (WW) for specifics.

When reports from a particular radar station do not appear on the chart, notations plotted at the radar site give the reason. Table 7-5 lists the notations and their meanings.

TABLE 7-5. Symbols indicating no echoes

Symbol	Meaning
NE	No echo (equipment operating but no echoes observed)
NA	Observation not available
OM	Equipment out for maintenance

USING THE CHART

The radar summary chart aids in preflight planning by identifying general areas and movement of precipitation and/or thunderstorms. Radar detects only drops or ice particles of precipitation size; it does not detect clouds and fog. Therefore, the absence of echoes does not guarantee clear weather. Furthermore, cloud tops may be higher than precipitation tops detected by radar. The chart must be used in conjunction with other charts, reports and forecasts.

Examine chart annotations carefully. Always determine location and movement of echoes. If echoes are anticipated near your planned route, take special note of echo intensity and trend. Echoes of light or moderate intensity may contain turbulence. Echoes of strong or greater intensity or echoes increasing in intensity may contain hazardous turbulence and hail. Echo tops also are often a clue to severity of thunderstorms. A good rule is to avoid echoes with tops of 35,000 feet or higher.

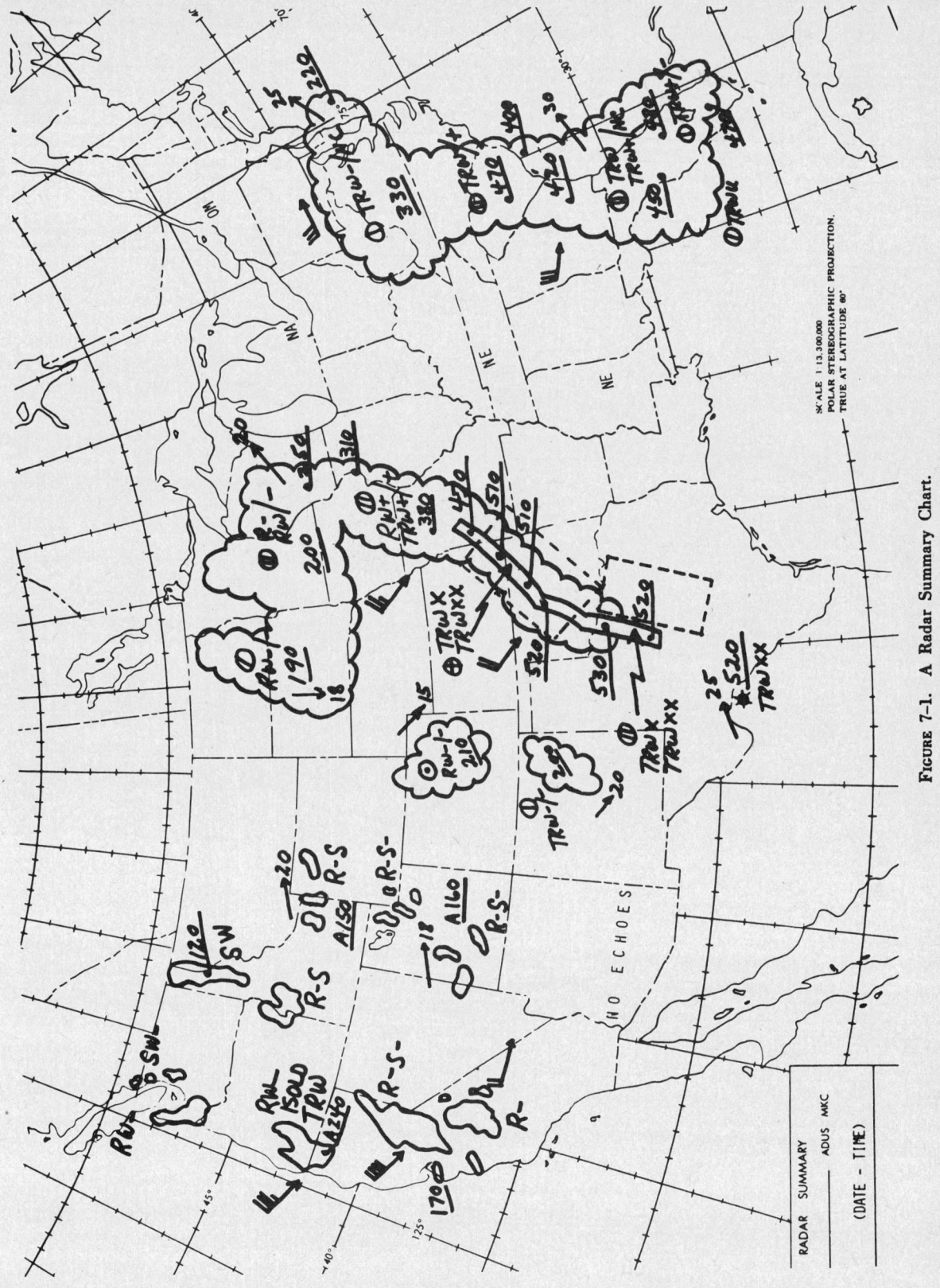

FIGURE 7-1. A Radar Summary Chart.

Radar Summary Chart

Suppose your proposed route will take you through an area of widely scattered thunderstorms with no increase anticipated. When these storms are separated by good VFR weather, you most likely can pick your way among them, visually sighting and circumnavigating the storms. However, widespread cloudiness may conceal the thundertorms. To avoid these embedded thunderstorms, you must either use airborne radar or detour the area. Echoes reported as broken or solid are difficult to circumnavigate either visually or by airborne radar. More details on avoiding hazards of thunderstorms are given in chapter 11, Aviation Weather.

Keep in mind that the chart is for preflight planning only. Once airborne, you must evade individual storms from inflight observations either by visual sighting or by airborne radar.

Chapter 20
SIGNIFICANT WEATHER PROGNOSTICS

Significant weather prognostic charts, called "progs" for brevity, portray forecast weather which may influence flight planning. Table 8-1 explains some symbols used on these charts. Significant weather progs are issued both for domestic and international flights.

TABLE 8-1. *Some standard weather symbols*

Symbol	Meaning
∧	Moderate turbulence
∧̂	Severe turbulence
ψ	Moderate icing
ψ̂	Severe icing
●	Rain
✱	Snow
❜	Drizzle
▽	Rain shower
✴	Snow shower
⎇	Thunderstorm
∽	Freezing rain
6	Tropical storm
⚲	Hurricane (typhoon)

NOTE: Character of precipitation is the manner in which it occurs. It may be intermittent or continuous. A single symbol denotes intermittent, a pair of symbols indicates continuous.

● ●	Continuous rain
✱	Intermittent snow
❜ ❜	Continuous drizzle

DOMESTIC FLIGHTS

Significant weather progs are prepared for the conterminous U.S. and adjacent areas. The U.S. low level significant weather prog is designed for domestic flight planning to 24,000 feet and a U.S. high level prog is for domestic flights above 24,000 feet to 45,000 feet. Chart legends include valid time in GMT.

U.S. Low Level Significant Weather Prog

The low level prog is a four-panel chart as shown in figure 8-1. The two lower panels are 12- and 24-hour surface progs. The two upper panels are 12- and 24-hour progs of significant weather from the surface to 400 millibars (24,000 feet). The charts show conditions as they are forecast to be *at* the valid time of the chart.

Surface Prog. The two surface prog panels use standard symbols for fronts and pressure centers explained in section 5. Movement of each pressure center is indicated by an arrow showing direction and a number indicating speed in knots. Isobars depicting forecast pressure pattern are included on some 24-hour surface progs.

The surface prog outlines areas of forecast precipitation and/or thunderstorms as shown in the lower panels of figure 8-1. Smooth lines enclose areas of expected continuous or intermittent precipitation; dash-dot lines enclose areas of showers or thunderstorms. Note that symbols indicate precipitation type and character. If precipitation will affect half or more of an area, that area is shaded; absence of shading denotes more sparse precipitation, specifically less than half areal coverage. Look at the lower left panel of figure 8-1. At 1200Z the forecast is for continuous snow affecting half or more of an area in portions of the northern Rocky Mountain States. Showers and thunderstorms affecting less than half the area are forecast on the same prog for the central Gulf Coast. Rain or drizzle over less than half the area is indicated for the Carolinas.

Significant Weather. The upper panels of figure 8-1 depict ceiling, visibility, turbulence, and freezing level. Note the legend near the center of the chart which explains methods of depiction.

Smooth lines enclose areas of forecast IFR weather; scalloped lines enclose areas of marginal

Significant Weather Prognostics

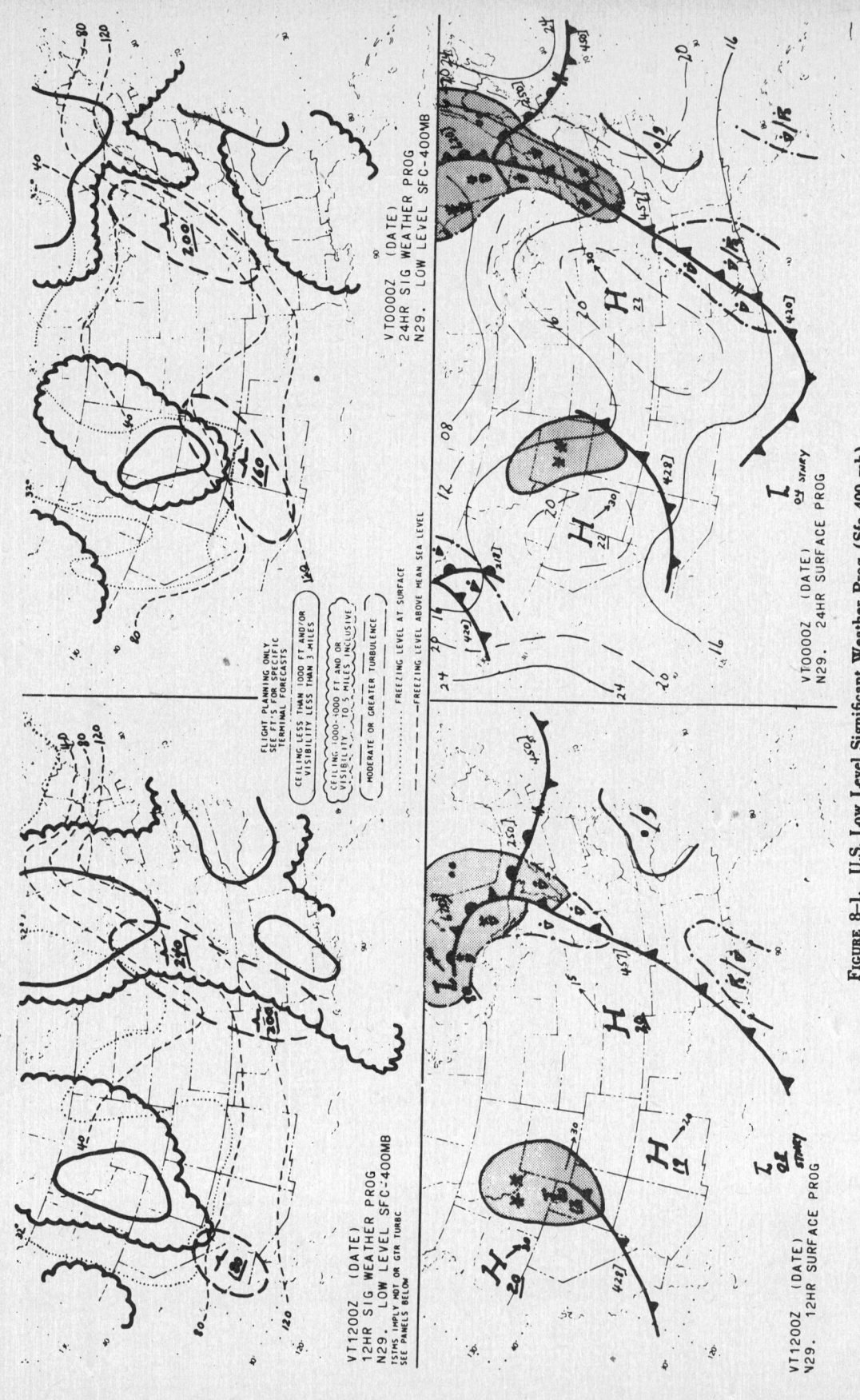

FIGURE 8-1. U.S. Low Level Significant Weather Prog (Sfc-400 mb).

weather (MVFR); VFR areas are not outlined. Recall that this is the same manner of depiction used on the weather depiction chart to portray ceiling and visibility.

Long-dashed lines enclose general areas of forecast moderate or greater turbulence. However, thunderstorms forecast on a surface prog always imply moderate or greater turbulence in the storms even though a general area of turbulence may not be outlined on the associated significant weather panel. (See legend, 12-hr sig weather prog, figure 8-1).

A symbol entered within a general area of forecast turbulence denotes intensity. Figures below and above a short line show expected base and top of the turbulent layer in hundreds of feet. Absence of a figure below the line indicates turbulence from the surface upward. No figure above the line indicates turbulence extending above the upper limit of the chart. Turbulence forecast from the surface to above 24,000 feet is indicated by the notation "SFC" below the line. In the upper left panel of figure 8-1, the annotation appearing in extreme southern California denotes moderate turbulence, surface to 18,000 feet.

Freezing level height contours for the uppermost freezing level are drawn at 4,000-foot intervals. The 4,000-foot contour terminates at the 4,000-foot terrain level along the Rocky Mountains. Contours are labelled in hundreds of feet MSL. Freezing level line at the surface is labelled "32°". An upper freezing level contour crossing the surface 32° line indicates multiple freezing levels due to layers of warmer air aloft.

The low level significant weather prog does not specifically outline areas of icing. However, icing is always implied in clouds and precipitation above the freezing level.

U.S. High Level Significant Weather Prog

The U.S. high level significant weather prog, figure 8-2, encompasses airspace from 400 to 150 millibars (24,000 feet to 45,000 feet pressure altitude). The prog outlines areas of forecast turbulence, continuous dense cirriform clouds, and cumulonimbus clouds. Table 8-2 interprets some examples of chart annotation.

Turbulence. Long-dashed lines enclose areas of probable moderate or greater turbulence. Symbols denote intensity, base, and top. Cumulonimbus clouds *imply* moderate or greater turbulence and icing.

Cirriform Clouds. Large-scalloped lines enclose areas of dense, continuous cirriform clouds of broken or overcast coverage. Expected base and top are given with the notation "LYR" meaning either single or multiple layers. A single digit preceding the notation "LYR" is coverage in "octas" or eighths, eight-eighths being overcast.

Cumulonimbus Clouds. Small-scalloped lines enclose areas of expected cumulonimbus development. The contraction "CB" denotes cumulonimbus; a digit preceding the contraction denotes coverage in octas. The notation "FEW CB" denotes less than one-eighth coverage.

Cumulonimbus coverage and heights represent an overall average for the forecast area. When a wide variation is expected within an area, separate CB amounts and heights may be indicated.

INTERNATIONAL FLIGHTS

Significant weather progs for international aviation cover large geographical areas. The areas covered extend from eastern Asiatic coastal areas across the Pacific; the Atlantic into Europe; northwestern Africa; and part of the Southern Hemisphere including northern South America and the South Pacific. These progs appear on both Mercator and polar stereographic projections, but methods of depiction are the same.

The only low level (surface to 400 millibars) international significant weather prog is for the North Atlantic area. All other international significant weather progs are high level progs from 400 millibars to 150 millibars.

North Atlantic Significant Weather Prog

Figure 8-3 is a North Atlantic low level significant weather prog. Note that it shows forecast positions of surface pressure centers and fronts using standard symbols. Scalloped lines depict areas of only broken or overcast layered clouds but any amount of cumulonimbus. Cloud cover is in octas, or eighths. Bases and tops are shown as on the U.S. significant weather progs. No figure above the line indicates tops above 24,000 feet. For example, note the lower left portion of figure 8-3. The annotations labelled "FEW CB" indicate less than one-eighth cumulonimbus with bases at 1,500 feet and tops above 24,000 feet. The area near the upper center of the chart labelled "8 LYR" denotes eight-eighths, or overcast, layered clouds with bases 1,500 feet and tops 18,000 feet.

Forecast cumulonimbus always implies turbulence and icing. Standard symbols and annotations indicate other areas of expected turbulence and icing. For example, on figure 8-3, the annotation at about 45°N and 50°W indicates icing from 11,000

Significant Weather Prognostics

TABLE 8-2. Depiction of clouds and turbulence on a High Level Significant Weather Prog

Depiction	Meaning
1. FEW CB 420 (in scalloped cloud outline)	Few (less than one-eighth coverage) cumulonimbus, tops 42,000 feet. Bases are below 24,000 ft.—the lower limit of the prog.
2. 3 CB ABV 450 (in scalloped cloud outline)	Three-eighths cumulonimbus, tops above 45,000 feet.
3. 6 LYR 330/280 (in cloud outline)	Six-eighths coverage (broken), layered cirriform clouds, base 28,000 and tops 33,000 feet.
4. ⋀ to ⋀ 330 (in dashed oval)	Moderate to severe turbulence from below lower limit of the prog (24,000 feet) to 33,000 feet. (Consult low-level prog for turbulence forecasts below 24,000 feet.)
5. ⋀ 390 (in dashed oval)	Severe turbulence from 39,000 feet to above upper limit of the prog (45,000 feet).

NOTES:

Base and top shown by figures below and above a short line respectively.

Cumulonimbus Clouds, Examples 1 and 2. Bases always below 400 millibars and are not shown. Tops above 150 millibars shown as "ABV 450".

Cirriform Clouds and Turbulence. Figure below the line omitted when base is below 400 millibars. Figure above omitted when top above 150 millibars.

feet to 16,000 feet within the "7 LYR" of clouds. Referring to the upper center of the chart, an area of moderate turbulence is forecast from the surface to 18,000 feet.

Freezing level on this prog is indicated only by a 10,000-foot MSL contour. Note on figure 8–3 the dashed line labelled "100". Freezing level north of this line is below 10,000 feet; south of the contour, freezing level is above 10,000 feet.

International High Level Significant Weather Progs

Figure 8–4 is an international high level significant weather prog. The legend "SIG WX (400–150MB)" in the upper right identifies the chart. This example on a Mercator projection covers the Pacific ocean from about Honolulu westward to the Asiatic coast.

Annotations on international high level significant weather progs are the same as on U.S. high level significant weather progs. In addition, the international prog shows surface positions of pressure centers, fronts, tropical storms, and hurricanes (typhoons in the western Pacific). Note on figure 8–4 near 28° N. and 138° E. the typhoon, "ELLEN". Also shown are tropical storm "DOT" near 31° N. and 127° E. and tropical depression "BILLIE", marked as a low pressure center at 37° N. and 121° E.

USING SIGNIFICANT WEATHER PROGS

Use the significant weather progs in planning your flight to avoid areas and/or altitudes of most probable significant icing and turbulence. You may also plan your flight to remain clear of extensive cloudiness. By comparing progs with analyses, you can determine expected movement and changes in weather patterns.

Significant Weather Prognostics

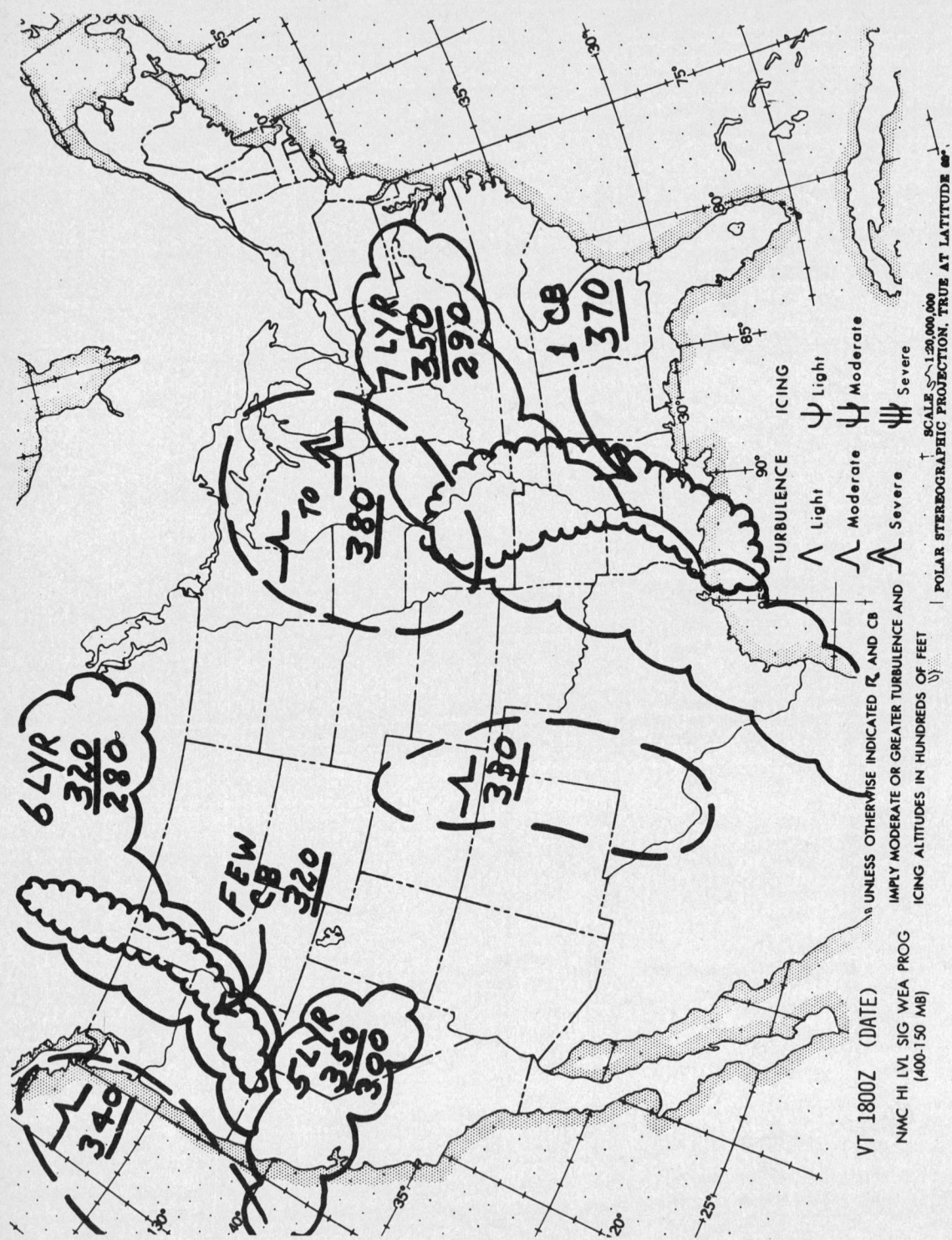

FIGURE 8-2. U.S. High Level Significant Weather Prog (400–150 mb).

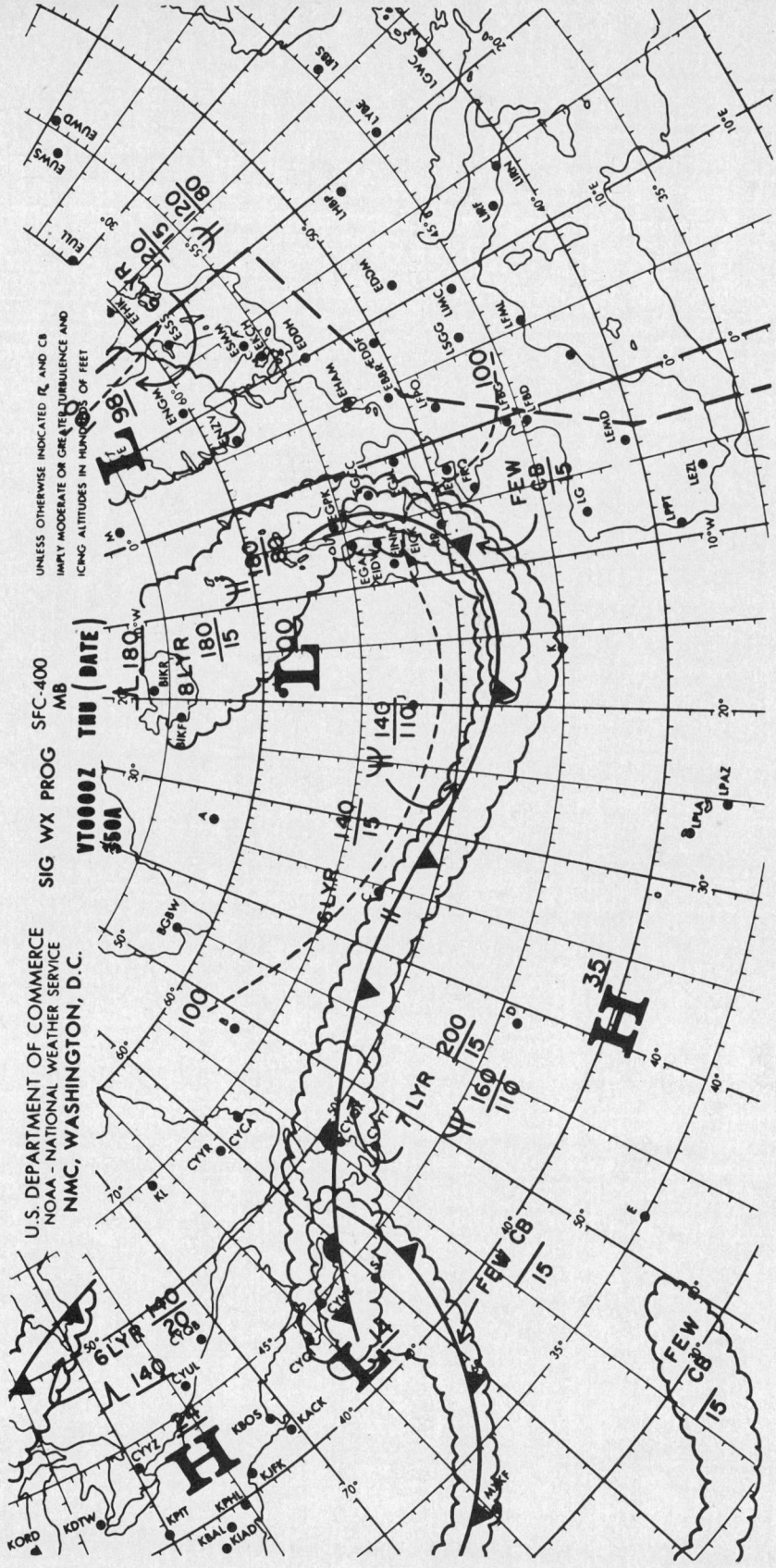

FIGURE 8-3. North Atlantic Low Level Significant Weather Prog (Sfc-400 mb).

Significant Weather Prognostics

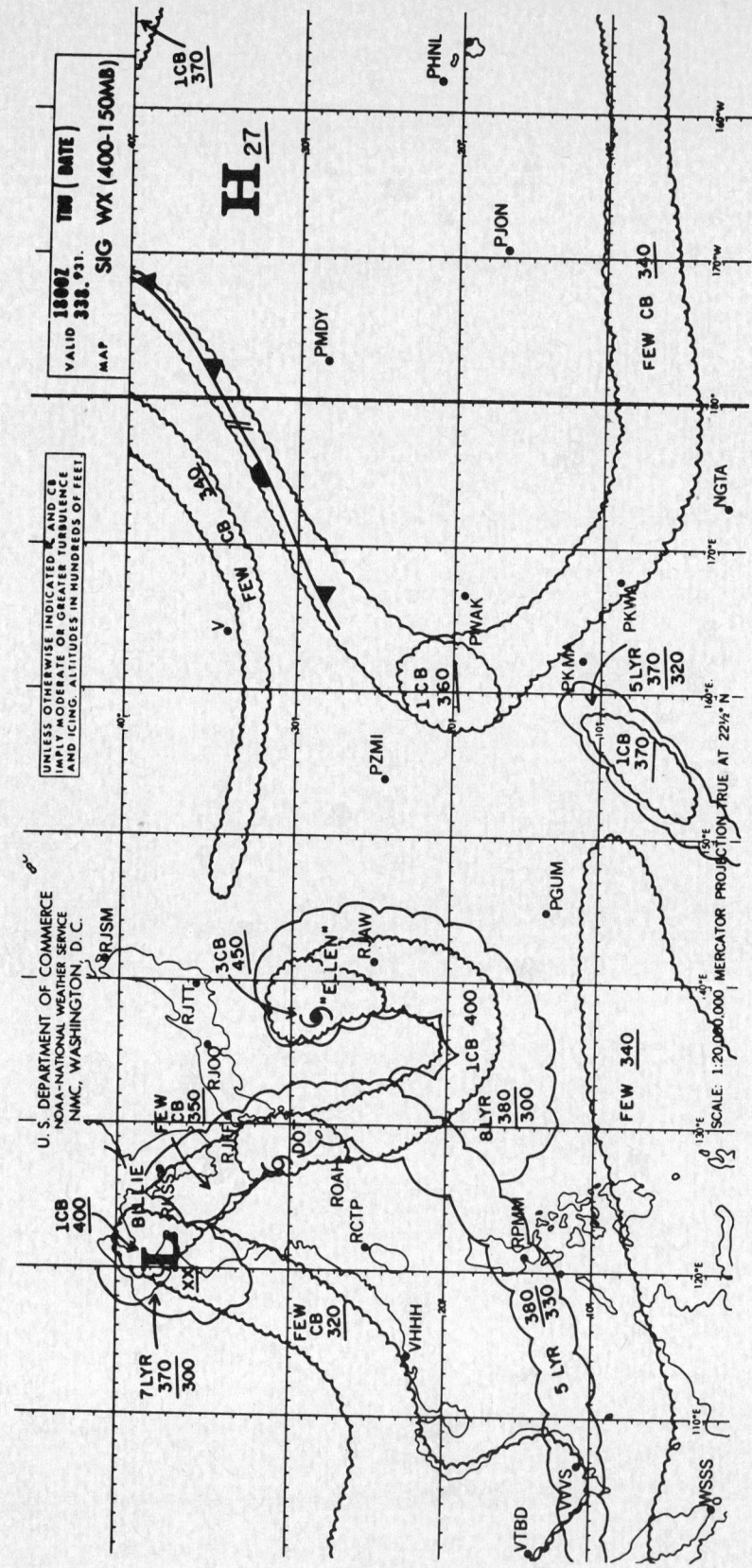

FIGURE 8-4. Pacific High Level Significant Weather Prog (400-150 mb). This Mercator projection extends from about Honolulu to eastern Asia.

Chapter 21
WINDS AND TEMPERATURES ALOFT

Winds aloft charts, both forecast and observed, are transmitted routinely by facsimile. The *forecast* winds aloft charts also contain forecast temperatures aloft. Forecast charts are computer prepared and observed charts are hand plotted.

FORECAST WINDS AND TEMPERATURES ALOFT (FD)

Forecast winds and temperatures aloft charts are prepared for eight levels on eight separate panels. A legend on each panel shows the valid time and the level of the chart. Levels below 18,000 feet are true altitudes; levels 18,000 and above are pressure altitudes or flight levels. Figure 9–1 is one panel of a winds and temperatures aloft forecast.

Temperature in °C for each forecast point is entered in two digits above the station circle. Arrows with pennants and barbs similar to those used on the surface map show wind direction and speed. Wind direction is drawn to the nearest 10 degrees, and the second digit of the coded direction is entered at the outer end of the arrow. First you determine the general direction and then use the digit to determine direction to the nearest 10°. For example, a wind in the northwest quadrant with a digit 3 indicates 330°. A calm or light variable wind is shown by "99" entered to the lower left of the station circle. Following are examples of plotted temepratures and winds with their interpretations:

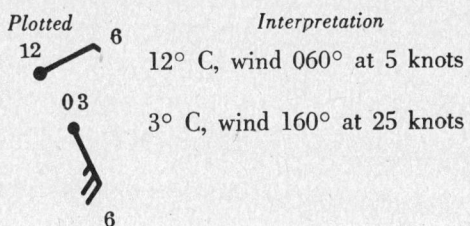

Plotted	Interpretation
12 / 6	12° C, wind 060° at 5 knots
03 / 6	3° C, wind 160° at 25 knots

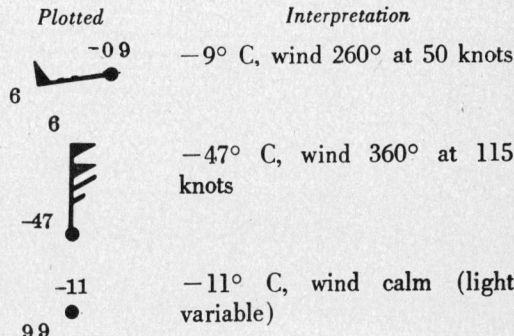

Plotted	Interpretation
−09 / 6	−9° C, wind 260° at 50 knots
−47 / 6	−47° C, wind 360° at 115 knots
−11 / 99	−11° C, wind calm (light variable)

OBSERVED WINDS ALOFT

Charts of observed winds are sent at 6-hour intervals for selected levels, each level on a separate panel. Wind direction and speed at each observing station is shown by arrows the same as on the forecast charts. The only difference is that a calm wind is shown by encircling the station. Figure 9–2 is a panel of the observed winds aloft chart.

USING THE CHARTS

The use of winds aloft charts seems obvious—to determine winds at a proposed flight altitude or to select the best altitude for a proposed flight. Temperatures also can be determined from the *forecast* charts. To determine winds and temperatures at a level between charted levels, interpolate between the charted levels.

Forecast winds are generally preferable to observed winds since they are more relevant to flight time. Observed winds are more than 2 hours old when received by facsimile, and their reliability diminishes with time.

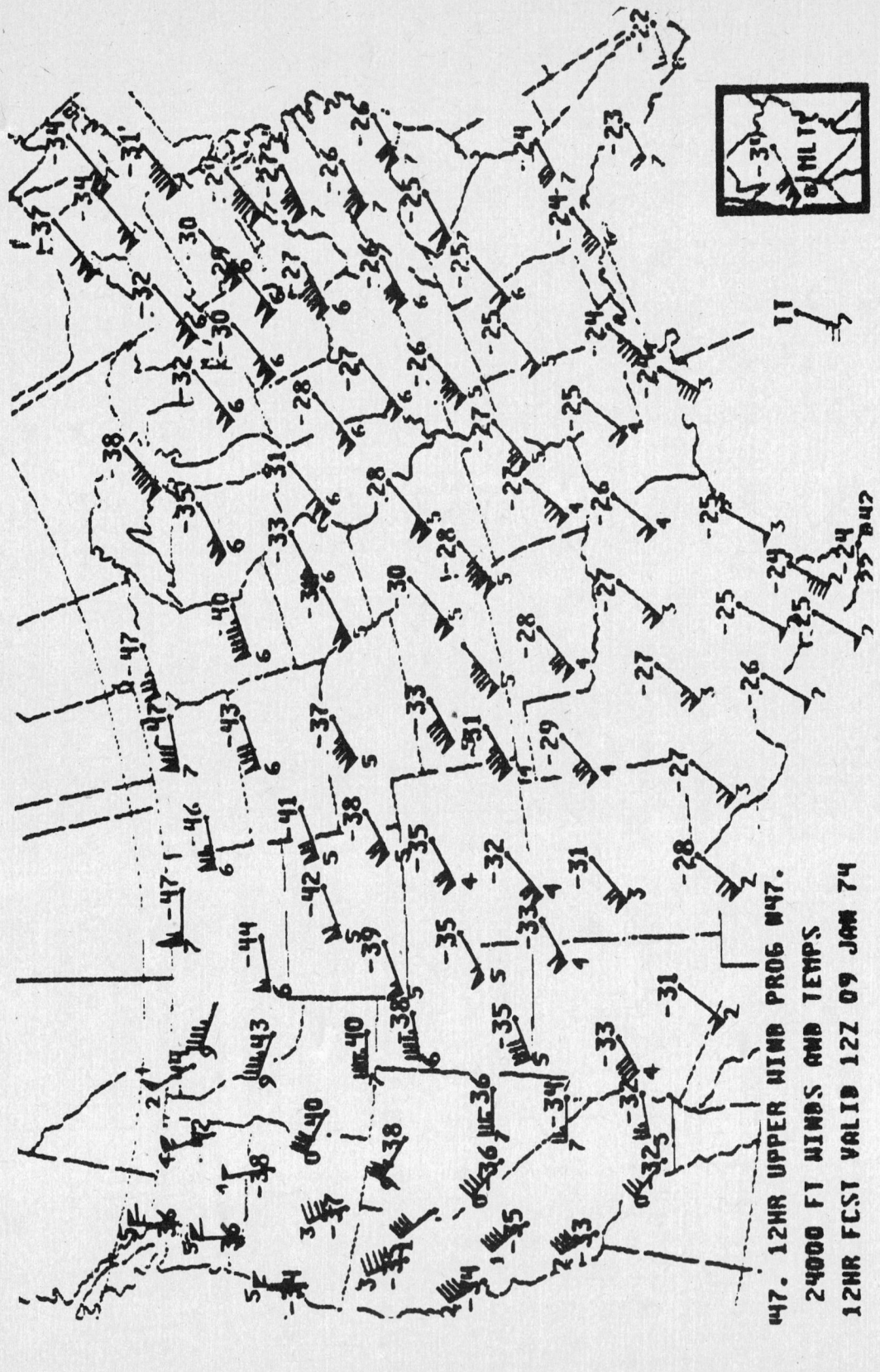

FIGURE 9-1. A panel of winds and temperatures aloft forecast for 24,000 feet pressure altitude.

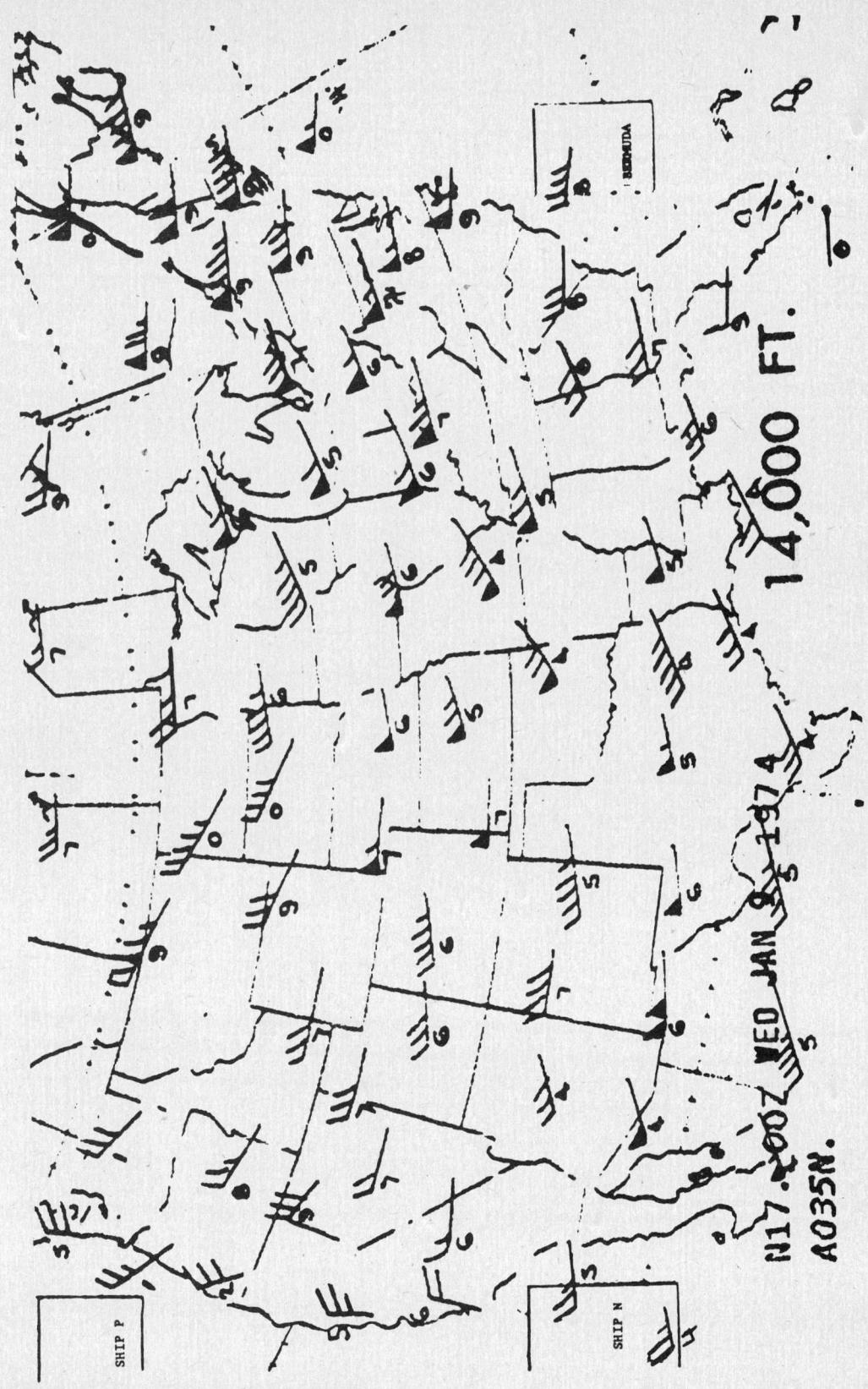

Figure 9-2. A panel of observed winds aloft for 14,000 feet.

Chapter 22
FREEZING LEVEL CHART

A freezing level chart, figure 10-1, is an analysis of observed freezing level data from upper air observations.

PLOTTED DATA

Table 10-1 explains plotting of freezing level data. Note that more than one entry denotes multiple crossings of the 0° C isotherm.

ANALYSIS

Solid lines are contours of the freezing level drawn for 4,000-foot intervals and labelled in hundreds of feet MSL. When a station reports more than one crossing of the 0° C isotherm, the lowest crossing is used in the analysis. This is in contrast to the low level significant weather prog on which the depicted forecast freezing level aloft is the highest freezing level. A dashed line shows the intersection of the freezing level with the surface.

USING THE CHART

The contour analysis shows an overall view of the lowest observed freezing level. Always plan for possible icing in clouds or precipitation above the freezing level—especially between temperatures of 0° C and −10° C.

Plotted multiple crossings of the 0° C isotherm at a station always show an inversion with warm air above subfreezing temperatures. This situation can produce very hazardous icing when precipitation is occurring.

Area forecasts show more specifically the areas of expected icing. Low level significant weather progs show anticipated changes in the freezing level.

TABLE 10-1. Plotting of freezing levels

Plotted	Interpreted as—
BF	Entire observation below freezing (0° C)
000	Surface temperature 0° C. Freezing level at surface
Three digits other than 000	Height of a freezing level aloft in hundreds of feet MSL, i.e.; 002, 200 feet MSL; 120, 12,000 feet MSL
110 051 BF	Below frezing from surface to 5,100 feet; above freezing from 5,100 feet to 11,000 feet; and below freezing above 11,000 feet
090 034 003	Lowest freezing level, 300 feet; below freezing from 300 feet to 3,400 feet; above freezing 3,400 to 9,000 feet; below freezing above 9,000 feet.

Freezing Level Chart

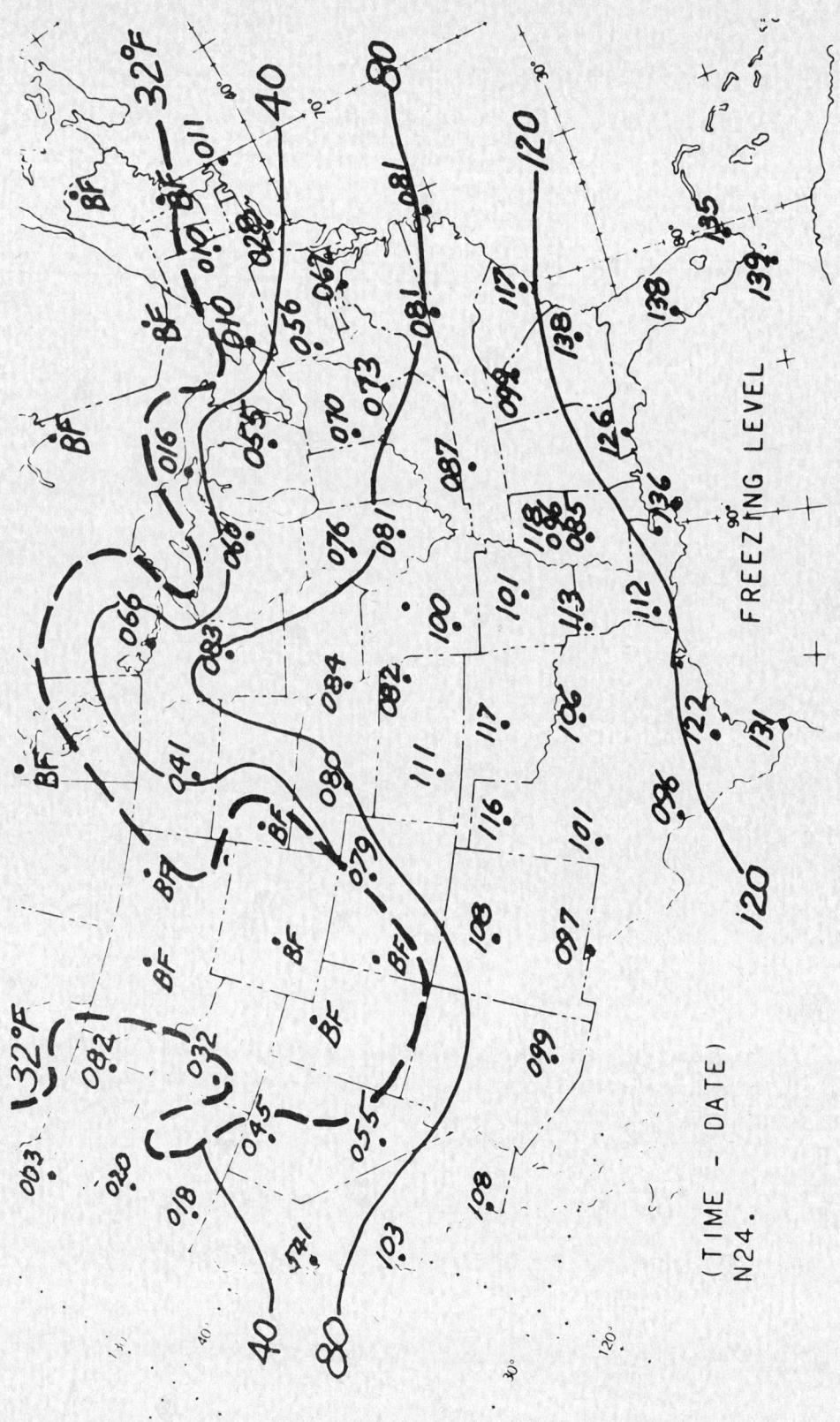

FIGURE 10-1. A freezing level chart.

205

Chapter 23
STABILITY CHART

The stability chart, figure 11-1, outlines areas of stable and unstable air. Two stability indices are computed for each upper air station; one is the *lifted index* and the other, the *K index*. At each station, lifted index is plotted above a short line, and the K index, below the line.

This section explains computation of the indices and the analysis and use of the chart. If you run into trouble with the discussion, you should review AVIATION WEATHER, chapter 6, "Stable and Unstable Air."

LIFTED INDEX

The lifted index is computed as if a parcel of air near the surface were lifted to 500 millibars. As the air is "lifted", it cools by expansion. The temperature the parcel would have at 500 millibars is then subtracted from the existing 500-millibar temperature. The difference is the lifted index—it may be positive, zero, or negative.

Positive Index

A positive index means that a parcel of air *if lifted* would be colder than existing air at 500 millibars. The air is stable. *Large positive (high) values indicate very stable air.*

Zero Index

A zero index denotes that air lifted to 500 millibars would attain the same temperature as the existing 500-millibar temperature. The air is neither stable nor unstable (neutrally stable).

Negative Index

A negative index means that the low-level air *if lifted* to 500 millibars would be warmer than existing air at 500 millibars. The air is unstable. *Large negative (low) values indicate very unstable air.*

K INDEX

The K index is primarily for the meteorologist; but a discussion is included for those who are interested. It combines moisture and stability but does not depend on lifting. It is computed using three terms as follows:

$$K = (850 \text{ mb temp} - 500 \text{ mb temp}) + (850 \text{ mb dew point}) - (700 \text{ mb temp-dew point spread})$$

The first term (850 mb temp -500 mb temp) is proportional to the mean lapse rate. A large temperature difference shows a steep or unstable lapse rate. The greater the difference, the more unstable the air and the higher the K value.

The second term (850 mb dew point) is a measure of low-level moisture. Since the dew point is added, high moisture content at 850 millibars increases the K value.

The third term (700 mb temp-dew point spread) is a measure of saturation at 700 millibars. The greater the spread, the drier is the air. Since the term is subtracted, it lowers the K value. However, moist air (small spread) lowers the value less than does dry air (large spread). Thus, the greater the degree of saturation at 700 millibars, the larger is the K value.

Putting the three together, we see that each of the following contributes to a large or high K index:

1. An unstable lapse rate
2. High moisture content at 850 millibars, and
3. A high degree of saturation at 700 millibars.

Thus, a large K index supports cloudiness and precipitation.

STABILITY ANALYSIS

The analysis is based on the lifted indices only. Solid lines are drawn for indices of $+4$, 0, and -4. A "U" identifies an area of instability; and an "S", an area of stability. In general, air with a lifted index between 0 and $+4$ is marginally stable. It may become unstable due to surface heating, upslope flow, frontal lifting, convergence, or inflow of cold air aloft.

Stability Chart

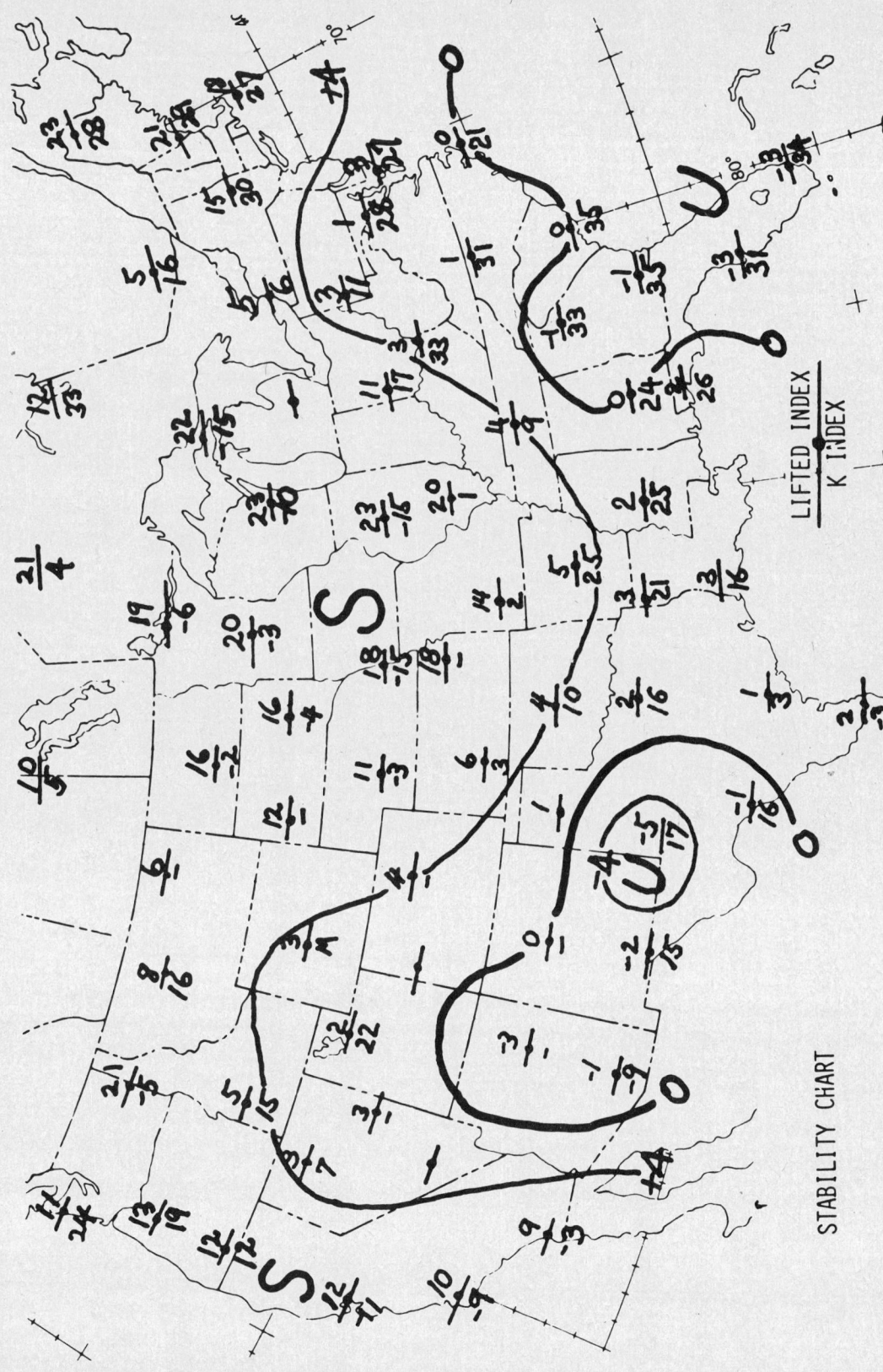

FIGURE 11-1. A stability chart. (This chart is keyed to table 11-1.)

207

Stability Chart

USING THE CHART

From the chart you can make a quick estimate of areas of probable convective turbulence. Those of you making use of the K index also can estimate most likely areas of clouds and precipitation. The K index should be compared with the lifted index. Table 11-1 shows the comparisons along with probable weather and operational impact. The table is keyed to the chart in figure 11-1.

Correlations between the indices and resulting weather vary with seasons and in different sections of the country. Also, extent of instability that develops depends on the degree of surface heating or on forced lifting such as by sloping terrain or a front. First estimates made from the chart are only preliminary to detailed flight planning.

TABLE 11-1. *Using the K and lifted indices*

Lifted Index	*K Index*	*Area in figure 11-1*	*Probable weather*	*Operational impact*
Zero or Negative (unstable)	High (wet)	Georgia Florida	Instability showers or thunderstorms	Turbulence; may be hazardous; soaring plagued by clouds
Zero or Negative (unstable)	Low (dry)	Southwest TX Southern NM Eastern AZ	Limited cumulus activity; little if any precipitation	Bumpy but not hazardous; good for thermal soaring
Positive (stable)	High (wet)	New England	Stratified cloudiness; steady precipitation	Smooth for IFR flight; may restrict VFR; no thermals
Positive (stable)	Low (dry)	Northern Plains, Calif. coast	Predominantly fair	Smooth flight; generally good VFR; weak thermals if any

Chapter 24
SEVERE WEATHER OUTLOOK CHART

The severe weather outlook is a preliminary 24-hour outlook presented in two 12-hour panels. Figure 12-1 is one panel of the outlook. The chart graphically delineates areas of general and severe thunderstorms and may show areas of possible tornadoes.

GENERAL THUNDERSTORMS

A line with an arrowhead delineates an area of probable thunderstorm activity. When you face in the direction of the arrow, activity is expected to the right of the line. An area labelled APCHG indicates probable thunderstorm activity *approaching* severe intensity.

SEVERE THUNDERSTORMS

A hatched area indicates severe thunderstorms; and the following notations show expected areal coverage:

Notation	Expected coverage
ISOLD	Extremely small number
FEW	Up to 15% coverage
SCTD	16% to 45% coverage
NMRS	More than 45% coverage

If an instability line is forecast, the chart may include a squall line symbol and expected time of development.

TORNADOES

Tornadoes are indicated only if a tornado watch is in effect at chart time. The watch area is cross-hatched. No areal coverage is specified.

USING THE CHART

The severe weather outlook is strictly for *advance* planning. It alerts all interests to the possibility of future storm development. As the time of severe weather approaches, the forecaster can more specifically delineate the time, extent, and nature of the weather and issue a severe weather watch (WW).

Severe Weather Outlook Chart

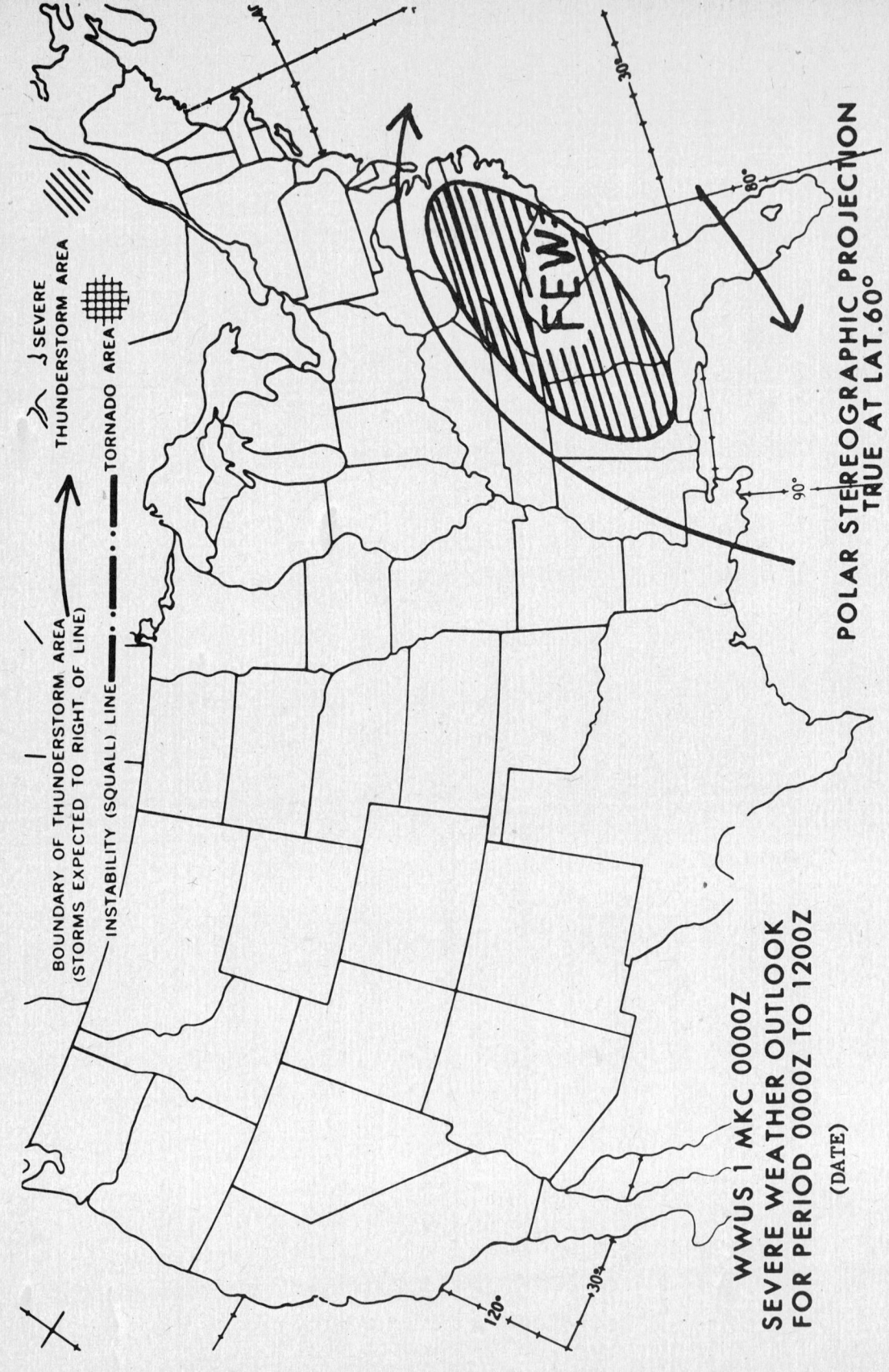

FIGURE 12-1. A panel of a severe weather outlook chart. If a severe weather watch is in effect when the chart is issued, the WW area is outlined on the chart and appropriately shaded as severe thunderstorms or tornadoes.

Chapter 25
CONSTANT PRESSURE CHARTS

A constant pressure chart is an upper air weather map at a specified pressure surface or flight level.* *Any surface of equal pressure in the atmosphere is a constant pressure surface, pressure altitude, and flight level.* Every 12 hours, computer prepared constant pressure charts are transmitted by facsimile for five pressure levels.

To the meteorologist, pressure is relevant to forecasting, so he labels the charts in millibars. To the pilot, pressure altitude or flight level of the chart is more relevant. Table 13–1 lists the approximate pressure altitude and height in meters of each chart along with other information pertinent to the analysis. Figures 13–2 through 13–6 are sections of each chart.

A constant pressure chart depicts highs, lows, troughs, and ridges aloft by the height contour patterns resembling isobars on a surface map. For a direct use of a constant pressure chart, assume

---
* *Flight level* is used here for all pressure altitudes, although in routine operations lower level flights are at indicated altitude called *cruising altitude*.

you are planning a flight at 10,000 feet. The 700-millibar chart is approximately 10,000 feet MSL. It is a source of observed temperature, temperature-dew point spread, and winds aloft for your flight.

PLOTTED DATA

Figure 13–1 illustrates and decodes the standard data plot. The format varies slightly depending on the pressure level and available data. For example, when air is too dry to measure the dew point, an "X" is plotted for the temperature-dew point spread on lower levels, but the entry is omitted on upper levels. A light variable wind or missing wind data are represented by "LV" and "M" respectively plotted to the lower right. Aircraft reports are used in analyses over areas of sparse data. A square in lieu of a station circle signifies an aircraft report. The flight level of the aircraft is plotted in feet; temperature and wind are at flight level of the aircraft (see figs. 13–5 and 13–6).

TABLE 13-1. Features of constant pressure charts—U.S.

Pressure (millibars)	Pressure altitude in feet (flight level)	Approximate height in meters	Temperature–dew point spread	Isotachs	Contour interval (meters)	Prefix to plotted value	Suffix to plotted value	Examples of height plotting and contour labelling	
								Plotted/labelled	Height
850	5,000	1,500	Yes	No	60	1	---	581	1,581
700	10,000	3,000	Yes	No	60	2 or 3*	---	882	2,882
500	18,000	5,500	Yes	No	60	----	0	578	5,780
300	30,000	9,000	Yes**	Yes	120	----	0	943	9,430
200	39,000	12,000	Yes**	Yes	120	1	0	217	12,170

* Prefix a "2" or "3" whichever brings the height closer to 3,000 meters.
** Omitted when air is too dry to measure dew point.
Flight level of an aircraft is plotted in lieu of height of constant pressure surface.

Constant Pressure Charts

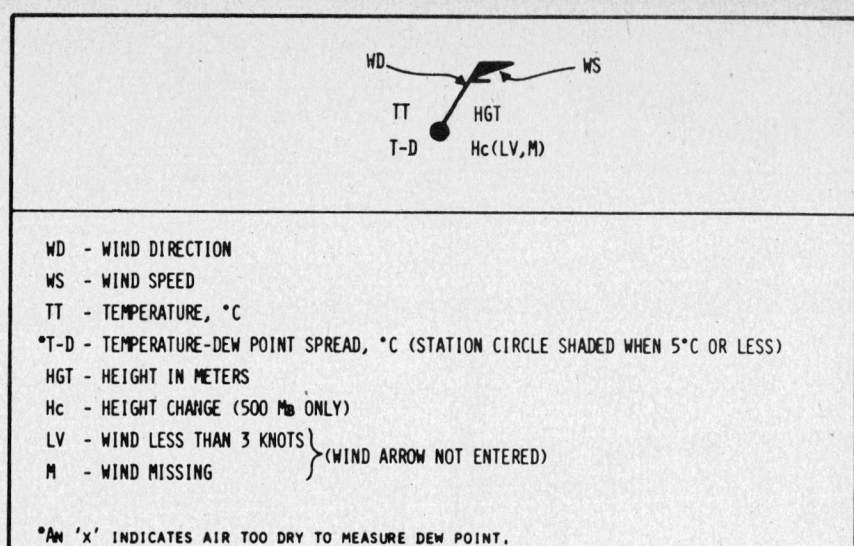

FIGURE 13-1. Standard station plotting model on constant pressure analyses. Entries vary slightly with levels and with data available. Note that when air is too dry to measure the dew point, an "X" is entered for temperature-dew point spread. The entry is usually omitted on the 200- and 300-millibar levels since air at these levels normally is too dry to measure dew point. Height change appears only on the 500-millibar chart.

The following examples show some station plots and how they are decoded. Meanings of the legend below each plot are: W, wind; T, temperature; T–DP, temperature-dew point spread; H, height of pressure surface in meters; Hc, 12-hour height change (500 millibars only):

(850 mb)
W: less than 3
T: 22° C
T–DP: 4° C
H: 1,479 m.

(700 mb)
W: 010° 20 kt
T: 9° C
T–DP: 17° C
H: 3,129 m.

(500 mb)
W: 210° 60 kt
T: −19° C
T–DP: Dry
H: 5,580 m.
Hc: +30 m.

(200 mb)
W: Missing
T: −60° C
T–DP: Dry
H: 11,910 m.

ANALYSIS

All charts contain *contours* and *isotherms*; some contain *isotachs*. Contours are lines of equal height; isotherms, lines of equal temperature; and isotachs, lines of equal windspeed.

Height Contours

A pattern of height contours resembles isobars on a surface weather map and we may compare a height analysis to a pressure analysis. A contour high, low, ridge, or trough is analogous to a pressure high, low, ridge, or trough. Since an upper air chart is above the surface friction layer, winds at these levels for practical purposes parallel the contours.

Refer to figures 13–2 through 13–4 and note the low aloft near northern California extending upward through 500 millibars. Figures 13–5 and 13–6 show the low opening into a trough at 300 and 200 millibars.

Isotherms

Isotherms drawn at 5° C intervals show horizontal temperature variations at chart altitude. Let's refer to the 850 millibar chart (5,000 feet pressure altitude), figure 13–2 and locate an isotherm. Note the dashed line extending from west to east through Utah and Colorado and labelled "00" at the eastern edge of Colorado. This is the 0° C isotherm. North of this isotherm, temperatures at 5,000 feet are below freezing. The −15° C isotherm extends across South Dakota. By inspecting isotherms, you can determine if your flight will be toward colder or warmer air. Subfreezing temperatures and a temperature-dew point spread of 5° C or less suggest possible icing.

Isotachs

Isotachs appear on only the 300- and 200-millibar charts. To aid in identifying areas of strong winds, hatching denotes windspeeds of 70 to 110 knots; a clear area within a hatched area indicates winds 110 to 150 knots, etc. Note the clear area in figures 13–5 and 13–6 extending eastward from northern New Mexico. On the 200-millibar chart, figure 13–6, the 130-knot isotach is over northern New Mexico and across the Texas Panhandle. Winds at 39,000 feet over Amarillo, Texas are from the west at 135 knots. Winds over northern California at this flight level are only 30 knots—note the dashed line labelled "30K".

THREE DIMENSIONAL ASPECTS

As established earlier, we may treat a height contour analysis as a pressure analysis. Closely spaced contours mean strong wind as do closely spaced isobars. Wind blows clockwise around a contour high and counterclockwise around a low.

Features on synoptic surface and upper air charts are related. However, a weak surface system often loses its identity in a large scale upper air pattern; while another system may be more evident on an upper air chart than on the surface chart. Many times weather is more closely associated with an upper air pattern than with features on the surface map.

We have learned as a general rule to regard a surface low as a producer of bad weather and a high as a producer of good weather. However, *widespread cloudiness and precipitation often develop in advance of an upper air low or trough which is not evident at the surface.* In contrast, an upper air high usually means good weather. An exception is an upper air high or ridge that has a stabilizing effect at low levels. Smoke, haze, dust, or even low stratus and fog may persist for extended periods; yet the surface map may show no cause for the restriction.

Highs and lows generally slope westward or northwestward into the upper atmosphere. Due to this slope, wind aloft with an upper system often blows across the associated surface system. Surface fronts, lows, and highs tend to move with the upper winds. For example, strong winds aloft across a surface front will cause the front to move rapidly; but if upper winds parallel a front, it moves slowly if at all.

An intense, cold low leans less than does a warm or weaker system. The low becomes almost vertical and is clearly evident on both surface and upper air maps. Upper winds encircle the surface low rather than blow across it. Thus, the storm moves very slowly and usually causes extensive and persistent cloudiness, precipitation, strong winds, and generally adverse flying weather. The term "cold low" describes such a system.

In contrast to the cold low is the "thermal low". A dry, sunny region becomes quite warm from intense surface heating resulting in a surface low pressure area. The warm air is carried to high levels by convection, but cloudiness is scant because of lack of moisture. Pressure decreases slowly with altitude in warm air; thus, the warm surface low often is "capped" by a high aloft. Unlike the cold low, the thermal low is relatively shallow with weak pressure gradients and no well defined cyclonic circulation. However, you must be alert for high density altitude, light to moderate convective turbulence, and isolated showers and thunderstorms which have high bases and generally produce light precipitation. The thermal low is a semipermanent feature of the desert regions in the southwestern United States and northern Mexico during warm weather.

These are only a few examples of associating weather with upper air features. They point out the need to view weather in three dimensions.

USING THE CHARTS

From the charts you can approximate the observed temperature, wind, and temperature-dew point spread along your proposed route. Usually you can select a constant pressure chart close to your planned altitude. For altitudes about midway between two charted surfaces, interpolate between the two charts.

Determine temperature from plotted data or the pattern of isotherms. To readily delineate areas of high moisture content, station circles are shaded

Constant Pressure Charts

indicating temperature-dew point spreads of 5° C or less. You can get the actual spread from plotted data. A small spread alerts you to cloudiness, precipitation, and icing. Determine windspeed for lower levels from plotted data; for the 300- and 200-millibar surfaces, determine speed from the isotach pattern. Wind direction parallels the contours.

As stated earlier, constant pressure charts often show the cause of weather and its movement more clearly than does the surface map. For example, the large scale wind flow around a low aloft may spread cloudiness, low ceilings, and precipitation far more extensively than indicated by the surface map alone.

Keep in mind that constant pressure charts are observed weather. Upper air prognostic charts (next section) show expected changes and may be more relevant at your flight time.

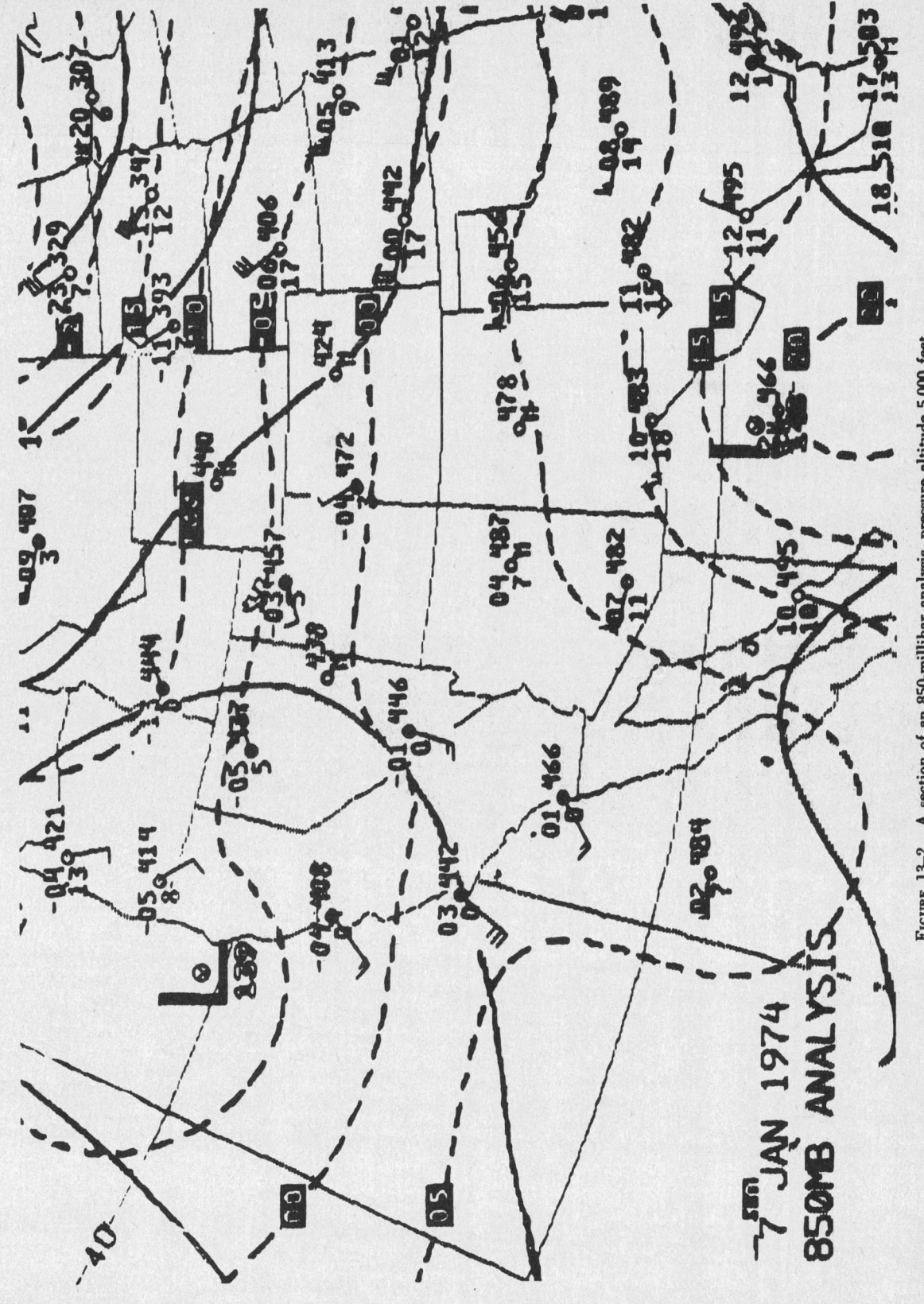

FIGURE 13-2. A section of an 850-millibar analysis, pressure altitude 5,000 feet.

Constant Pressure Charts

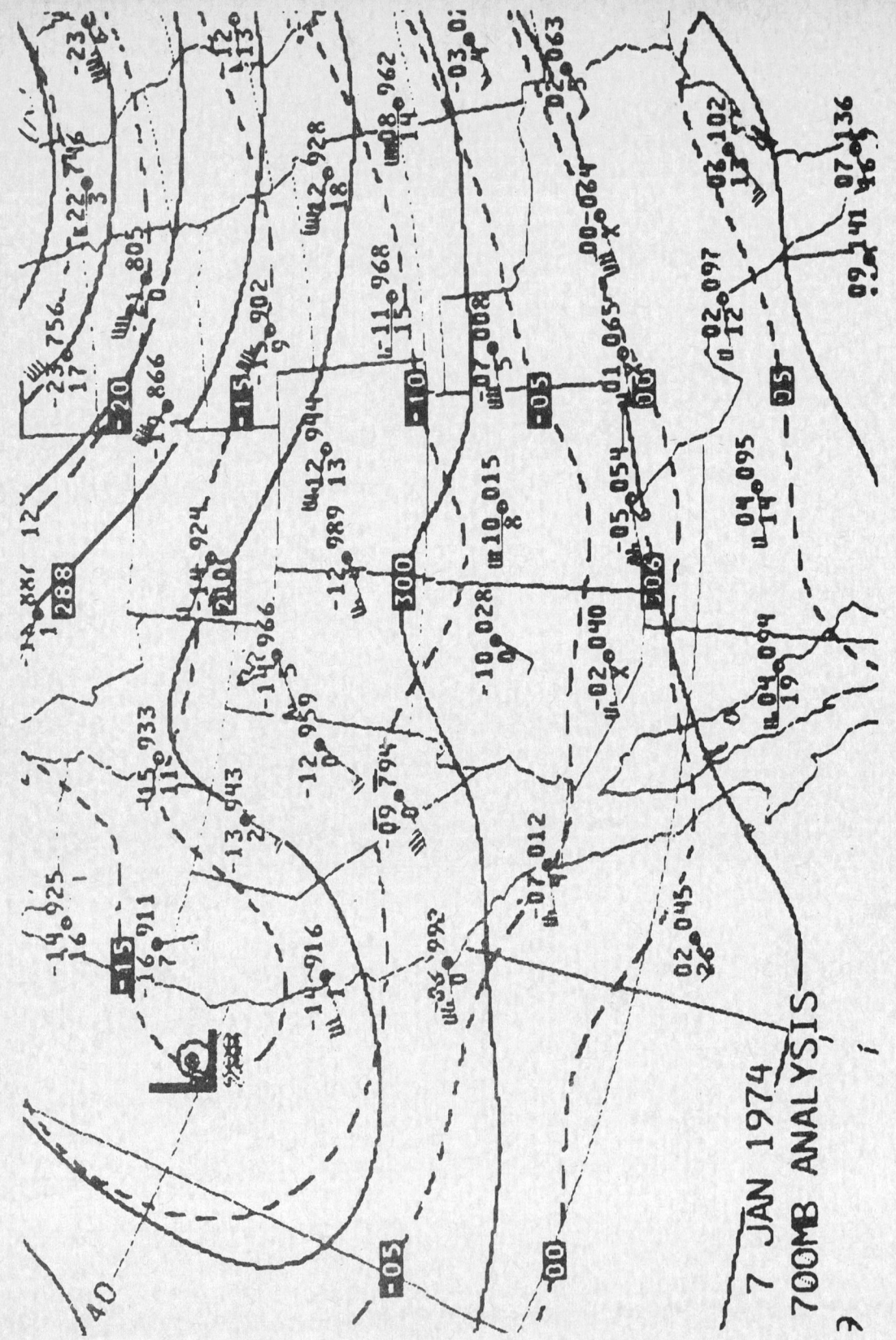

FIGURE 13-3. A section of a 700-millibar analysis, pressure altitude 10,000 feet.

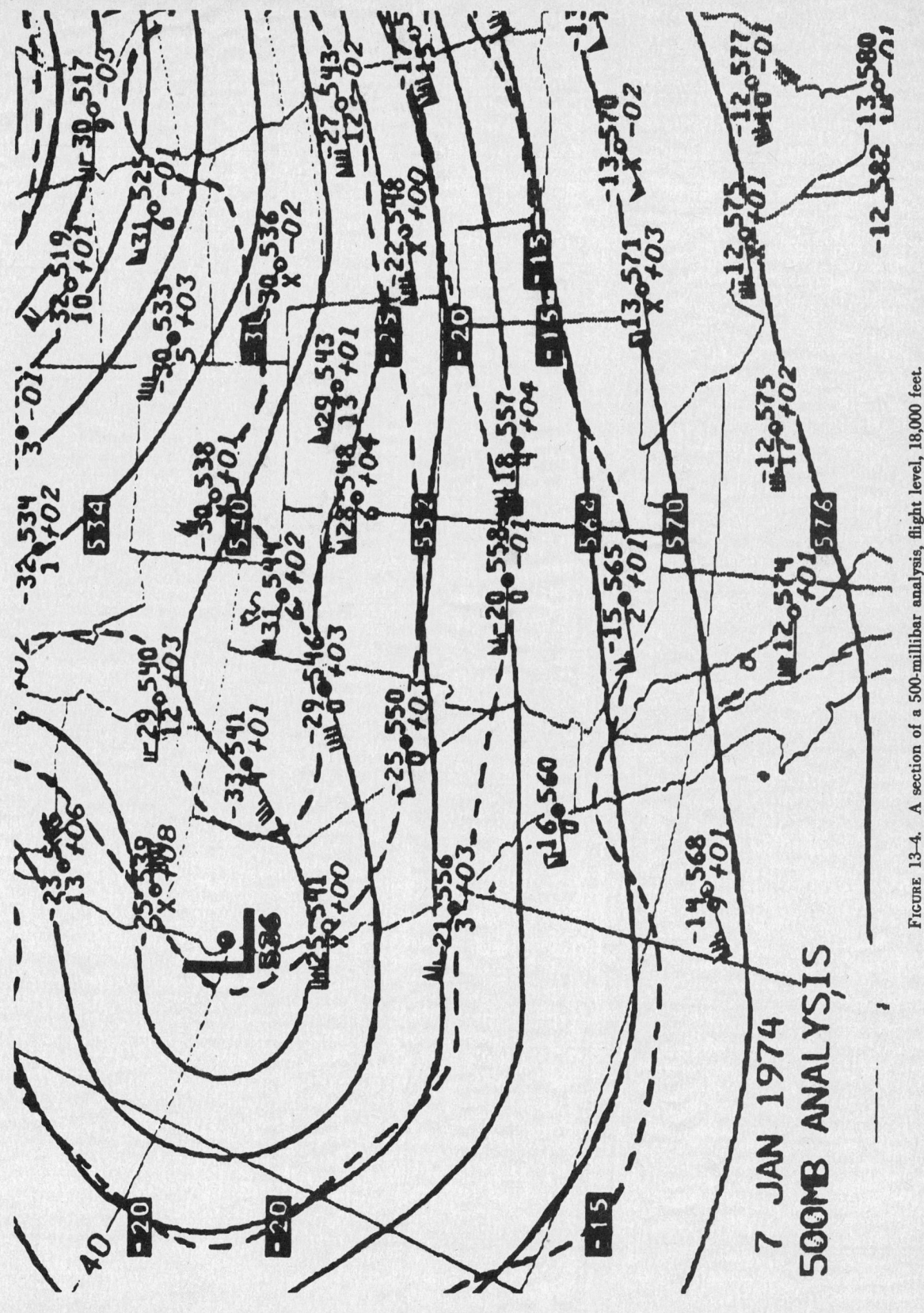

FIGURE 13-4. A section of a 500-millibar analysis, flight level, 18,000 feet.

Constant Pressure Charts

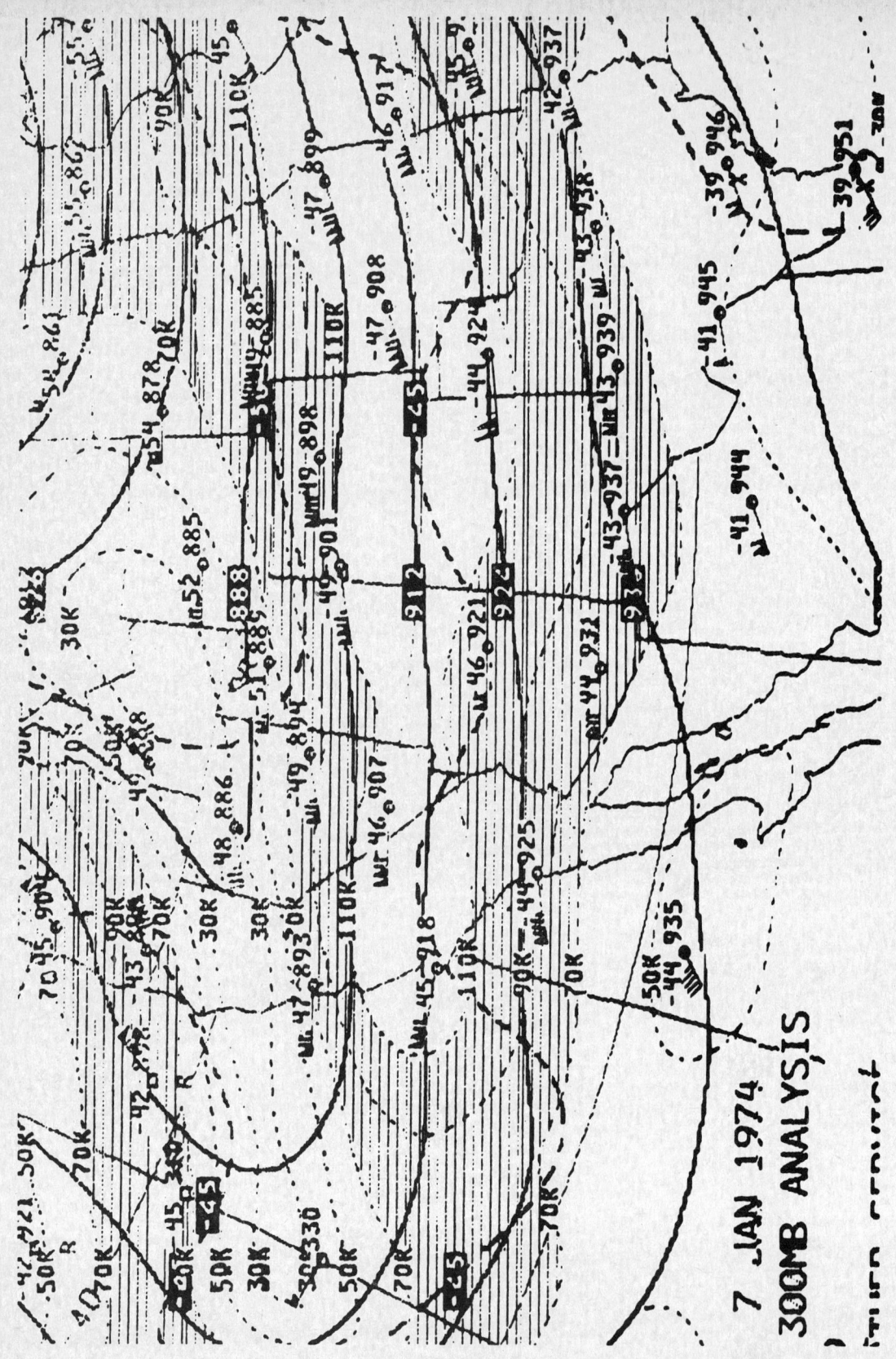

FIGURE 13-5. A section of a 300-millibar analysis, flight level, 30,000 feet.

Constant Pressure Charts

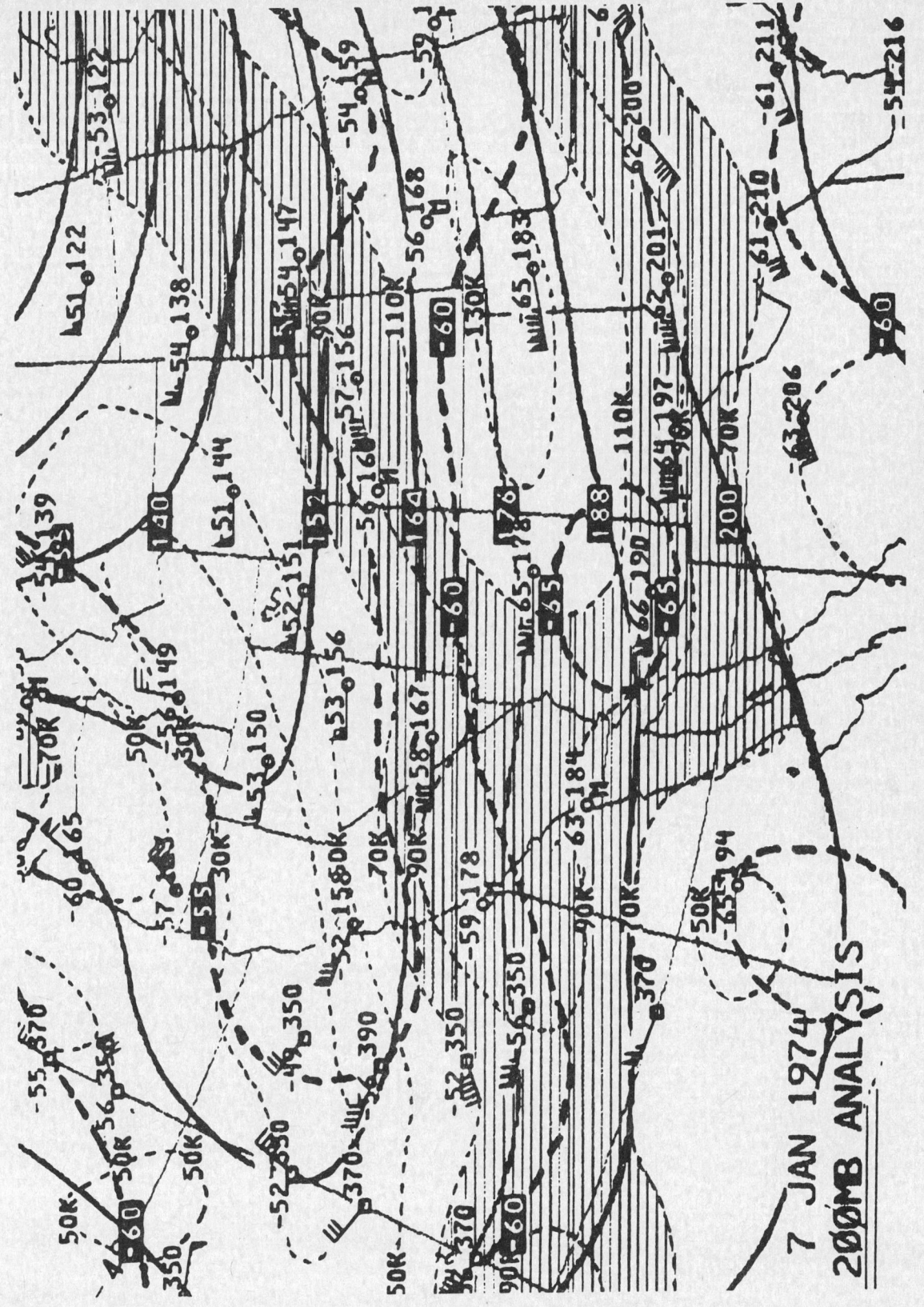

Figure 13-6. A section of a 200-millibar analysis, flight level, 39,000 feet.

Chapter 26
CONSTANT PRESSURE PROGNOSTICS

Constant pressure prognostic charts are computer prepared for the same standard levels as the analyses discussed in section 13. Progs are issued for large geographical areas. Some are primarily for the meteorologist while others are specifically for international flight documentation. Format varies slightly with different computer techniques. 850- and 700-millibar progs normally are not prepared for flight documentation, but a briefer may use them on occasion during preflight briefing. A legend on each prog identifies pressure surface and valid time; and on some charts, a legend along the side shows a few of the parameters depicted on the chart.

HEIGHT CONTOURS/STREAMLINES

Each chart has either height contours or streamlines. These solid lines show forecast positions of upper highs, lows, troughs, and ridges. Charts on polar stereographic projections have contours with heights labelled in meters. Charts on Mercator projections have streamlines (lines parallel to wind direction); streamlines are not labelled.

TEMPERATURE

The forecast temperature field is indicated either by isotherms at 5° C intervals or by encircled "spot" temperatures in sufficient density to delineate the thermal field. Some hemispheric progs do not show the temperature field.

WINDSPEED

Windspeed is shown by isotachs drawn for 20-knot intervals for speeds from 10 knots to 130 knots and at 40-knot intervals for speeds stronger than 130 knots. On hemispheric progs, areas of windspeed from 70 knots to 110 knots are hatched to aid in identifying strong winds. A clear area inside an area of hatching indicates speeds stronger than 110 knots. No hatching is used on larger scale charts prepared for flight documentation.

In addition to isotach windspeed at the constant pressure surface, some 500- and 300-millibar progs also show "spot" winds at levels other than the chart level:

Chart Level	Winds at
500 mb	400 mb
300 mb	250 mb

Thus, forecast winds are also presented for FL240 and FL340. These winds are depicted by arrows and flags as on winds aloft charts.

FORMATS

A brief look at some different formats will help you use the charts. When given a constant pressure prog, you should be able to interpret the chart using this discussion and the legend on the chart.

Figure 14-1 is a section of a 500-millibar hemispherical prog. This format is used on some hemispherical 300- and 200-millibar progs also. Note that the chart has only contours and isotachs. Contours and high and low centers are labelled in meters as shown in table 14-1. Isotachs are labelled in knots. Note also that areas of wind greater than 70 knots are shaded.

Figure 14-2 illustrates the format of a constant pressure prog on a polar stereographic projection. This format is used on 500- and 300-millibar progs prepared for international flight documentation. Open figures are used to label contours and high and low centers in meters as shown in table 14-1. Isotachs are labelled in knots using dark squares with white figures. Spot temperatures at the pressure surface are circled. Figure 14-3 is a section of a 300-millibar prog in this format. This particular chart has no spot winds for 250 millibars, but some 300-millibar progs do.

Figure 14-4 illustrates the format of a 200-millibar prog on a Mercator projection. Light solid lines are streamlines and are unlabelled. Heavy solid lines are height contours of the tropopause and are labelled in flight levels. Note the heavy line labelled "FL450" meaning the tropopause along this line is at flight level 45,000 feet (150 millibars). Temperatures at the 200-millibar surface are circled and temperatures of the tropopause are in rectangles. Spot winds on the 200-millibar prog are also for 200 millibars and conform to the streamline and isotach patterns. Figure

Constant Pressure Prognostics

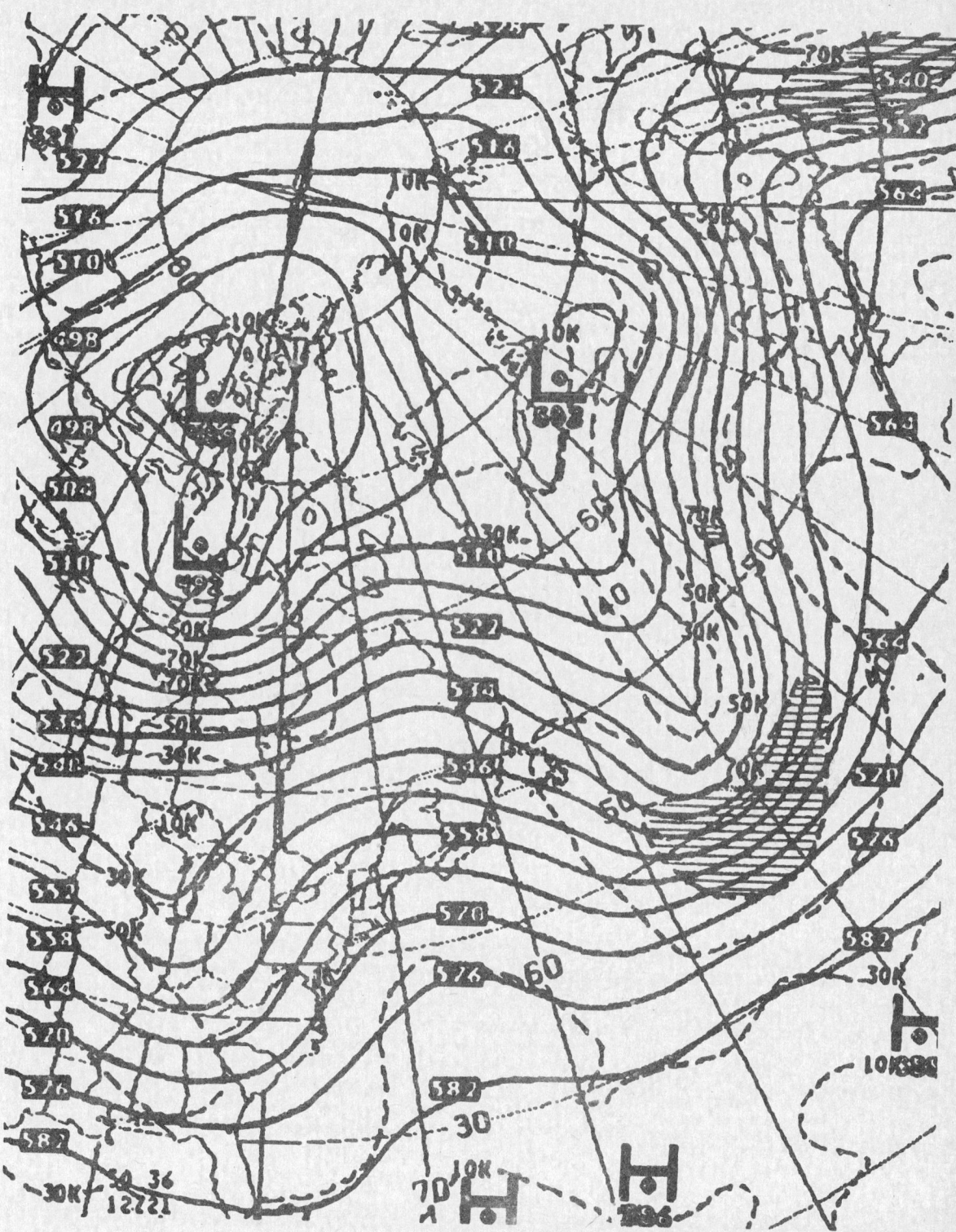

FIGURE 14-1. A section of a hemispheric 500-millibar prog with contours and isotachs but no temperature field. This format is used on hemispheric 300- and 200-millibar charts also. Hatching shows wind between 70 and 110 knots.

Constant Pressure Prognostics

TABLE 14–1. Height labelling of contours and of high and low centers

Pressure in millibars	Flight level in feet	Approx. height in meters	Prefix to labelled value	Suffix to labelled value	Examples	
					Labelled	Height in meters
850	5,000	1,500	1	---	444	1,444
700	10,000	3,000	2 or 3*	---	288	3,288
500	18,000	5,500	---	0	582	5,820
300	30,000	9,000	---	0	948	9,480
200	39,000	12,000	1	0	205	12,050

* Prefix a "2" or "3" whichever brings the height closer to 3,000 meters.

14–5 is a section illustrating a 200-millibar prog prepared for international flight documentation.

A combination prog is prepared for the North Pacific between the Western U.S. coast and eastern Asia. It includes a 300-millibar prog and a tropopause wind shear prog (explained in section 15). The progs are in the same formats, but are sent as a single transmission in two panels.

USING THE CHARTS

Use a constant pressure prog primarily to extract forecast wind and temperature for flights within plus or minus 6 hours of the valid time of the chart. A constant pressure prog shows expected movements of weather systems; and you may compare it with any type of prog having the same valid time to get a three dimensional picture of expected weather.

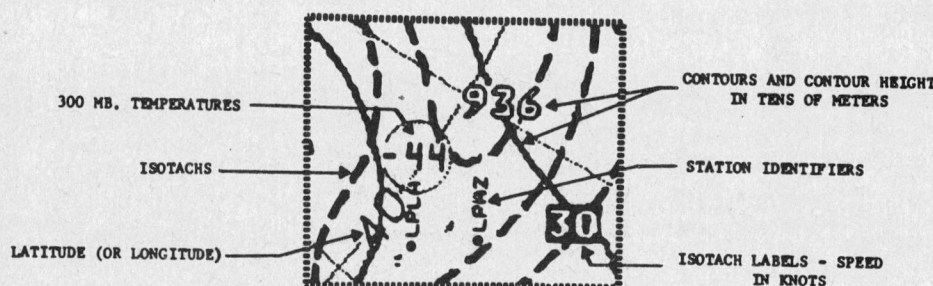

FIGURE 14–2. Format of a 300-millibar prog prepared on a polar stereograph projection. Contours are labelled in meters (open figures). Isotachs are labelled in knots. Temperatures are circled. The same format is used on some 500-millibar progs.

Constant Pressure Prognostics

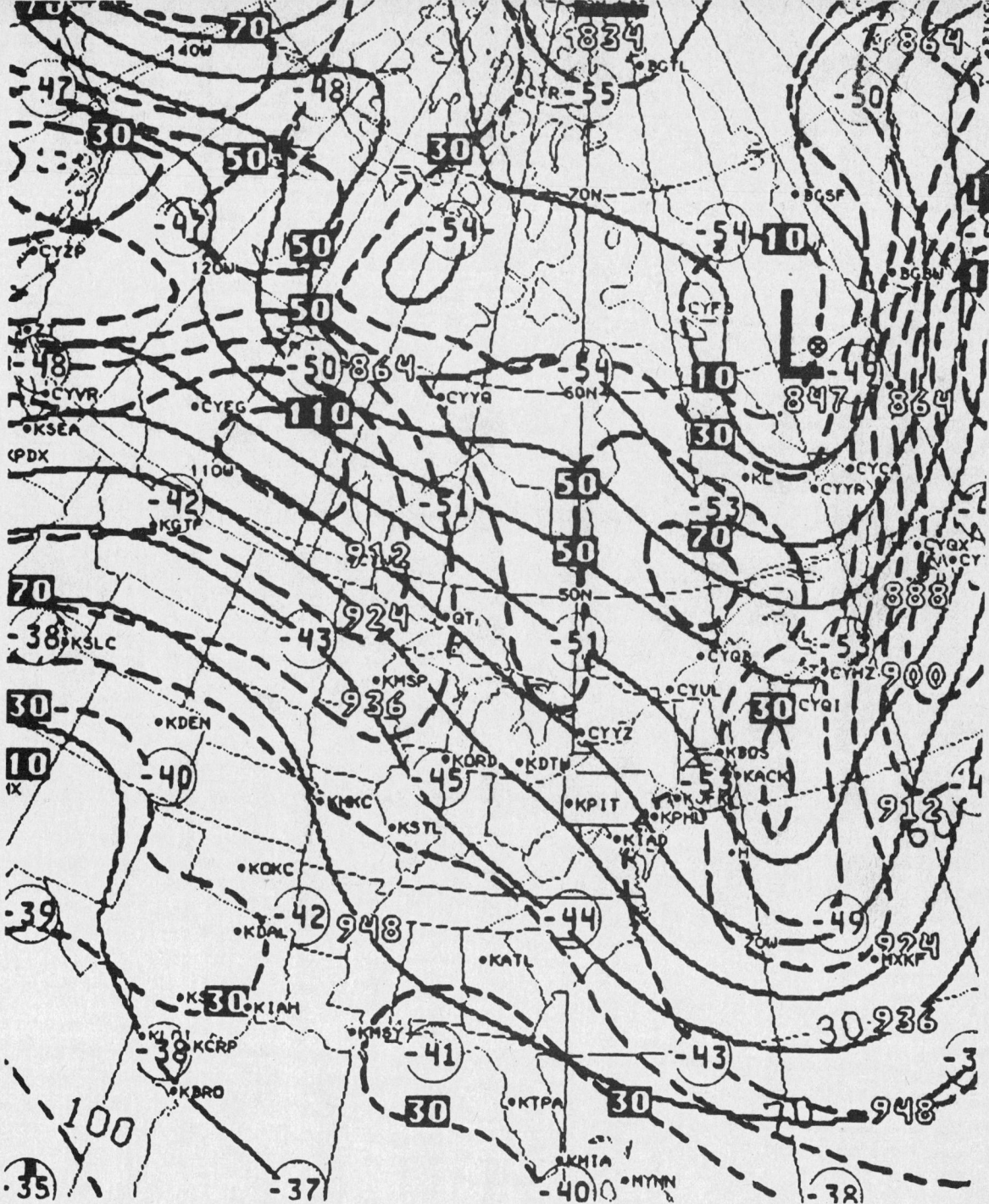

FIGURE 14-3. A section of a 300-millibar prog for international flight documentation. See figure 14-2 for explanation. Some 300-millibar progs show spot winds for 250 millibars.

Constant Pressure Prognostics

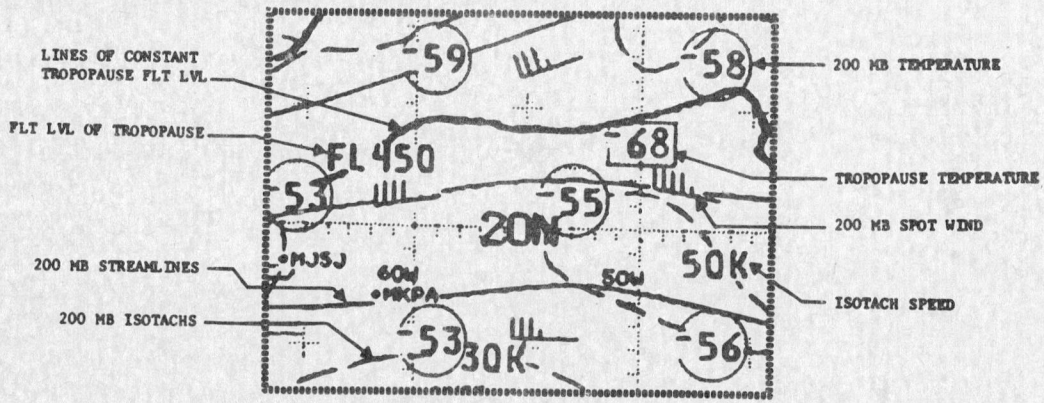

FIGURE 14-4. Format of a 200-millibar prog prepared on a Mercator projection. Heavy lines are lines of constant flight level of the tropopause. Light solid lines are streamlines and are unlabelled. 200-millibar temperatures are circled. Tropopause temperatures are in rectangles. Spot winds are also for the 200-millibar surface.

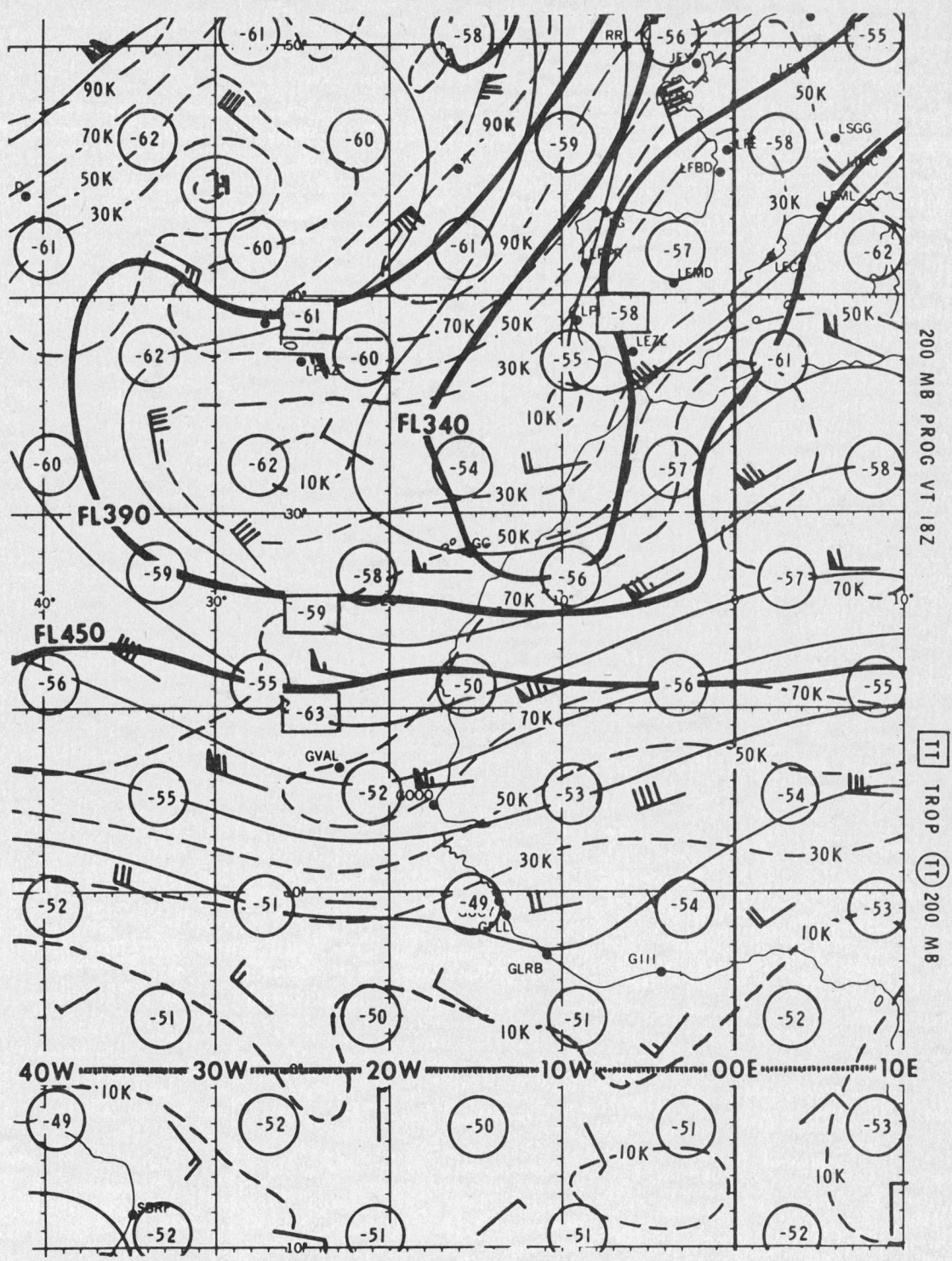

FIGURE 14–5. A section of a 200-millibar prog for international flight documentation.

Chapter 27
TROPOPAUSE, MAX WIND, AND WIND SHEAR CHARTS

A chart of the observed tropopause is prepared for the contiguous 48 States. A maximum wind prog and vertical wind shear prog, both associated with the tropopause, are prepared also for the 48 States. Tropopause/vertical wind shear progs are prepared for international flights.

OBSERVED TROPOPAUSE CHART

The observed tropopause chart shows for each upper air observing station the pressure, temperature, and wind at the tropopause. Figure 15-1 shows the chart with Albuquerque, New Mexico (ABQ) identified to aid in explaining the station model. Decode the plotted data at Albuquerque as follows: tropopause wind, 240° at 115 knots; tropopause temperature, −61° C; tropopause pressure, 200 millibars.

Using the Chart

Maximum wind occurs near the tropopause, so this chart is essentially a chart of observed maximum winds. A close inspection of the chart reveals a jet stream from central New Mexico across southeastern Kansas, central Missouri, and southern Illinois to West Virginia. The reason wind data are missing over Oklahoma, eastern Kansas, and Illinois is that strong winds carried the radiosonde instruments too far from observing stations to obtain reliable wind data. This area of missing wind data is actually the area of strongest winds in the jet stream.

From the chart you can determine wind and temperature at the tropopause. You can then use constant pressure progs or the FD winds and temperatures aloft forecast to interpolate for a flight level between a constant pressure level and the tropopause.

DOMESTIC TROPOPAUSE WIND AND WIND SHEAR PROGS

Forecast parameters at the tropopause over the contiguous 48 States and some adjoining oceanic and North American areas are shown on two charts—the *tropopause winds* and the *tropopause height/vertical wind shear progs.*

Tropopause Winds

The tropopause winds prog, figure 15-2, depicts wind direction by streamlines—solid lines. Streamlines have no dimensions and are unlabelled; they are in sufficient density to show the direction field. Direction of the streamline basically is from west to east in mid latitudes. A high or low may be encircled by a closed streamline; you can readily determine whether it is a high or low, and you know the circulation around these systems.

Windspeed is shown by isotachs at 20-knot intervals—dashed lines in figure 15-2. They are labelled in knots. Areas of windspeeds between 70 and 110 knots are hatched as are windspeeds between 150 and 190 knots. The shading criteria are the same as used on selected constant pressure analyses and progs.

Tropopause Height/Vertical Wind Shear

The tropopause height/vertical wind shear prog, figure 15-3, depicts height of the tropopause in millibars and vertical wind shear in knots per 1,000 feet. Solid lines trace intersections of the tropopause with standard constant pressure surfaces. The intercepts are labelled in millibars by white letters in dark rectangles. Since a line on a constant pressure surface is also a line of constant pressure altitude, the isopleths are also lines of constant flight level (see chapter 3, AVIATION WEATHER, discussion of Pressure Altitude). Following is a listing of pressures and corresponding flight levels:

Millibars	Flight Level
500	18,000
450	21,000
400	24,000
350	27,000
300	30,000
250	34,000
200	39,000
150	45,000

Vertical wind shear is in knots per 1,000 feet depicted by dashed lines at 2-knot intervals. Wind shear is averaged through a layer from about 8,000 feet below to 4,000 feet above the tropopause.

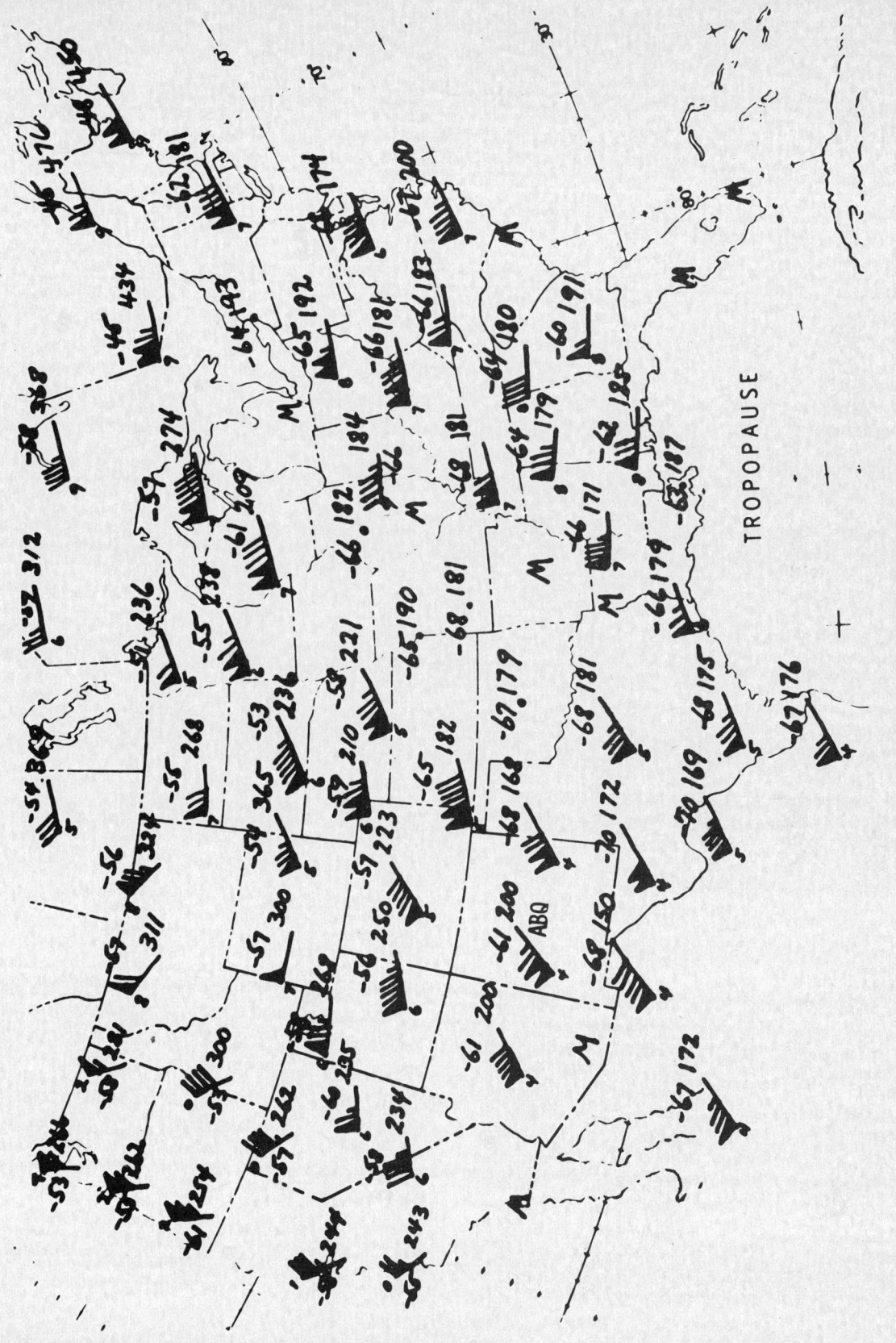

Figure 15-1. An observed tropopause chart.

Tropopause, Max Wind, and Wind Shear Charts

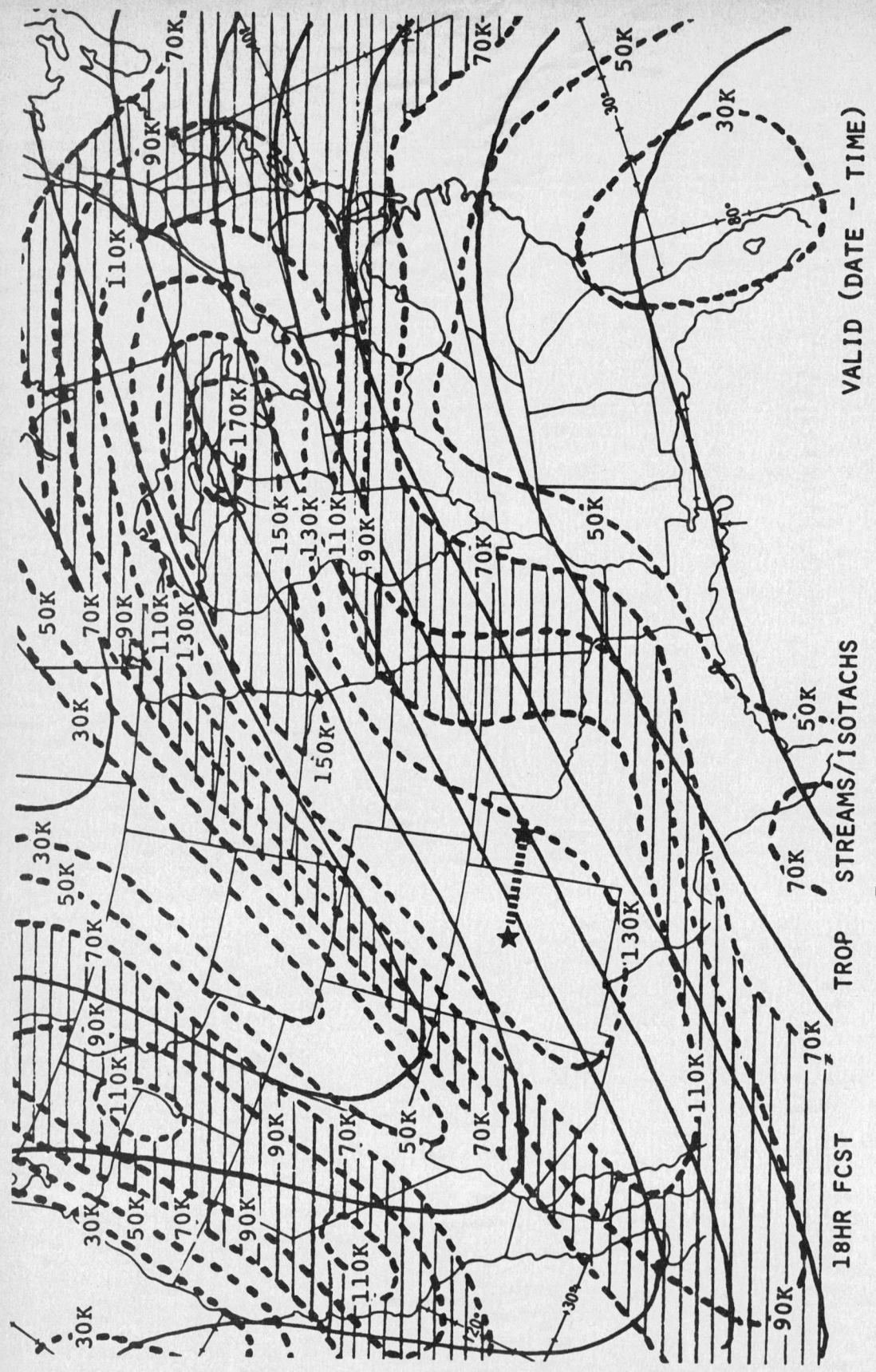

FIGURE 15-2. Section of a tropopause wind prog.

Tropopause, Max Wind, and Wind Shear Charts

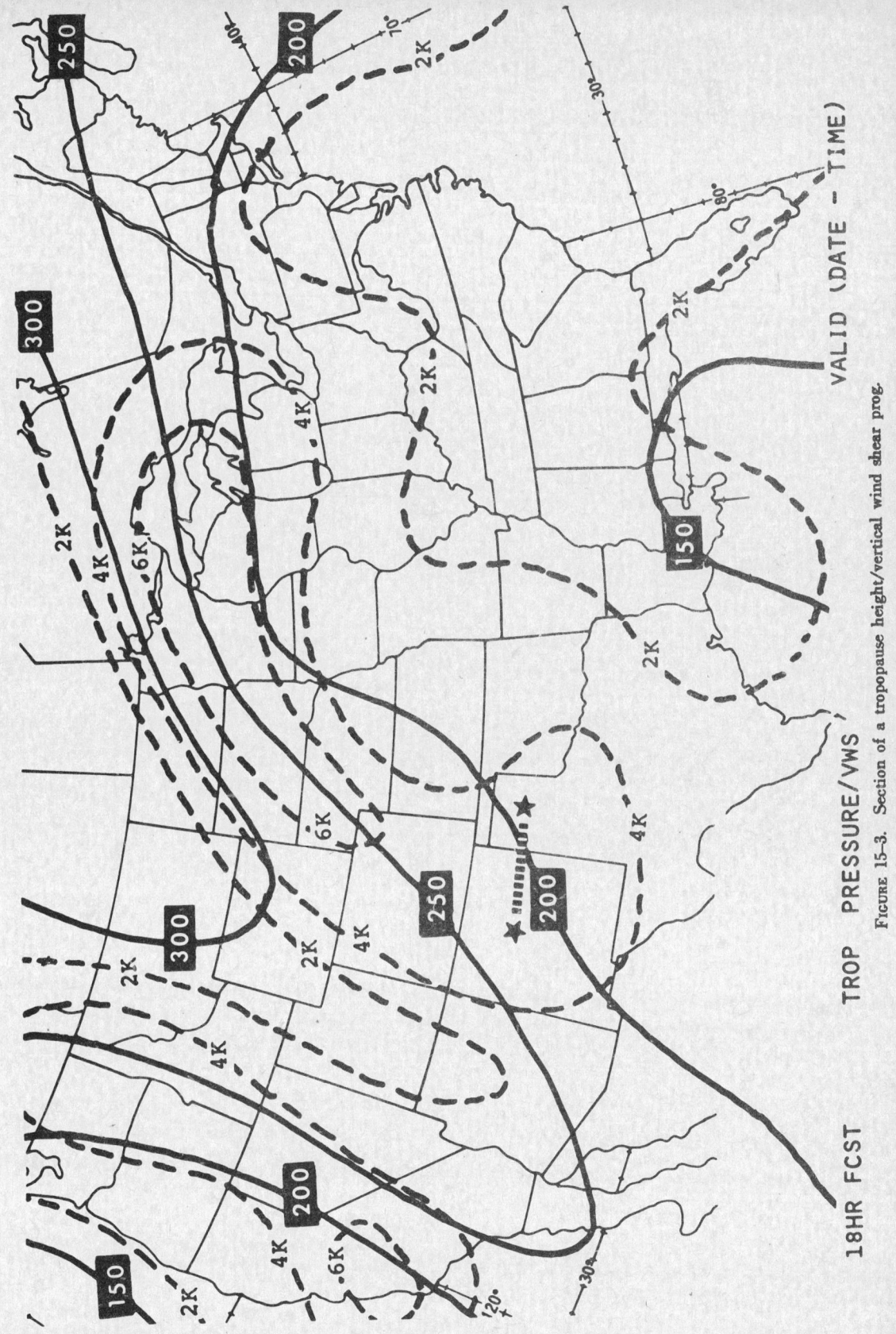

FIGURE 15-3. Section of a tropopause height/vertical wind shear prog.

Using the Charts

The progs are issued twice daily and may be used for a period up to plus or minus 6 hours from the valid time. The charts may be used to determine vertical and horizontal wind shears as clues to probable wind shear turbulence. They also may be used to determine winds for high level flight planning.

Although neither chart depicts the jet stream, locating the jet is not difficult. It passes through isotach and vertical shear maxima. It is well to examine both charts to get better accuracy than is likely if you use either chart alone. Examine figures 15–2 and 15–3; note a jet maximum from eastern Washington and Oregon extending southward and slightly westward through central California. It reappears near the southwest corner of the charts, enters the U.S. near the Arizona-New Mexico border, extends northeastward across central Nebraska, and then swings more easterly through the central Great Lakes and southern New England.

Horizontal wind shear can be determined from the spacing of isotachs. The horizontal wind shear critical for turbulence is 40 knots per 150 miles (see chapter 13, AVIATION WEATHER, discussion on clear air turbulence). 150 nautical miles is 2½° latitude.

Refer to figure 15–2 and measure 2½° latitude by laying a pencil along a meridian in the Atlantic. Move the pencil perpendicular to the isotachs across north central Montana, and you can see that the horizontal shear—the difference in windspeed—is about 40 knots along this distance. This spacing represents the wind shear critical for probable wind shear turbulence. The strong wind shear from southwestern Arizona to northwestern Minnesota suggests a probability of turbulence due to horizontal wind shear.

Vertical wind shear can be determined directly from the dashed lines in figure 15–3. The vertical shear critical for probable turbulence is 6 kt/1,000 ft. You find this critical value in central California and from western Nebraska to the Great Lakes. An area of extremely high probability of moderate or even hazardous turbulence is the three-State junction of the Dakotas and Minnesota where horizontal shear is about 80 kt/150 mi., and vertical shear is in excess of 6 kt/1,000 ft.

Wind direction and speed at the tropopause flight level may be read directly from the streamlines and isotachs. To determine wind at a flight level below or above the tropopause, first determine direction and speed at the tropopause. Wind direction changes very little within several thousand feet of the tropopause, so this direction may be used throughout the layer for which vertical wind shear is computed. Next determine wind shear and the number of thousands of feet the desired flight level differs from flight level of the tropopause. Multiply the shear by the thousands of feet and subtract this value from the speed at the tropopause.

As an example, let's assume a westbound flight wants the probability of turbulence and the wind for a leg from Amarillo to Albuquerque. Note from the charts in figures 15–2 and 15–3 that horizontal wind shear is negligible. Vertical wind shear is interpolated between the 4- and 6-knot shear lines and is about 5 kt/1,000 ft. Widespread significant turbulence is unlikely. (You should also refer to the high-level significant weather prog and pilot reports for further clues to turbulence.)

Wind direction along the route determined from the streamlines is about 230°—a quartering headwind. Speed is strongest at the tropopause; so for a westbound flight, choose a flight level as far as practical above or below the tropopause. Height of the tropopause determined from figure 15–3 is flight level 39,000 feet (200 millibars). Let's further assume that the high-level significant weather prog shows cirrus prevalent below the tropopause and you wish to stay clear of clouds. You would like to check the wind at 43,000 feet. From figure 15–2, you determine tropopause windspeed to be on the high side of the 130-knot isotach but quite a distance from the 150-knot isotach. Let's interpolate the speed as 135 knots. The flight level, 43,000 feet, is 4,000 feet above the tropopause. Multiply the 5-knot shear by 4, and you get a difference of 20 knots. Subtract 20 knots from 135, the speed at the tropopause, and you get a speed of 115 knots. Wind at FL430 is 230° at 115 knots.

INTERNATIONAL TROPOPAUSE AND WIND SHEAR PROGS

Tropopause and wind shear progs prepared for international flight documentation cover several areas to serve flights across the North Atlantic, the North and South Pacific, and the Caribbean. These progs are computer prepared for direct transmission by facsimile. They show flight level of the tropopause, mean vertical wind shear, and temperatures. Figure 15–4 shows a section of and figure 15–5 the format of a tropopause wind shear prog.

Tropopause, Max Wind, and Wind Shear Charts

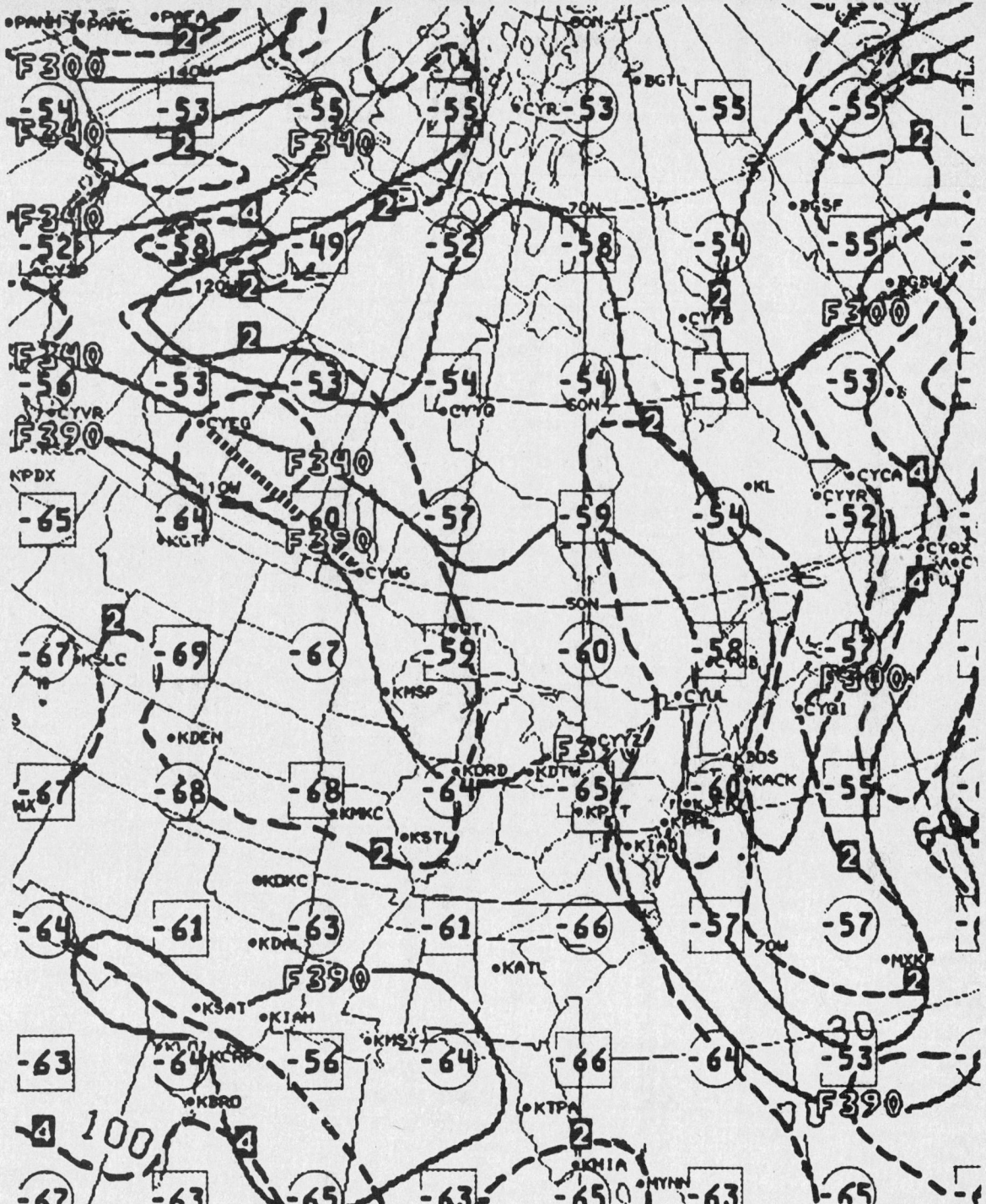

FIGURE 15-4. Western section of an international tropopause wind shear prog for the Atlantic. This section was chosen to show a geographical area with which the reader is most likely familiar.

Tropopause, Max Wind, and Wind Shear Charts

Tropopause Flight Levels

Tropopause heights are in flight levels. Solid lines trace intersections of the tropopause with constant pressure surfaces. They are labelled as flight levels in hundreds of feet preceded by the letter "F". The figures are "open" to distinguish them from other labels. Note the line in figure 15–5 labelled "F390" and the several flight level labels along the upper left edge of figure 15–4.

Mean Vertical Wind Shear

Vertical wind shear is shown by dashed lines. Shear is in knots per 1,000 feet with dashed lines drawn at 2-knot intervals. Labels are dark squares with white numbers. Note in figure 15–5 the dashed shear line labelled "2" meaning a shear of 2 knots/1,000 feet. Wind shear is averaged through a layer from about 8,000 feet below to 4,000 feet above the tropopause.

Temperatures

Spot temperatures are entered for both the tropopause and for 150 millibars—flight level 45,000 feet (FL450). Tropopause temperatures are enclosed in squares; temperatures at FL450 are circled. Note how these are indicated in figure 15–5.

Using the Chart

When you are planning to fly at a level in the vicinity of the tropopause, you can often use the tropopause and wind shear prog together with constant pressure progs to compute winds and temperatures for your planned flight level. You can also use the chart when planning a flight above FL390, the highest constant pressure or flight level prog.

Both the rate of change of windspeed with height and the temperature lapse rate change abruptly at the tropopause. Therefore, you cannot interpolate *through* it; you must restrict each interpolation to airspace either *above* or *below* it. When both your flight level and the tropopause fall between two constant pressure prog levels, you must use the wind shear prog for interpolation.

When using the domestic wind shear prog, you worked upward or downward from the tropopause maximum wind. Since for this prog you have no maximum wind, you must work downward or upward from a constant flight level prog. To illustrate, let's assume you are returning from Alaska to Chicago O'Hare airport via Edmonton (CYEG) and Winnipeg (CYWG), Canada at FL370. Let's compute wind and temperature for the leg from CYEG to CYWG. This segment is shown by the heavy dotted line on figure 15–4.

The tropopause over CYEG is at FL340 and over CYWG, near FL390. Obviously at FL370, you will fly through the tropopause during this leg of your flight. Figure 15–6 is a cross section of the route to assist you in following the computation.

The constant pressure prog nearest your flight level is 200 millibars (FL390), 2,000 feet above your flight level. We begin in our computation by extracting wind direction and speed from the FL390 prog at CYEG, at CYWG, and at a point about midway between where you will fly through the tropopause. Assume we determined the 39,000-foot (200 millibar) winds to be:

CYEG   280° at 75 knots,
Midpoint 290° at 95 knots,
CYWG   300° at 100 knots.

Next we get wind shear for the three points from figure 15–4. Note a line of constant wind shear near CYEG and encircling the first half of the route. The line is unlabelled. However, the steeply sloping tropopause through the area suggests a maximum wind shear. Also, to the north of CYEG, we see a 4-knot shear on the same side of the 2-knot line. This unlabelled line, therefore, has to be 4 knots/1,000 ft. and is the shear at both CYEG and midpoint. CYWG lies about half way between the 4- and 2-knot shear lines, so we interpolate shear over CYWG at 3 knots/1,000 ft.

Now let's compute wind at FL370 for each point starting with CYEG. Look at figure 15–6 and note that your flight level is between the tropopause and FL390. Therefore, expect stronger wind at FL370 than at FL390. Multiplying the shear (4 knots) by the altitude difference in thousands of feet (2), we get a speed difference of 8 knots. *Adding* this to wind at FL390, we get a wind over CYEG at FL370 of 93 knots from 280°. In the same manner we get a wind at midpoint of 290°, 103 knots. From figure 15–4, we see that the tropopause over CYWG for all practical purposes is at FL390, and this is also shown on the cross section of figure 15–6. Therefore, wind is at a maximum at FL390 and becomes less at FL370 as we get farther from the tropopause. Multiplying the shear (3 knots) by the altitude difference, we get a windspeed difference of 6 knots. *Subtracting* this difference from the 39,000 foot windspeed, we get a wind over CYWG at FL370 of 94 knots from 300°.

Now let's interpolate temperatures at FL370 for the same three points. Look again at figure 15–4 and note tropopause temperatures enclosed in

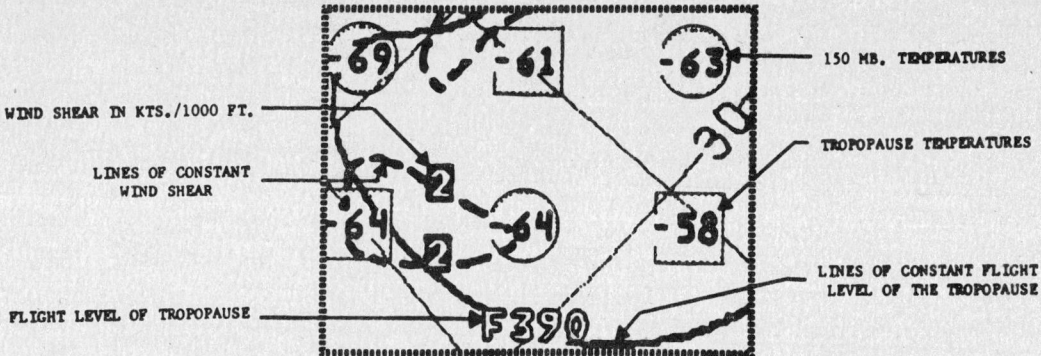

FIGURE 15-5. Format of the international tropopause wind shear prog.

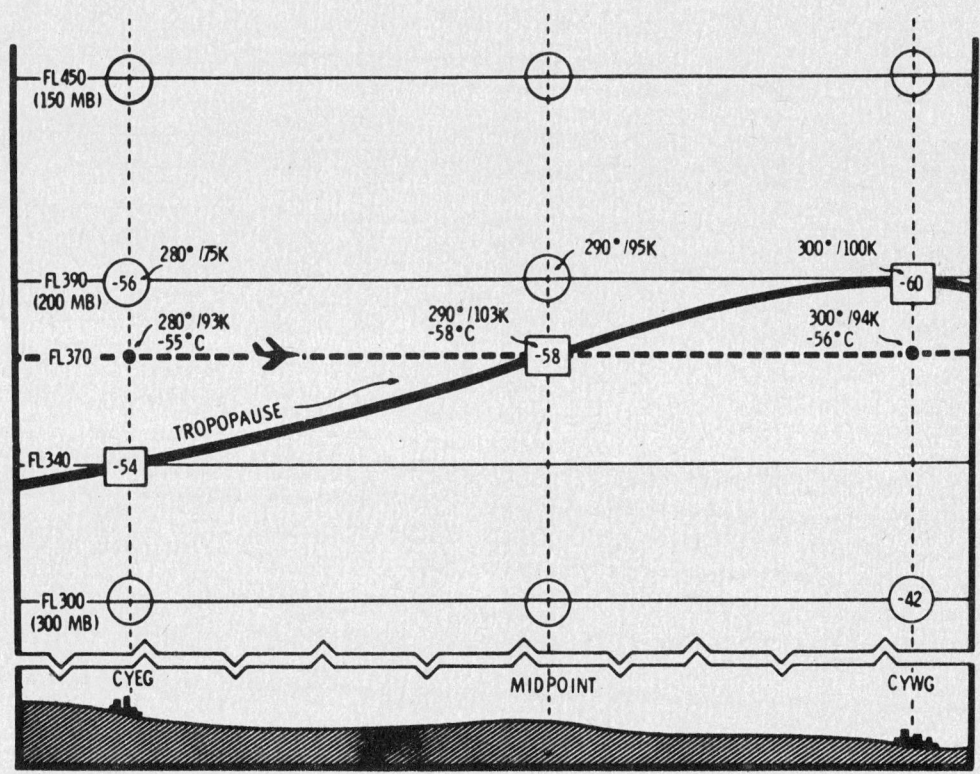

FIGURE 15-6. Cross section of a route from Edmonton, Canada to Winnipeg, Canada based on the prog in figure 15-4. It is used in figuring wind and temperature at flight level 37,000. Forecast temperatures are available at flight levels 30,000 and 39,000 from constant pressure progs and at 45,000 from the tropopause wind shear prog. Only the temperatures used in this computation are entered.

squares. Just northwest of CYEG, temperature is −53° C; and just northwest of CYWG, −60° C. From these values, we estimate tropopause temperatures as shown below and as entered in squares on the cross section in figure 15–6:

CYEG   −54° C,
Midpoint −58° C,
CYWG   −60° C.

Remember that lapse rate changes at the tropopause, so we cannot interpolate *through* it. Note in figure 15–6 how it is necessary to choose temperature at FL390 over CYEG and at FL300 over CYWG. These temperatures would come from the 200- and 300-millibar progs. At midpoint, you will be at the tropopause, so temperature at your flight level is the tropopause temperature, −58° C. Interpolating for FL370 at the other two points we get −55° C over CYEG and −56° C over CYWG.

Winds and temperatures for FL370 at:
CYEG   280° at 93 knots, −55° C,
Midpoint 290° at 103 knots, −58° C,
CYWG   300° at 94 knots, −56° C.

This example is complex since the flight is through the tropopause. In most cases, you can compute an average wind and temperature for a relatively long segment of a route.

Chapter 28
TABLES AND CONVERSION GRAPHS

This section provides graphs and tables you can use operationally in decoding weather messages during preflight and inflight planning and in transmitting pilot reports. Information included covers:

1. Icing intensities and reporting.
2. Turbulence intensities and reporting.
3. Locations of probable turbulence by intensity versus weather and terrain features.
4. Standard temperature, speed, and pressure conversions.
5. Density altitude computations.
6. Selected contractions and acronyms.

The table of *Icing Intensities* classifies each intensity according to its operational effects on aircraft.

The table of *Turbulence Intensities* classifies each intensity according to its effects on aircraft control and structural integrity and on articles and occupants within the aircraft.

The table of *Locations of Probable Turbulence* lists each turbulence intensity along with terrain and weather features conducive to turbulence of that intensity.

The graph for *Density Altitude Computations* provides a means of computing density altitude, either on the ground or aloft, using the aircraft altimeter and outside air temperature.

Contractions are used extensively in surface, radar, and pilot reports and in forecasts. Most of them are known from common usage or can be deciphered phonetically. The list of *Selected Contractions* contains only those most likely to give you difficulty. Acronyms used in this manual are defined in the list of *Acronyms*.

TABLE 16–1. Icing intensities, airframe ice accumulation, and pilot report

Intensity	Airframe ice accumulation	Pilot report
Trace	Ice becomes perceptible. Rate of accumulation slightly greater than rate of sublimation. It is not hazardous even though deicing/anti-icing equipment is not used unless encountered for an extended period of time—over one hour.	Aircraft identification, location, time (GMT), intensity and type of icing,* altitude/FL, aircraft type, IAS
Light	The rate of accumulation may create a problem if flight is prolonged in this environment (over one hour). Occasional use of deicing/anti-icing equipment removes/prevents accumulation. It does not present a problem if the deicing/anti-icing equipment is used.	
Moderate	The rate of accumulation is such that even short encounters become potentially hazardous and use of deicing/anti-icing equipment or diversion is necessary.	*Example of pilot's transmission:* Holding at Westminister VOR 1232. Light Rime Icing. Altitude six thousand, Jetstar IAS 200 kt
Severe	The rate of accumulation is such that deicing/anti-icing equipment fails to reduce or control the hazard. Immediate diversion is necessary.	

* Icing may be rime, clear, or mixed.
    Rime ice: Rough milky opaque ice formed by the instantaneous freezing of small supercooled water droplets.
    Clear ice: A glossy, clear or translucent ice formed by the relatively slow freezing of large supercooled water droplets.
    Mixed ice: A combination of rime and clear ice.

TABLE 16-2. Turbulence reporting criteria

Intensity	Aircraft reaction	Reaction inside aircraft	Reporting term-definition
Light	Turbulence that momentarily causes slight, erratic changes in altitude and/or attitude (pitch, roll, yaw). Report as **Light Turbulence;*** or Turbulence that causes slight, rapid and somewhat rhythmic bumpiness without appreciable changes in altitude or attitude. Report as **Light Chop.**	Occupants may feel a slight strain against seat belts or shoulder straps. Unsecured objects may be displaced slightly. Food service may be conducted and little or no difficulty is encountered in walking.	Occasional — Less than 1/3 of the time. Intermittent — 1/3 to 2/3. Continuous — More than 2/3.
Moderate	Turbulence that is similar to Light Turbulence but of greater intensity. Changes in altitude and/or attitude occur but the aircraft remains in positive control at all times. It usually causes variations in indicated airspeed. Report as **Moderate Turbulence;*** or Turbulence that is similar to Light Chop but of greater intensity. It causes rapid bumps or jolts without appreciable changes in aircraft altitude or attitude. Report as **Moderate Chop.**	Occupants feel definite strains against seat belts or shoulder straps. Unsecured objects are dislodged. Food service and walking are difficult.	**NOTE** 1. Pilots should report location(s), time (GMT), intensity, whether in or near clouds, altitude, type of aircraft and, when applicable, duration of turbulence. 2. Duration may be based on time between two locations or over a single location. All locations should be readily identifiable. EXAMPLES: a. Over Omaha, 1232Z, Moderate Turbulence, in cloud, Flight Level 310, B707. b. From 50 miles south of Albuquerque to 30 miles north of Phoenix, 1210Z to 1250Z, occasional Moderate Chop, Flight Level 330, DC8.
Severe	Turbulence that causes large, abrupt changes in altitude and/or attitude. It usually causes large variations in indicated airspeed. Aircraft may be momentarily out of control. Report as **Severe Turbulence.***	Occupants are forced violently against seat belts or shoulder straps. Unsecured objects are tossed about. Food service and walking are impossible.	
Extreme	Turbulence in which the aircraft is violently tossed about and is practically impossible to control. It may cause structural damage. Report as **Extreme Turbulence.***		

* High level turbulence (normally above 15,000 feet ASL) not associated with cumuliform cloudiness, including thunderstorms, should be reported as CAT (clear air turbulence) preceded by the appropriate intensity, or light or moderate chop.

Locations of Probable Turbulence by Intensities Versus Weather and Terrain Features

LIGHT TURBULENCE

1. In hilly and mountainous areas even with light winds.

2. In and near small cumulus clouds.

3. In clear-air convective currents over heated surfaces.

4. With weak wind shears in the vicinity of:
   a. Troughs aloft.
   b. Lows aloft.
   c. Jet streams.
   d. The tropopause.

5. In the lower 5,000 feet of the atmosphere:
   a. When winds are near 15 knots.
   b. Where the air is colder than the underlying surfaces.

MODERATE TURBULENCE

1. In mountainous areas with a wind component of 25 to 50 knots perpendicular to and near the level of the ridge:
   a. At all levels from the surface to 5,000 feet above the tropopause with preference for altitudes:
      (1) Within 5,000 feet of the ridge level.
      (2) At the base of relatively stable layers below the base of the tropopause.
      (3) Within the tropopause layer.
   b. Extending outward on the lee of the ridge for 150 to 300 miles.

2. In and near thunderstorms in the dissipating stage.

3. In and near other towering cumuliform clouds.

4. In the lower 5,000 feet of the troposphere:
   a. When surface winds exceed 25 knots.
   b. Where heating of the underlying surface is unusually strong.
   c. Where there is an invasion of very cold air.

5. In fronts aloft.

6. Where:
   a. Vertical wind shears exceed 6 knots per 1,000 feet, and/or
   b. Horizontal wind shears exceed 18 knots per 150 miles.

SEVERE TURBULENCE

1. In mountainous areas with a wind component exceeding 50 knots perpendicular to and near the level of the ridge:
   a. In 5,000-foot layers:
      (1) At and below the ridge level in rotor clouds or rotor action.
      (2) At the tropopause.
      (3) Sometimes at the base of other stable layers below the tropopause.
   b. Extending outward on the lee of the ridge for 50 to 150 miles.

2. In and near growing and mature thunderstorms.

3. Occasionally in other towering cumuliform clouds.

4. 50 to 100 miles on the cold side of the center of the jet stream, in troughs aloft, and in lows aloft where:
   a. Vertical wind shears exceed 6 knots per 1,000 feet, and
   b. Horizontal wind shears exceed 40 knots per 150 miles.

EXTREME TURBULENCE

1. In mountain wave situations, in and below the level of well-developed rotor clouds. Sometimes it extends to the ground.

2. In growing severe thunderstorms (most frequently in organized squall lines) indicated by:
   a. Large hailstones (3/4 inch or more in diameter).
   b. Strong radar echoes, or
   c. Almost continuous lightning.

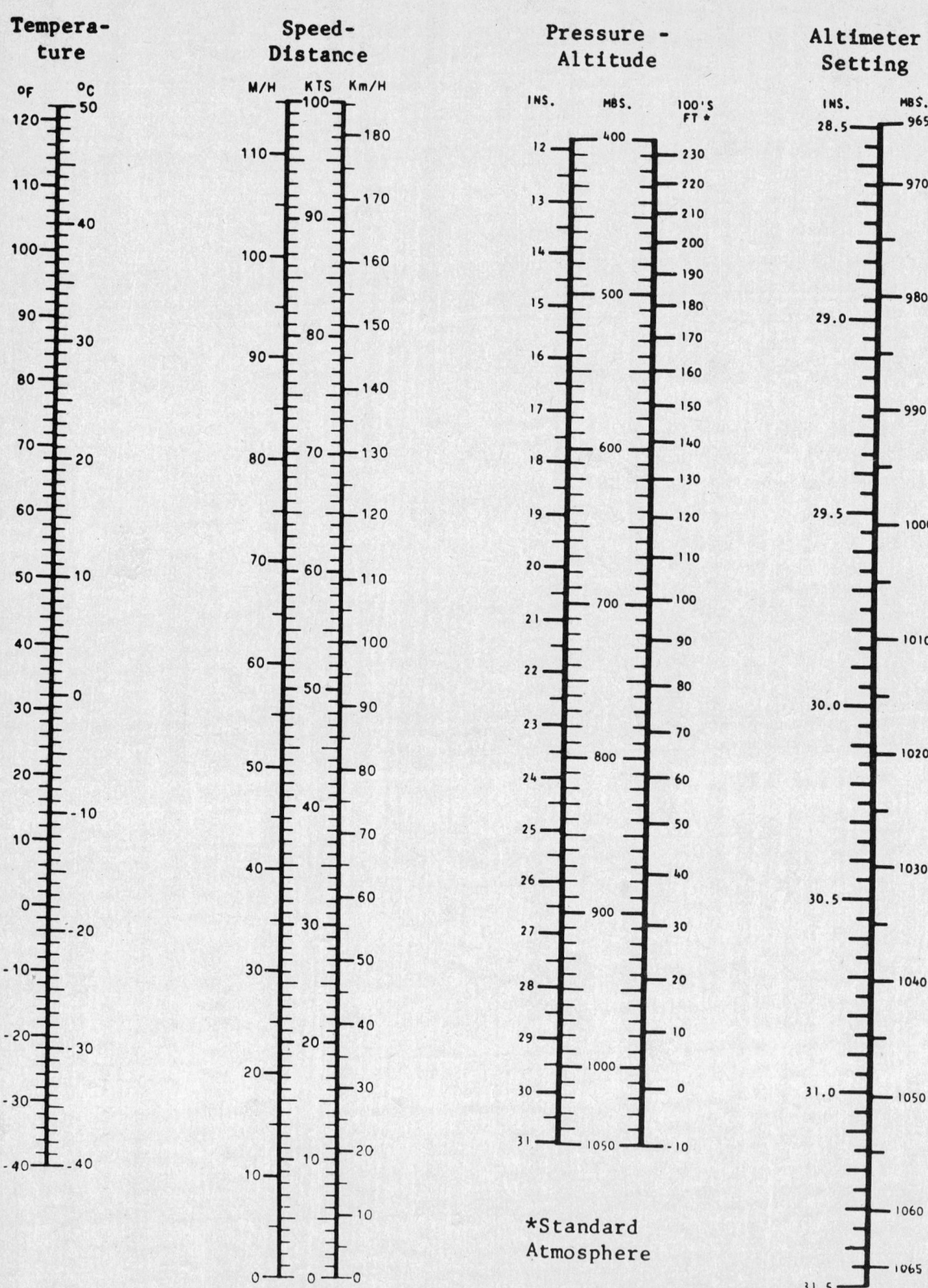

Density Altitude Computation

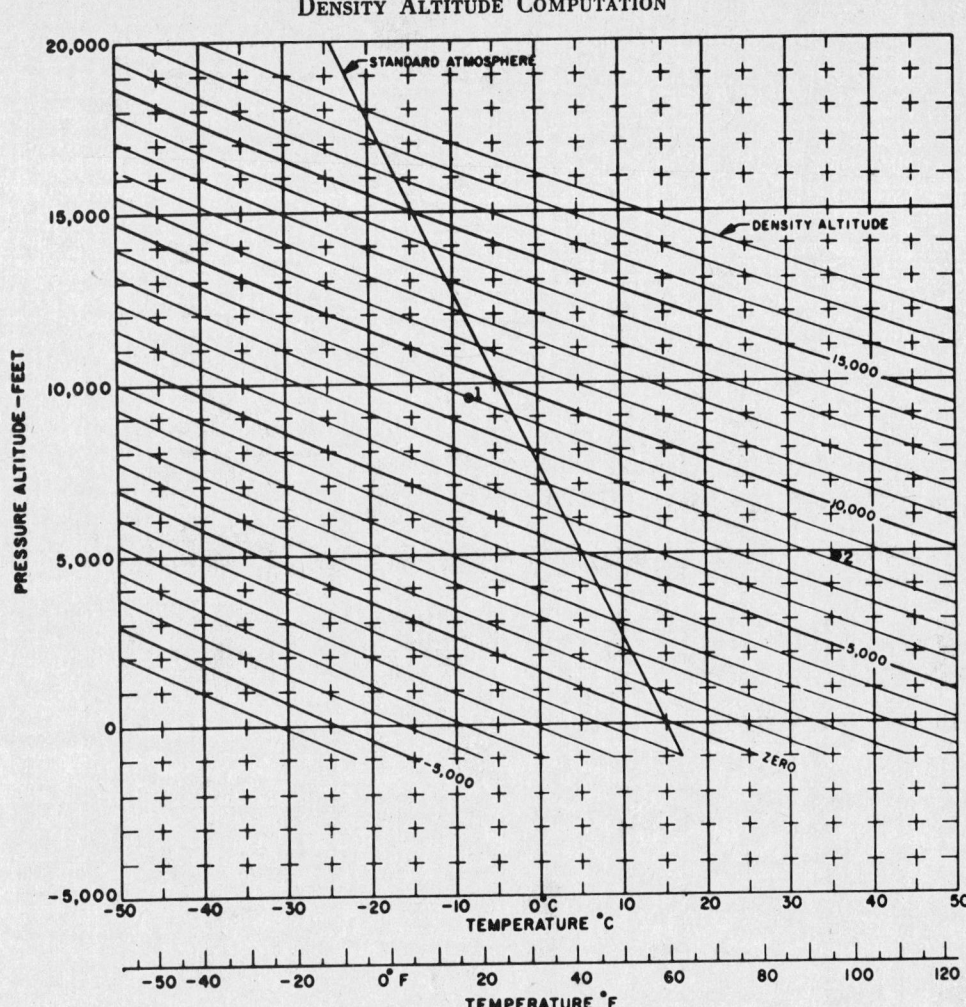

Use this graph to find density altitude either on the ground or aloft. Set your altimeter at 29.92 inches; it now indicates pressure altitude. Read outside air temperature. Enter the graph at your pressure altitude and move horizontally to the temperature. Read density altitude from the sloping lines.

Example 1. Find density altitude in flight. Pressure altitude is 9,500 feet; and temperature, −8° C. Find 9,500 feet on the left of the graph and move across to −8° C. Density altitude is 9,000 feet (marked "1" on the graph).

Example 2. Find density altitude for take-off. Pressure altitude is 4,950 feet; and temperature, 97° F. Enter the graph at 4,950 feet and move across to 97° F. Density altitude is 8,200 feet (marked "2" on graph). Note that in the warm air, density altitude is considerably higher than pressure altitude.

Tables and Conversion Graphs

Selected Contractions

A

ACLD	above clouds
ACSL	standing lenticular altocumulus
ACYC	anticyclonic
AFDK	after dark
ALQDS	all quadrants
AC	altocumulus
ACCAS	altocumulus castellanus
AS	altostratus
AOA	at or above
AOB	at or below

B

BCKG	backing
BFDK	before dark
BINOVC	breaks in overcast
BL	between layers
BLZD	blizzard
BOVC	base of overcast
BRKHIC	breaks in higher overcast

C

CBMAM	cumulonimbus mamma
CC	cirrocumulus
CCSL	standing lenticular cirrocumulus
CFP	cold frontal passage
CI	cirrus
CLRS	clear and smooth
CRLCN	circulation
CS	cirrostratus
CU	cumulus
CUFRA	cumulus fractus
CYC	cyclonic

D

DFUS	diffuse
DNSLP	downslope
DP	deep
DTRT	deteriorate
DURGC	during climb
DURGD	during descent
DWNDFTS	downdrafts

E

EMBDD	embedded

F

FNTGNS	frontogenesis (front forming)
FNTLYS	frontolysis (front decaying)
FROPA	frontal passage

G

GFDEP	ground fog estimated _____ feet deep

H

HDEP	haze layer estimated _____ feet deep
HLSTO	hailstones
HLYR	haze layer aloft

I

ICG	icing
ICGIC	icing in clouds
ICGICIP	icing in clouds and precipitation
ICGIP	icing in precipitation
INTMT	intermittent
INVRN	inversion
IPV	improve
ISOLD	isolated

K

KDEP	smoke layer estimated _____ feet deep
KLYR	smoke layer aloft
KOCTY	smoke over city

L

LTG, LTNG	lightning
LTGCC	lightning cloud-to-cloud
LTGCCCG	lightning cloud-to-cloud, cloud-to-ground
LTGCW	lightning cloud-to-water
LTGIC	lightning in cloud
LTNG, LTG	lightning

M

MEGG	merging
MLTLVL	melting level
MNLD	mainland
MOGR	moderate or greater
MRGL	marginal
MSTR	moisture

Tables and Conversion Graphs

N

NCWX	no change in weather
NPRS	non persistent
NRW	narrow
NS	nimbostratus

O

OAOI	on and off instruments
OAT	outside air temperature
OCFNT	occluded front
OCLD	occlude
OFP	occluded frontal passage
OFSHR	off shore
OI	on instruments
OMTNS	over mountains
ONSHR	on shore
OTAS	on top and smooth
OVRNG	overrunning

P

PDW	priority delayed weather
PRESFR	pressure falling rapidly
PRESRR	pressure rising rapidly
PRIND	present indications are
PRST	persist

Q

QSTNRY	quasistationary
QUAD	quadrant

R

RGD	ragged
RTD	routine delayed weather

S

SC	stratocumulus
CLR, SKC	sky clear
SNOINCR	snow depth increase in past hour
SNRS, SR	sunrise
SNST, SS	sunset
SNWFL	snowfall
SQAL	squall
SQLN	squall line
SR, SNRS	sunrise
SS, SNST	sunset
ST	stratus
STFRA	stratus fractus
STFRM	stratoform
STM	storm

T

TCU	towering cumulus
TOVC	top of overcast
TROP	tropopause
TWRG	towering

U

UDDF	up and down drafts
UPDFTS	updrafts
UPSLP	upslope

V

VLNT	violent
VR	veer

W

WDSPRD	widespread
WFP	warm frontal passage
WK	weak
WRMFNT	warm front
WSHFT	wind shift
WV	wave

Acronyms

AC—Convective Outlook Bulletin; identifies a forecast of probable convective storms.

AIRMET—Airman's Meteorological Information; an inflight advisory forecast of conditions possibly hazardous to light aircraft or inexperienced pilots.

ARTCC—Air Route Traffic Control Center, FAA.

FA—Area Forecast; identifies a forecast of general aviation weather over a relatively large area.

FAA—Federal Aviation Administration, Department of Transportation.

FD—Winds and Temperatures Aloft Forecast; a forecast identifier.

FSS—Flight Service Station, FAA.

FT—Terminal Forecast; identifies a forecast in the U.S. forecast code.

ICAO—International Civil Aviation Organization.

IFSS—International Flight Service Station, FAA.

LAWRS—Limited Aviation Weather Reporting Station; usually a control tower; reports fewer weather elements than a complete SA.

NESS—National Environmental Satellite Service, National Oceanic and Atmospheric Administration, Department of Commerce; serves NWS with satellite weather observations.

NHC—National Hurricane Center, NWS.

NMC—National Meteorological Center, NWS.

NOAA—National Oceanic and Atmospheric Administration, Department of Commerce.

NSSFC—National Severe Storms Forecast Center, NWS.

NWS—National Weather Service, National Oceanic and Atmospheric Administration, Department of Commerce.

PATWAS—Pilot's Automatic Telephone Weather Answering Service; a self-briefing service.

PIREP—Pilot Weather Report.

RAREP—Radar Weather Report.

SA—Surface Aviation Weather Report; a message identifier.

SAWRS—Supplemental Aviation Weather Reporting Station; usually an airline office at a terminal not having NWS or FAA facilities.

SIGMET—Significant Meteorological Information; an inflight advisory forecast of weather hazardous to aircraft.

TAF—Terminal Aviation Forecast; identifies a terminal forecast in the ICAO code.

TWEB—Transcribed Weather Broadcast; a self-briefing radio broadcast service.

UA—Teletypewriter identifier of a pilot weather report (PIREP).

WA—Teletypewriter identifier of an AIRMET valid for a specified period.

WAC—Teletypewriter identifier of an AIRMET continuing in effect until cancelled.

WH—Hurricane Weather Advisory; identifies a hurricane advisory forecast specifically for aviation.

WS—Teletypewriter identifier of a SIGMET.

WSFO—Weather Service Forecast Office, NWS.

WSO—Weather Service Office, NWS.

WW—Severe Weather Watch; identifies a forecast of probable severe thunderstorms or tornadoes.

Chapter 29
WEATHER INFORMATION

Flight Service Stations (FSS) and Combined Station/Tower (CS/T) provide information on airport conditions, radio aids and other facilities, and process flight plans. CS/T personnel are not certificated pilot weather briefers; however, they provide factual data from weather reports and forecasts. Airport Advisory Service is provided at the pilot's request on 123.6 by FSSs located at airports where there are not control towers in operation.

The telephone area code number is shown in parentheses. Each number given is the preferred telephone number to obtain flight weather information. Automatic answering devices are sometimes used on listed lines to given general local weather information during peak workloads. To avoid getting the recorded general weather announcement, use the selected telephone number listed.

● **FAST FILE FLIGHT PLAN SYSTEM**

Some Flight Service Stations have inaugurated this system for pilots who desire to file IFR/VFR flight plans with or without a weather briefing. Pilots may call the discrete telephone numbers listed and file flight plans in accordance with prerecorded taped instructions. IFR flight plans will be extracted from the recorder and subsequently entered into the appropriate ARTCC computer. VFR flight plans will be transcribed; and both IFR/VFR flight plans will be filed in the FSS. This equipment is designed to automatically disconnect after 8 seconds of no transmission, so pilots are instructed to speak at a normal speech rate without lengthly pauses between flight plan elements. Pilots are urged to file flight plans into this system at least 30 minutes in advance of proposed departure. The system may be used to close and cancel flight plan.

Preflight weather briefing services remain available through regular telephone numbers.

★ Indicates Pilot's Automatic Telephone Weather Answering Service (PATWAS) or telephone connected to the Transcribed Weather Broadcast (TWEB) providing transcribed aviation weather information.

◆ Indicates a restricted number, use for aviation weather information.

■ Call FSS for "one call" FSS–WSO briefing service.

✻ Automatic Aviation Weather Service (AAWS).

● §§ Indicates Fast File telephone number for pre-recorded and transcribed flight plan filing only.

Location and Identifier		Area Code	Telephone
ALABAMA			
Anniston ANB	FSS	(205)	831-2303
Birmingham BHM	FSS	(205)	595-6151 ■
	FSS	(205)	595-2101 ★
Dothan DHN	FSS	(205)	983-3551
Huntsville	WS	(205)	772-3521 ◆
Mobile MOB (Bates)	FSS	(205)	344-3610 ■
Montgomery MGM (Dannelly)	FSS	(205)	832-7516 ■
Muscle Shoals MSL	FSS	(205)	383-6541 ■
	FSS	(205)	381-2500 ★
Tuscaloosa TCL	FSS	(205)	758-3628
ARIZONA			
Douglas DUG (Bisbee-Douglas)	FSS	(602)	364-8458
Flagstaff	WS	(602)	774-2851
		(602)	774-1424
		(602)	774-0475
Phoenix PHX (Sky Harbor)	FSS	(602)	261-4295 ■
E bound		(602)	267-7239 ✻
W bound		(602)	267-1181 ✻
Prescott PRC	FSS	(602)	445-2160
Grand Canyon		(602)	638-2943
Kingman		(602)	753-5659

Location and Identifier		Area Code	Telephone
ARIZONA (Can't.)			
Tusocn	FSS	(602)	792-6359 ■
			294-2635 ✻
Tuscon-Hermosillo, Mexico Route		(602)	898-8549 ★
			(0500-2200)
E Bound		(602)	294-7441 ✻
W Bound		(602)	294-8263 ✻
Winslow	WS	(602)	289-3592
Yuma YUM	FSS	(602)	726-2601 ■
ARKANSAS			
El Dorado ELD (Goodwin)	FSS	(501)	863-5128
Fayetteville FYV (Drake)	FSS	(501)	HI 2-8277
Ft. Smith	WS	(501)	646-7885 ◆
Harrison HRO	FSS	(501)	EM 5-3433
Jonesboro JBR	FSS	(501)	WE 5-3471
		(0600-2200	Other hrs. Memphis)
Little Rock	FSS	(501)	376-0721
		(501)	835-7626
Texarkana TXK	CS/T	(501)	774-4151 ■
CALIFORNIA			
Arcata ACV	FSS	(707)	839-1545

Weather Information

Location and Identifier		Area Code	Telephone
CALIFORNIA (Con't.)			
Bakersfield BFL (Meadows)	FSS	(805)	399-1787■
(No wea bcst avbl 2300-0500 lcl time)			
Bishop	WS	(714)	873-3213
			(0545-1915)
Blythe BLH	FSS	(714)	922-6151
Crescent City CEC (McNamara Fld)	FSS	(707)	464-2514
(0600-2200 other hrs Arcata)			
Daggett DAG	FSS	(714)	254-2958
			254-2959
Eureka	WS	(707)	442-2171◆
Fresno FAT (Air Terminal)	FSS	(209)	251-8269■
Imperial IPL	FSS	(714)	352-8740
Lancaster	FSS	(805)	948-5385
Long Beach	WS	(213)	429-0337
Los Angeles LAX (International)	FSS	(213)	776-2727■
North of LAX		(213)	466-4116§§
South of LAX		(213)	263-6776§§
		(213)	670-1000■
LAX Basin Forecast		(213)	776-8803★
Route Forecast		(213)	776-1640★
Van Nuys		(213)	781-5213■
LAX Basin Forecast		(213)	787-6580★
Route Forecast		(213)	787-4911★
Burbank		(213)	841-3904■
LAX Basin Forecast		(213)	843-6911★
Route Forecast		(213)	841-0034★
Long Beach		(213)	639-2618■
	WS	(213)	429-0337
LAX Basin Forecast		(213)	639-4200★
Rourte Foecast		(213)	639-2647★
Orange County		(714)	542-3585■
LAX Basin Forecast		(714)	546-1610★
Route Forecast		(714)	546-0595★
El Monte		(213)	728-9957■
LAX Basin Forecast		(213)	442-3113★
Route Forecast		(213)	442-7800★
Marysville MYV (Yuba Co.)	FSS	(916)	742-8852
Montague SIY (Siskiyou Co.)	FSS	(916)	459-3003
(Other hrs. Red Bluff)			(0615-2145)
Mt. Shasta	WS	(916)	926-2227
			(0630-1500)
Needles EED	FSS	(714)	326-3511
Oakland OAK (International)	FSS	(415)	562-7807■
		(415)	569-0313★
Concord		(415)	933-8990■
Fremont		(415)	656-5093■
Palo Alto		(415)	326-2941■
South San Francisco		(415)	588-8623■
		(415)	589-6711★
San Jose		(408)	248-8912■
		(408)	263-0123★
San Mateo		(415)	342-8626■
Redwood City		(415)	364-2828★
Ontario ONT	FSS	(714)	983-2618
		(714)	986-2006 ⚹

Location and Identifier		Area Code	Telephone
CALIFORNIA (Con't)			
Colton		(714)	825-0749
Corona		(714)	734-0280
El Monte		(714)	728-9957
Hemet		(714)	925-9230
Santa Ana		(714)	836-0776
Paso Robles PRB	FSS	(805)	238-2448
Paso Robles San Luis Obispo-Arroyo Grande route			544-6323
Red Bluff RBL (Bidwell)	FSS	(916)	527-0242■
		(916)	527-8310★
			(0500-2200)
Redding Muni		(916)	246-1556★
			(0500-2200)
Sacramento SAC	FSS	(916)	428-6500■
			428-4027★
Salinas SNS	FSS	(408)	422-4723
San Diego SAN (Lindbergh)	FSS	(714)	291-6381■
			291-0750 ⚹
			(0600-2200)
			San Diego ⚹
			Santa Barbara ⚹
			Mexican Border ⚹
			Inland over Coastal and Tehachapi Mtns. ⚹
San Francisco (See Oakland FSS)			
Santa Barbara SBA	FSS	(805)	967-2305
Santa Maria	WS	(805)	925-0246
			(0600-2200)
Santa Rosa STS (Sonoma County)	WS	(707)	545-3724
Stockton SCK	FSS	(209)	982-4284■
Thermal TRM	FSS	(714)	399-5155
		(714)	345-1612
Ukiah UKI	FSS	(707)	462-8877
COLORADO			
Akron AKO	FSS	(303)	345-2271
Alamosa	WS	(303)	589-2547
			(0400-2000)
Colorado Springs	WS	(303)	596-0553◆
(Denver FSS)	FSS	(303)	634-3127
Denver DEN (Stapleton INTL)	FSS	(303)	321-0031■
	FSS	(303)	388-3653★
			(0500-2200)
Denver-Cheyenne-North Platte-Akron area			398-3967 ⚹
Denver-Colorado Springs-Pueblo-La Junta area			398-3967 ⚹
Denver-Grand Junction route			398-5391 ⚹
Denver-Salt Lake City			398-5392
Denver-Billings route			398-5393 ⚹
Denver-Kansas City			398-5394 ⚹
Eagle EGE	FSS	(303)	328-6575
Grand Junction GJT (Walker Field)	FSS	(303)	242-1801■
La Junta LHX	FSS	(303)	384-4311
Pueblo	CS/T	(303)	948-3301
	WS	(303)	948-3376◆
Trinidad TAD	FSS	(303)	846-2623

Weather Information

Location and Identifier		Area Code	Telephone
CONNECTICUT			
Bridgeport	WS	(203)	378-2344
Windsor Locks BDL (Bradley Field)	FSS	(203)	623-2416■
DELAWARE			
Wilmington	WS	(302)	571-6360◆
(Millville)	FSS	(302)	652-3479
DISTRICT OF COLUMBIA			
Washington Dulles Intl (toll)	WS	(703)	661-8526◆
(For International Flight Briefing)			
Washington National DCA	FSS	(202)	DI 7-4040■
Washington:			
Local Area		(202)	347-4950★
Northerly Routes		(202)	920-4000★
Southerly Routes		(202)	920-3603★
IFR Flight Plans Only		(202)	521-7333
Baltimore:			
Local Area		(301)	766-0757★
Northerly Routes		(301)	768-6510★
Southerly Routes		(301)	768-6650★
IFR Flight Plans Only		(301)	521-7333
FLORIDA			
Apalachicola	WS	(904)	653-3171
Crestview CEW	FSS	(904)	682-2795
Daytona Beach	WS	(904)	252-3112
		(904)	253-6131◆
Ft. Myers FMY (Page Field)	FSS	(813)	936-1857■
	WS	(813)	332-5595
		(0700-1700 Mon-Fri.)	
		(0700-1500 Sat-Sun.)	
Gainesville GNV	FSS	(904)	376-7515/6
Jacksonville JAX (Craig Muni)	FSS	(904)	641-8333■
	FSS	(904)	641-8055★
Key West EYW (International)	FSS	(305)	296-2042
	WS	(305)	296-2741
Lakeland	WS	(813)	682-4221
		(0600-2400 Nov 15-Mar 15)	
		(0800-1700 Mar 16-Nov 14)	
Melbourne MLB (Melbourne Rgnl)	FSS	(305)	723-6151
			783-7833
			269-2022
Miami MIA	IFSS	(305)	233-2600■
	FSS	(305)	233-2616★
Orlando ORL (Herndon)	FSS	(305	894-0861
Pensacola PNS	FSS	(904)	438-4390
	FSS	(904)	432-3037
	WS	(904)	453-2488
St. Petersburg PIE (Clearwater Intl)	FSS	(813)	531-1495/6/7
	FSS	(813)	531-8200★
Tallahassee TLH	FSS	(904)	576-3141
	WS	(904)	575-1811◆
			576-6318
Tampa	WS	(813)	229-1708★
	WS	(813)	879-3907◆

Location and Identifier		Area Code	Telephone
FLORIDA (Con't.)			
Vero Beach VRB	FSS	(305)	562-2321/2
		(305)	464-1817
		(305)	287-8021
W. Palm Beach	WS	(305)	683-3032◆
GEORGIA			
Albany ABY	FSS	(912)	435-6201
Alma AMG	FSS	(912)	632-4422
(0600-2200 EST Other hrs. Brunswick)			
Athens	WS	(404)	548-7318◆
Atlanta ATL (Fulton Co.)	FSS	(404)	691-2240■
	FSS	(404)	755-6608★
		(404)	691-0282§§
Augusta	WS	(404)	793-6610◆
Brunswick SSI (Malcom McKinnon)	FSS	(912)	638-8641
Columbus	WS	(404)	322-1793◆
		(0630-1845)	
Macon MCN (Lewis B. Wilson)	FSS	(912)	788-5064■
Rome	WS	(912)	232-6801
		(Mon-Fri 0645-1445)	
Savannah SAV (Travis)	FSS	(912)	964-7730■
Valdosta VLD	FSS	(912)	CH 4-2361
IDAHO			
Boise BOI (Air Terminal)	FSS	(208)	343-2525■
	FSS	(208)	345-6163/4★
Burley BYI (Burley)	FSS	(208)	678-8361/2
Idaho Falls IDA (Fanning Field)	FSS	(208)	522-9024
Lewistown	WS	(208)	743-3841
		(0330-1930)	
Pocatello	WS	(208)	233-0143◆
(0500-2100 lcl time)			
ILLINOIS			
Chicago CHI (Du Page)	FSS	(312)	626-8266
	FSS	(231)	584-5830★
	FSS	(312)	626-8629★
	FSS	(312)	584-5010
	WS	(312)	686-2155◆
Decatur DEC	FSS	(217)	429-2311
Moline (Davenport, Iowa)	WS	(319)	326-1322
Quincy UIN (Baldwin)	FSS	(217)	885-3251
Rockford RFD (Greater Rockford)	FSS	(815)	965-6758■
Springfield	WS	(217)	575-3867
INDIANA			
Evansville	WS	(812)	426-2987◆
Ft. Wayne FWA (Baer Field)	FSS	(219)	747-3139
Indianapolis IND (Weir-Cook)	FSS	(317)	244-3316■
	FSS	(317)	247-2209★
Lafayette LAF (Purdue University)	FSS	(317)	743-1802/3
South Bend SBN (St. Joseph Co.)	FSS	(219)	232-5858
Terre Haute (Hulman)	FSS	(812)	232-0984

Weather Information

Location and Identifier		Area Code	Telephone
IOWA			
Burlington BRL	FSS	(319)	753-1626
Cedar Rapids CID	FSS	(319)	364-0244
Davenport (via Moline, Ill.)	WS	(319)	326-1322
Des Moines DSM	FSS	(515)	285-4640 ▄
Dubuque	WS	(319)	582-3171
			(0545-2105)
Mason City MCW	FSS	(515)	423-7512
Ottumwa OTM	FSS	(515)	682-3492
Sioux City SUX	WS	(712)	255-3944 ♦
Waterloo ALO	WS	(319)	234-1602 ♦
KANSAS			
Chanute CNU	FSS	(316)	431-4450
Concordia	WS	(913)	243-3141
Dodge City DDC	FSS	(316)	225-0218/9 ▄
Emporia EMP	FSS	(316)	DI 2-7475
Garden City GCK (New Municipal)	FSS	(316)	275-9208
Goodland GLD	FSS	(913)	899-7154 ▄
Hill City HLC	FSS	(913)	674-5642
		(0600-2200 Other hrs. Goodland)	
Manhattan MHK	FSS	(913)	539-4606
		(0600-2200 other hrs. Salina)	
Russell RSL	FSS	(913)	483-2165
Salina SLN	FSS	(913)	825-0506/7
Wichita ICT	FSS	(316)	942-2261/2 ▄
	FSS	(316)	942-3284 ★
KENTUCKY			
Bowling Green BWG (Warren County)	FSS	(502)	843-1152
Erlanger	WS	(606)	371-6681 ♦
London LOZ	FSS	(606)	878-6122
		(606)	254-2743
		(Lexington Ecxhange)	
		(606)	679-6159
		(Somerset Exchange)	
Louisville LOU (Bowman)	FSS	(502)	451-5344
	WS	(502)	451-5344 ▄
Paducah PAH (Barkley)	FSS	(502)	442-6828
LOUISIANA			
Alexandria ESF	FSS	(318)	445-3663 ▄
Lafayette LFT	FSS	(318)	233-4952
Lake Charles LCH	FSS	(318)	477-1784 ▄
Monroe MLU	FSS	(318)	322-3157
New Orleans NEW	FSS	(504)	241-2935 ▄
	FSS	(504)	241-2351 ★
Shreveport SHV (Downtown)	FSS	(318)	221-2211
(Greater Shreveport)	WS	(318)	631-3558 ♦
	WS	(318)	635-7769 ★
MAINE			
Augusta AUG (State)	FSS	(207)	622-6491
Bangor BGR	FSS	(207)	947-4028
Caribou	WS	(207)	489-3377 ♦

Location and Identifier		Area Code	Telephone
MAINE (Con't.)			
Houlton HUL (International)	FSS	(207)	532-2475
Portland	WS	(207)	775-3071 ♦
MARYLAND			
Baltimore	WS	(301)	761-1333 ♦
(Washington FSS)	FSS	(301)	766-0757 ★
	FSS	(301)	766-3420
Salisbury SBY (Salisbury-Wicomico)	FSS	(301)	742-8719
MASSACHUSETTS			
Boston BOS (Logan)	FSS	(617)	223-6447 ▄
	FSS	(617)	567-7420 ▄
	FSS	(617)	569-1773 ★
Worcester ORH	CS/T	(617)	PL 5-6083
	WS	(617)	798-3815 ♦
			(0600-2200)
MICHIGAN			
Alpena	WS	(517)	354-8733 ♦
Battle Creek BTL (Kellogg)	CS/T	(616)	962-7878
Detroit DET (City)	FSS	(313)	372-3737
	FSS	(313)	372-1711 ★
	WS	(313)	729-2111 ♦
Flint	WS	(313)	234-3987 ♦
Grand Rapids	WS	(616)	949-2580 ✕
	(75 NM radius)		(0600-2400)
	WS	(616)	456-2268
Houghton CMX (Houghton County)	FSS	(906)	482-0380
Houghton Lake	WS	(517)	366-5392
Jackson JXN (Reynolds Mun.)	FSS	(517)	782-0355
Lansing LAN (Capital City)	FSS	(517)	371-1150
Marquette MQT	FSS	(906)	GR 5-4197
	WS	(906)	226-8642 ▄
			(0500-2100)
Pellston PLN (Emmet County)	FSS	(616)	LE 9-8401
Saginaw MBS (Tri City)	FSS	(517)	695-2511
Saulte Ste Marie SSM	FSS	(906)	635-1551
			(1000-1800)
(Traverse City)		(906)	635-1381
			(1800-1000)
	WS	(906)	632-7751
Traverse City TVC	FSS	(616)	947-5056
Ypsilanti	WS	(313)	729-2111 ♦
MINNESOTA			
Alexandria AXN	FSS	(612)	763-6593
Duluth	WS	(218)	722-7982 ▄
Hibbing HIB (Chisholm-Hibbing)	FSS	(218)	262-3826
International Falls	WS	(218)	285-5151
	WS	(612)	283-8425 ♦
Minneapolis MSP St. Paul Intl. (Wold-Chamberlain)	FSS	(612)	726-1130 ▄
	FSS	(612)	726-1104 ✕
	(50 NM radius)		
	FSS	(612)	726-9494 ★

Weather Information

Location and Identifier		Area Code	Telephone
MINNESOTA (Con't)			
Redwood Falls RWF	FSS	(507)	637-8530
Rochester RST	FSS	(507)	288-7576∎
St. Cloud	WS	(612)	253-2540
			(0400-2200)
MISSISSIPPI			
Greenwood GWO	FSS	(601)	GL 3-2631
Jackson JAN (Thompson Field)	FSS	(601)	939-5212∎
McComb MCB (Pike County)	FSS	(601)	684-7070
Meridian MEI (Key Field)	FSS	(601)	482-1243
	WS	(601)	483-5270◆
MISSOURI			
Cape Girardeau CGI	FSS	(314)	334-2803
Columbia COU	FSS	(314)	449-3836∎
Joplin JLN	FSS	(417)	MA 3-6868
Kansas City MKC	FSS	(816)	471-7565
	FSS	(816)	471-2131★
IFR Flight Plans only		(816)	471-7570§§
50 NM radius		(816)	421-0919★
Easterly		(816)	421-1940★
Southerly		(816)	421-1941★
St. Louis STL (Spirit of St. Louis)	FSS	(314)	532-1011∎
		(314)	532-1041★
50 NM Radius		(314)	532-1238
Route NE-bnd to Chicago		(314)	532-1449
Route W-bnd to Kansas City		(314)	532-1324
IFR flights plan only		(314)	532-1321§§
Springfield SGF	FSS	(417)	862-3588∎
Vichy VIH (Rolla National)	FSS	(314)	299-3911
MONTANA			
Billings BIL (Logan Field)	FSS	(406)	259-4545∎
Bozeman BZN (Gallatin Field)	FSS	(406)	388-4242
Butte BTM (Silver Bow)	FSS	(406)	494-3004
Cut Bank CTB	FSS	(406)	938-4522
Dillon DLN	FSS	(406)	683-5651
Glasgow	WS	(406)	228-4042
Great Falls GTF (International)	FSS	(406)	761-7110∎
Havre	WS	(406)	265-6424
Helena HLN	CS/T	(406)	442-9902
	WS	(406)	442-7312
Kalispell	WS	(406)	756-4829
Lewistown LWT	FSS	(406)	538-3639
Livingston LVM (Mission Field)	FSS	(406)	222-2411
Miles City MLS	FSS	(406)	232-1503
Missoula MSO (Missoula City)	FSS	(406)	542-2230∎
NEBRASKA			
Chadron CDR	FSS	(308)	432-3153
Grand Island GRI	FSS	(308)	382-5196∎
Lincoln LNK (Muni/AFB)	FSS	(402)	477-3929∎

Location and Identifier		Area Code	Telephone
NEBRASKA (Con't.)			
North Platte LBF (Lee Bird)	FSS	(308)	532-4034∎
Omaha OMA (Eppley)	FSS	(402)	422-6866
	FSS	(402)	342-3603★
Scottsbluff BFF	FSS	(308)	635-2615∎
Sidney SNY	FSS	(308)	254-3130
Valentine	WS	(402)	376-3442
NEVADA			
Elko EKO	FSS	(702)	738-7222
Ely ELY (Yelland Fld)	FSS	(702)	289-3051∎
		(0500-2100 Other hrs. Elko)	
Las Vegas LAS (McCarren)	FSS	(702)	736-1573/4∎
		(702)	739-7863/4/5✳
Lovelock LOL (Derby Fld)	FSS	(702)	273-2448
Reno RNO	FSS	(702)	784-5414∎
			786-7787★
Tonopah TPH (Nye Co.)	FSS	(702)	482-6421
Winnemuca	WS	(702)	623-2203
NEW HAMPHSIRE			
Concord CON	FSS	(603)	224-7474∎
Lebanon LEB (Regional)	FSS	(603)	298-8360
NEW JERSEY			
Atlantic City	WS	(609)	645-2345◆
Millville MIV	FSS	(609)	825-1173
		(609)	825-1983
Newark	WS	(201)	624-7272★
			(Via New York)
Teterboro TEB	FSS	(201)	288-9092
		(212)	898-5256
		(914)	352-4535
NEW MEXICO			
Albuquerque ABQ (Sunport/Kirtland)	FSS	(505)	243-7831∎
	FSS	(505)	242-2661★
Carlsbad CNM	FSS	(505)	TU 5-2042
Clayton	WS	(505)	374-9511
		(0430-1130 Mon-Fri) (0430-0930 Sat)	
Deming DMN	FSS	(505)	546-2726
Farmington FMN	CS/T	(505)	327-4479
			(Answered in Gallup)
Gallup (GUP)	FSS	(505)	722-4308
Hobbs HOB (Lea Co.)	CS/T	(505)	393-6143
Las Vegas LVS	FSS	(505)	425-7411
Roswell ROW	FSS	(505)	347-5400∎
Santa Fe SAF	CS/T	(505)	982-3871
			(Answered in Albuquerque)
Truth or Consequences TCS	FSS	(505)	894-3277
Tucumcari TCC	FSS	(505)	461-2900

247

Weather Information

Location and Identifier		Area Code	Telephone
NEW YORK			
Albany ALB (Albany County)	FSS	(518)	869-9225
	FSS	(518)	UN 9-9173 ∎
Binghamton	WS	(607)	797-0784 ◆
Buffalo BUF (Greater Buffalo)	FSS	(716)	842-5790 ∎
	FSS	(716)	632-5042 ★
Elmira ELM (Chemung Co.)	FSS	(607)	739-2471
Glens Falls GFL (Warren Co.)	FSS	(518)	RX 3-2593
New York ISP (Islip MacArthur)	FSS/IFSS	(516)	737-3535/6/7
	FSS	(914)	723-4330
	FSS	(212)	995-8657
From New York City		(212)	656-5988/9 §§
From Westchester Co.		(914)	723-3862/3 §§
From Long Island		(516)	737-3617/8 §§
		(516)	737-3535/6/7
		(516)	737-3595/6
From New York City		(212)	995-8657/8
From Westchester County		(914)	723-4330-4
International Flight Planning:			
From New York City		(212)	656-8558
		(212)	995-8659
Massena MSS	FSS	(315)	RO 9-2033
New York	WS	(212)	639-5690
	WS	(212)	476-5950 ★
Local New York City Area			
New York City		(212)	476-8800 ✳
No. New Jersey		(201)	288-3100 ✳
Routes Northbound			
New York City		(212)	426-8300 ✳
No. New Jersey		(201)	288-5570 ✳
Routes South and Westbound			
New York City		(212)	426-9300 ✳
No. New Jersey		(201)	288-9250 ✳
Poughkeepsie POU (Dutchess Co.)	FSS	(914)	462-3400
Rochester ROC (Monroe County)	CS/T	(716)	325-3320
	WS	(716)	328-7361 ◆
Syracuse SYR (Hancock)	WS	(315)	455-1214 ◆
Utica UCA (Oneida) Local	FSS	(315)	736-9023
From Rome		(315)	337-0115
From Syracuse		(315)	475-9904/5
Other		(315)	962-5667
Watertown ART	FSS	(315)	639-6228
NORTH CAROLINA			
Asheville	WS	(704)	684-3136 ◆
Cape Hatteras	WS	(919)	995-2321
Charlotte	WS	(704)	399-6000 ◆
Elizabeth City CG (Coast Guard Air Station)	FSS	(919)	338-3808
(0700-2200 Other hrs. New Bern)			
Greensboro	WS	(919)	294-4800 ◆
	WS	(919)	668-0789 ◆
Hickory HKY	FSS	(704)	328-5656
New Bern EWN (Simmons-Nott)	FSS	(919)	638-3133

Location and Identifier		Area Code	Telephone
NORTH CAROLINA (Con't.)			
Raleigh-Durham RDU	FSS	(919)	755-4306 ∎
			596-2446 ∎
			(Durham Exchange)
Greensboro		(919)	273-8660
Rocky Mount RWI (Rocky Mt-Wilson)	FSS	(919)	442-7171
Wilmington	WS	(919)	763-8331
Winston Salem	WS	(919)	725-6882 ◆
NORTH DAKOTA			
Bismarck	WS	(701)	223-0920 ◆
Dickinson DIK	FSS	(701)	225-2989
Fargo	WS	(701)	232-1584 ◆
Grand Forks GFK (Intl.)	FSS	(701)	772-7201
Jamestown JMS	FSS	(701)	252-4350
Minot MOT (Intl.)	FSS	(701)	852-3696
Williston	WS	(701)	572-3198 ◆
OHIO			
Akron	WS	(216)	896-2246 ◆
Cincinnati LUK (Lunken)	FSS	(513)	871-8220
	FSS	(513)	871-6200 ★
	WS	(606)	371-6681 ∎
Cleveland CLE (Hopkins)	FSS	(216)	267-3700 ∎
	FSS	(216)	267-3410 ★
Columbus CMH	FSS	(614)	237-7461 ∎
	FSS	(419)	526-2132
Dayton DAY	FSS	(513)	898-3692 ∎
Findlay FDY	FSS	(419)	422-6176/6177
Mansfield MFD	CS/T	(419)	526-2132
			(Columbus Exchange)
	WS	(419)	522-7070
			(0600-1800)
Toledo	WS	(419)	865-8859 ◆
Youngstown YNG	FSS	(216)	539-5121
	FSS	(216)	759-2117
	FSS	(216)	856-1993
	WS	(216)	545-1755
Zanesville ZZV	FSS	(614)	453-0649
OKLAHOMA			
Gage GAG	FSS	(405)	923-2601
Hobart HBR	FSS	(405)	726-5234
McAlester MLC	FSS	(918)	GA 3-4091
Oklahoma City OKC (Wiley Post)	FSS	(405)	787-9323 ∎
	FSS	(405)	787-9060/1 ★
Ponca City PNC	FSS	(405)	RO 5-5485
Tulsa TUL	FSS	(918)	836-3505 ∎
	FSS	(918)	835-2364 ★
(Bartlesville Exchange)	FSS	(918)	336-5833
(Muskogee Exchange)	FSS	(918)	683-1204

Weather Information

Location and Identifier		Area Code	Telephone
OREGON			
Astoria	WS	(503)	861-2722
Baker BKE	FSS	(503)	523-2961
Eugene EUG (Mahlon-Sweet)	CS/T	(503)	688-8411
	WS	(503)	687-6407 ◆
Klamath Falls (Kingsley Field)	CS/T	(503)	882-4641
	WS	(503)	882-9474
			(0745-1545)
Medford MFR	CS/T	(503)	779-3241
	WS	(503)	773-1525 ◆
North Bend OTH	FSS	(503)	756-4916
Portland PDX (International)	FSS	(503)	222-1699 ■
(Hillsboro)	FSS	(503)	648-2111 ■
Redmond RDM (Roberts Field)	FSS	(503)	548-2522
Salem	WS	(503)	363-9829
The Dalles DLS	FSS	(509)	767-1187
PENNSYLVANIA			
Allentown ABE (Allentown-Bethechen	WS	(215)	264-1944
Altoona AOO (Blair)	FSS	(814)	793-3113
Bradford BFD (Bradford-McKean)	FSS	(814)	362-8860
Du Bois DUJ	FSS	(814)	328-2231
Erie ERI (Port Erie)	FSS	(814)	833-1345
	(0600-2200 Other hrs. Du Bois)		
	WS	(814)	838-1010 ◆
Harrisburg HAR (Capital City)	FSS	(717)	782-3777
	FSS	(717)	774-3626 ★
	WS	(717)	782-3775 ◆
Johnstown JST (Cambria)	FSS	(814)	535-3088
Philadelipha PNE (N. Philadelphia)	FSS	(215)	677-0744 ★
		(215)	464-6699 ★
	WS	(215)	365-7218 ★
Philipsburg PSB (Mid-State)	FSS	(814)	342-0830
Pittsburgh AGC (Allegheny)	FSS	(412)	462-3707
	FSS	(412)	462-5585/6 ★
Pittsburgh (Grtr Pittsburgh)	WS	(412)	644-2887 ◆
Wilkes-Barre AVP (Wilkes-Barre/Scranton)	FSS	(717)	346-4512
	(717)		982-4301
	WS	(717)	457-5650 ◆
			(toll call)
Williamsport IPT (Williamsport/Lycoming)	FSS	(717)	368-8547
	(Sunbury Exchange)		
	FSS	(717)	286-2770
	WS	(717)	368-1866 ◆
RHODE ISLAND			
Providence PVD (Green)	WS	(401)	737-3171 ◆
SOUTH CAROLINA			
Anderson AND	FSS	(803)	224-2573/4
Charleston CHS	FSS	(803)	747-5293 ■
	FSS	(803)	747-5778 ★
Columbia	WS	(803)	794-2593 ◆
	WS	(803)	796-8710/11 ★

Location and Identifier		Area Code	Telephone
SOUTH CAROLINA (Con't.)			
Florence FLO	FSS	(803)	662-8197
Greer GSP (Greenville)	FSS	(803)	271-8930 ■
Myrtle Beach CFE (Crescent Beach)	FSS	(803)	272-6903
	(0700-2100 other hrs. Florence)		
SOUTH DAKOTA			
Aberdeen ABR	FSS	(605)	225-5264 ■
	(0600-2200 Other hrs. Huron)		
Huron HON (W. W. Howes)	FSS	(605)	352-3806 ■
Pierre PIR	FSS	(605)	224-5894
Rapid City RAP	FSS	(605)	342-2302 ■
Watertown ATY	FSS	(605)	TU 6-4581
TENNESSEE			
Chattanooga	WS	(615)	892-6302 ◆
Crossville CSV (Crossville-Cumberland)	FSS	(615)	484-9541
Dyersburg DYR	FSS	(901)	285-4842
	(0600-2200 Others hrs. Jackson)		
Jackson MKL (McKeller)	FSS	(901)	423-0252
Knoxville TYS (McGhee Tyson)	FSS	(615)	577-6651 ■
(from Maryville)		(615)	983-4000
Memphis MEM	FSS	(901)	398-9268 ■
	FSS	(901)	398-2347 ★
Nashville BNA (Metro)	FSS	(615)	749-5378
	FSS	(615)	361-0737 ★
Tri City TRI (Bristol)	FSS	(615)	323-6204
			(0600-2200)
TEXAS			
Abilene ABI	FSS	(915)	677-4336/7
Alice ALI	FSS	(512)	664-0184 ■
Amarillo AMA (Air Terminal)	FSS	(806)	335-1608 ■
Austin AUS (Robert Mueller)	FSS	(512)	GR 8-6695
Beaumont BPT (Jefferson Co.)	FSS	(713)	722-0288
	WS	(713)	722-7011 ◆
			(0630-2130)
Brownsville BRO (Brownsville Intl.)	CS/T	(512)	LI 6-6421
			GA 5-1115
	(Harlingen Exchange)		
	WS	(512)	542-8231 ◆
Childress CDS	FSS	(817)	WE 7-3892
College Station CLL (Easterwood)	FSS	(713)	VI 6-8784/5
Corpus Christi	WS	(512)	888-8061 ◆
Cotulla COT	FSS	(512)	TR 9-2417
Dalhart DHT	FSS	(806)	CH 9-2006
Dallas DAL (Love Field)	FSS	(214)	350-3311 ■
	FSS	(214)	357-4343 ★
	Routes West and North of Dallas		
		(214)	357-4344 ★
	Routes East and South of Dallas		
Del Rio	WS	(512)	775-2115
El Paso ELP (Intl.)	FSS	(915)	778-6448 ■
	FSS	(915)	778-4487 ★

Weather Information

Location and Identifier		Area Code	Telephone
TEXAS (Con't.)			
Fort Worth FTW (Meacham)	FSS	(817)	624-8471 ■
	FSS	(817)	626-3071/2 ★
Galveston GLS (Scholes)	FSS	(713)	SH 4-3255
(0600-2200 other hrs. Houston)			
	WS	(713)	765-5448 ◆
Gregg County GGG (Longview)	CS/T	(214)	643-2266/7
Houston HOU (Hobby)	FSS	(713)	644-8361 ■
Local Area	WS	(713)	641-3000 ★
Houston-New Orleans	WS	(713)	641-3001 ★
Houston-Dallas	WS	(713)	641-3002 ★
Houston-Midland	WS	(713)	641-3003 ★
Lubbock LBB	FSS	(806)	762-0511
Lufkin LFK (Angelena Co.)	FSS	(713)	634-3319
McAllen MFE (Miller Fld)	FSS	(512)	MU 2-2878/9
Midland MAF (Air Terminal)	FSS	(915)	563-2611 ■
Mineral Wells MWL	FSS	(817)	FA 5-5922
Palacios PSX	FSS	(512)	972-2559
Port Arthur	WS	(713)	722-0476 ◆
San Angelo SJT (Mathis)	CS/T	(915)	944-1538
	WS	(915)	944-3322 ◆
San Antonio SAT (Intl.)	FSS	(512)	826-9561 ■
Tyler TYR (Pounds)	CS/T	(214)	597-8051
Victoria	WS	(512)	575-3182 ◆
Waco ACT	WS	(817)	754-3126 ◆
Wichita Falls SPS (Sheppard AFB/Wichita Falls Air Trml)	FSS	(817)	855-5574 ■
Wink INK	FSS	(915)	LA 7-3351
UTAH			
Bryce Canyon BCE	FSS	(801)	834-5311
Cedar City CDC	FSS	(801)	586-3806
Salt Lake City SLC (Intl.)	FSS	(801)	524-5183 ■
	FSS	(801)	364-5571 ★
Salt Lake to Denver Routes		(801)	531-8445 ★
Salt Lakes to La Vegas and Reno Routes		(801)	531-8523 ★
Salt Lake to Boise and Idaho Falls Routes		(801)	531-8554 ★
VERMONT			
Burlington	WS	(802)	862-9883 ◆
Montpelier MPV (Barre-Montpelier)	FSS	(802)	223-2376
VIRGINIA			
Bristol	WS	(615)	323-8242 ◆
Charlottesville CHO (Charlottesville/Albermarle)	FSS	(804)	973-4316
Danville DAN	FSS	(804)	793-1163
(0600-2200 Other hrs. Roanoke)			
Lynchburg	WS	(804)	239-5811 ◆
			(0600-1930)
Newport News PHF (Partick Henry)	FSS	(804)	877-0209 ■
Norfolk	WS	(804)	855-3029 ■
Richmond RIC (Byrd International)	FSS	(804)	222-7203 ■
Roanoke ROA (Woodrum)	FSS	(703)	362-1668 ■

Location and Identifier		Area Code	Telephone
WASHINGTON			
Bellingham BLI	FSS	(206)	734-6400
Ephrata EPH	FSS	(509)	SK 4-2361
Hoquiam HQM (Bowerman)	FSS	(206)	533-3432
Olympia	WS	(206)	357-6169 ◆
Seattle SEA (Boeing)	FSS	(206)	767-2726 ■
	FSS	(206)	767-4002 ★
Spokane SFF (Felts)	FSS	(509)	456-4546 ■
Toledo TDO (Toledo-Winlock)	FSS	(206)	864-2371
(0600-2200) (other hrs. Portland)			
Walla Walla ALW (City-County)	FSS	(509)	529-1413
Wenatchee EAT (Pangborn)	FSS	(509)	884-6656
Yakima YKM	WS	(509)	453-8975
WEST VIRGINIA			
Beckley	WS	(304)	252-3171 ◆
Bluefield BLF (Mercer County)	FSS	(304)	325-6521
Charleston CRW (Kanawha)	FSS	(304)	343-8919 ■
Elkins EKN (Randolph)	FSS	(304)	636-0810 ■
Huntington HTS (Tri-State)	FSS	(304)	453-3951 ■
Martinsburg MRB	FSS	(304)	AM 3-9353
Morgantown MGW	FSS	(304)	292-9489
Parkersburg PKB (Wood Co.)	FSS	(304)	485-6421
Wheeling HLG (Ohio Co.)	CS/T	(304)	277-1252
WISCONSIN			
Eau Clare EAU	FSS	(715)	835-2269
Green Bay GRB (Austin Straubel)	FSS	(414)	494-7417 ■
LaCrosse LSE	FSS	(608)	784-3170
Winona, Mo.		(507)	452-1046
Lone Rock LNR (Tri-Co.)	FSS	(608)	583-2661
(0800-1600 other hrs. LaCrosse or (608) 583-5011)			
Milwaukee MKE (Gen. Mitchell)	FSS	(414)	481-1060 ■
	FSS	(414)	744-7810 ★
Wausau AUW	FSS	(715)	845-7396
WYOMING			
Casper CPR (Air Terminal)	FSS	(307)	235-1555 ■
Cheyenne	WS	(307)	638-6437 ◆
Denver	FSS	(307)	635-4187
Lander	WS	(307)	332-2718
Laramie LAR (Gen. Brees Fld)	FSS	(307)	745-4845
Rawlins RWL	FSS	(307)	324-3241
Rock Springs RKS	FSS	(307)	362-2121
Sheridan SHR (Sheridan Co.)	FSS	(307)	674-7426 ■
Worland WRL	FSS	(307)	347-4122
PUERTO RICO			
San Juan SJU (Puerto Rico Intl.)	IFSS		791-1780
	WS		791-3490

Part III
Aviation Weather—Review Questions for FAA Written Test

REVIEW QUESTIONS

Following are 100 review questions. If you are able to answer all these questions satisfactorily and understand the principles behind them, you should have no trouble with your certification examination.

1. The visibility at the control tower level is 3/4 mile. However, at the usual observation point, the visibility is 4 miles. Prevailing visibility and required remarks (if any) are:

    a. 4 and no remark required
    b. 3/4 and a remark SFC VSBY 4
    c. 3/4 and no remark required
    d. 4 and a remark TWR VSBY 3/4

2. Tower visibility is entered in column 4a only when:

    a. Tower visibility is less than 4
    b. Surface visibility is less than 4
    c. Tower visibility is prevailing
    d. Obscuring phenomena is below the level of the tower

3. Fog 50 feet deep covers the area surrounding the station. Visibility at the usual point of observation is 1/8 mile. Above the fog, tower visibility is 15 miles. Prevailing visibility and remarks are:

    a. 15 and a remark FDEP 50
    b. 1/8 and a remark TWR VSBY 15 FDEP 50
    c. 1/8 and a remark F 50 FT DEEP
    d. 15 and a remark SFC VSBY 1/8 FDEP 50

4. Variable visibility describes a condition in which the prevailing visibility rapidly increases and decreases by one or more reportable values during the period of the observation. The prevailing visibility is reported as:

    a. The lowest of the values
    b. The average of all observed values
    c. The highest of the values
    d. The average of the extremes

5. The sky is partially obscured by 0.3 fog. The fog merges into a layer of low clouds. The correct entries for column 5 (weather and obstructions to vision) and column 13 (remarks) are:

    a. F and a remark GF3
    b. GF and a remark F3
    c. F and a remark F3
    d. GF and a remark GF3

6. Visibility by sectors is N=6, E=6, S=3, W=3. The prevailing visibility is reported as:
   a. 6
   b. 3
   c. 4
   d. 5

7. Visibility is rapidly varying between 1 and 4 miles, and the average of all observed values during the observation is 1 1/2. What is the prevailing visibility?
   a. 1
   b. 3
   c. 1 1/2
   d. 2 1/2

8. Visibility by sectors is N=6, E=5, S=4, W=3. Visibility is reported as follows:
   a. 4 1/2 and a remark VSBY W3
   b. 4 and a remark VSBY W3
   c. 5
   d. 4

9. Visibility is a term that denotes the greatest distance at which
   a. All objects can be seen and recognized
   b. Selected objects can be seen and identified
   c. Objects can be detected but not identified
   d. All objects can be detected but not identified

10. Visibility by sectors is N=7/8, E=1, S=2 1/2, W=1 1/2. Prevailing visibility is:
    a. 2 1/2
    b. 1
    c. 1 1/2
    d. 7/8

11. For determination of visibility during daylight hours, the preferred choice of markers should be confined to:
    a. Objects which subtend a smaller angle than approximately 0.3°
    b. Light objects appearing against a terrestrial background
    c. Dark or nearly dark objects against the horizon sky
    d. Light colored objects

12. Visibility by sectors is N=3, E=2, S=1, W=1 1/2. Prevailing visibility and appropriate visibility remarks are:
    a. 3 and remark VSBY S1
    b. 2 and a remark VSBY 1V3
    c. 1 and a remark VSBY N3E2W11/2
    d. 2 and a remark VSBY S1W11/2

13. Which one of the following would *not* be entered as a visibility value on MF1-10C?

    a. 13
    b. 0
    c. 95
    d. 2 3/4

14. Prevailing visibility is 3 miles, but is varying rapidly between 2 and 4 miles. This is reported as:

    a. 3V and a remark VSBY 2V4
    b. 3 and no remark
    c. 3V and no remark
    d. 3 and a remark VSBY 2V4

15. A special observation is *not* required when visibility of:

    a. 10 miles becomes 3 miles
    b. 1 1/4 miles becomes 2 1/2 miles
    c. 1 mile becomes 2 miles
    d. 2 miles becomes 3 miles

16. Visibility of 3 1/2 miles is reported as:

    a. 3.5
    b. 3 1/2
    c. 4
    d. 3

17. The prevailing visibility is 3/4 mile but varies rapidly between 1/2 and 2 miles. This is reported as:

    a. 3/4 and no remark
    b. 3/4V and a remark VSBY 1/2V2
    c. 1 1/4V and a remark VSBY 1/2V2
    c. 1/2V2 and no remark

18. Visibility at the usual point of observation is 3/4 mile in Fog and the Tower Visibility is 1/4 mile. Report prevailing visibility, and remarks if appropriate, as:

    a. 3/4 mile and a remark TWR VSBY 1/4
    b. 1/4 mile and a remark SFC VSBY 3/4
    c. 1/4 mile and no remark
    d. 3/4 mile and no remark

19. When the prevailing visibility is exactly halfway between two reportable values, select:

    a. The higher value only if the preceding digit is odd
    b. The lower value
    v. The higher value
    d. The even number

255

20. Which of the following would *not* be entered as a visibility value on MF1-10C?

   a. 15
   b. 7/8
   c. 100
   d. 3 1/2

21. An observer estimates prevailing visibility is 25 miles. The farthest visible visibility marker is at 7 miles. Which of the following visibilities is reported?

   a. 15
   b. 10
   c. 25
   d. 7

22. The prevailng visibility is 1 mile in fog. During the past 10 minutes the RVR on Runway 24 Center has remained above 6000 feet. Encode Runway Visual Range as:

   a. R24CVR60+
   b. RVR24C60+
   c. CR24VR60+
   d. Not encoded

23. Which type of light may *not* be used as a visibility marker at night?

   a. Building lights
   b. Focused lights
   c. Taxiway lights
   d. Obstruction lights

24. Visibility by sectors is NE=6, SE=2 1/2, SW=3, NW=5. Prevaiilng visibility and required remarks are:

   a. 3 and no remark
   b. 5 and a remark VSBY NE6SE2 1/2SW3
   c. 5 and a remark VSBY NE6SE2 1/2SW3NW5
   d. 5 and a remark VSBY SE2 1/2

25. Which of the following groups contain a visibility value not authorized for use in aviation weather reports?

   a. 1/16, 1 3/4, 100
   b. 78, 2, 75
   c. 0, 7/16, 3
   d. 3/8, 2 1/4, 2 1/2

26. Drizzle or snow grains usually fall from:

   a. Nimbostratus clouds
   b. Cirrus clouds
   c. Low stratus clouds
   d. Cumulus clouds

27. The sky is hidden by 0.5 clouds (all opaque) at an estimated 5000 feet, and 0.5 at an estimated 18,000 feet (all opaque). This is reported as:

    a. E50 SCT E180 OVC
    b. E180 OVC 50 SCT
    c. 50 SCT E180 OVC
    d. E50 BKN 180 OVC

28. The intensity of which type of precipitation may be determined by the visibility?

    a. Ice pellets
    b. Freezing rain
    c. Hail
    d. Snow

29. Blowing dust reduces the visibility to 4 miles, except in the south where it is 1 mile. The correct remark to be included in the report is:

    a. None should be included
    b. VSBY S1
    c. VSBY 4S1
    d. VSBY N4E4S1W4

30. Ceiling values of 2500, 2100, and 2200 feet are determined during the period of observation. This should be reported as:

    a. M23V and a remark CIG 21V25
    b. M25V and a remark CIG 21V25
    c. M21V and a remark CIG 21V25
    d. M22V and a remark CIG 21V25

31. Which of the following changes does *not* require a special observation?

    a. S— begins
    b. T ends
    c. ZR— begins
    d. T to T+

32. Which of the following sky conditions is a correct report?

    a. M40 BKN E300 —OVC
    b. 40 SCT E300 —OVC
    c. M40 BKN 300 —OVC
    d. M40 SCT 300 —OVC

33. Errors in entries in columns 1-13 on MF1-10C are erased and corrected if discovered:

    a. Before the report is transmitted
    b. Never
    c. If a correction is sent
    d. Within 5 minutes after transmission time

34. Which of the following sky conditions is a correct report?

   a. E7 X
   b. 15 −SCT E100 −BKN
   c. 5 −X E50 OVC
   d. W2 X

35. Which of the following sky conditions is a correct report?

   a. −X
   b. W8 OVC
   c. 5 SCT M25 −BKN
   d. 5 SCT 25 −BKN V 50 −OVC

36. If the ceiling is classified "W" (indefinite):

   a. The cloud or obscuration covers 0.9 or less
   b. The height of the cloud or obscuration is changing rapidly
   c. The base of the cloud or obscuration is ragged
   d. The base of an obscuration is at the surface

37. Temperature is recorded in column 7 of MF1-10C to the nearest:

   a. Half of a degree
   b. Whole degree
   c. Quarter of a degree
   d. Tenth of a degree

38. The intensity of drizzle occurring alone must be determined by:

   a. Total acccumulation of water
   b. Degree to which the drizzle affects visibility
   c. Rate of accumulation of water
   d. Size of the drops

39. A layer of smoke aloft (all opaque) at an estimated 1000 feet covers 0.2 of the sky. Clouds (all opaque) cover 0.5 of the sky at an estimated 4000 feet. The correct entries for sky cover, and remarks if required are:

   a. −X E40 BKN and a remark K2
   b. E40 BKN and a remark K10 SCT
   c. 10 SCT E40 BKN and a remark K10 SCT
   d. 10 SCT E40 BKN and no remarks

40. The symbol "−X" is used to indicate:

   a. That the sky is completely hidden by precipitation and/or an obstruction to vision
   b. That visibility is not being reported
   c. That 0.1 to 0.9 of the sky is hidden by precipitation and/or an obstruction to vision
   d. That the sky cover is doubtful

41. An entry for weather and/or obstructions to vision must be made when the prevailing visibility is:

    a. 15 miles or less
    b. 10 miles or less
    c. 6 miles or less
    d. Less than 10 miles

42. Which of the following is reported in the correct order?

    a. TRW−A
    b. S+ZR−
    c. IPZL−
    d. T+S−R

43. An opaque layer of 0.2 clouds at an estimated 1500 feet with an additional opaque layer of 0.8 clouds at a measured 6900 feet will be reported as:

    a. E15 SCT M69 BKN
    b. 15 SCT M70 OVC
    c. 15 SCT M69 OVC
    d. 15 SCT M70 BKN

44. Time entries on MF1-10C, Column 2, are always:

    a. True solar time
    b. Greenwich civil time
    c. Local daylight time when in effect
    d. Local standard time

45. A ceiling is classified as "M" (measured) when the ceiling is determined from:

    a. The known heights of unobscured portions of objects within 1 1/2 nautical miles of a runway
    b. A pilot while flying within 1 1/2 miles of any runway
    c. The maximum penetration of ceiling light during fog
    d. The time at which a ceiling balloon completely disappears

46. The station anemometer is inoperative but the observer estimates wind speed at 25 knots. Wind vane indicates direction from 290°. Encode wind as:

    a. 2925E
    b. 29E25
    c. M
    d. E2925

47. Ground fog obscures 0.2 of the sky. A layer of clouds (0.2 opaque) at an estimated 9000 feet covers 0.7 of the sky.

    Correct entries are:

    a. −X E90 BKN and a remark GF2
    b. −X E90 −BKN and a remark GF2
    c. −X 90 −BKN and a remark F2
    d. −X 90 SCT and a remark F2

48. When the sky is obscured by surface-based obscuring phenomena, and a ceilometer or ceiling balloon is used as a guide in determining vertical visibility, the appropriate ceiling designator is:

   a. M or W
   b. W
   c. E or M
   d. E or W

49. A thunderstorm is considered to be in progress when:

   a. Lightning is seen at the time of observation
   b. Lightning is seen within 15 minutes of the time of observation
   c. Thunder is heard within 15 minutes of the time of observation
   d. Cumulonimbus clouds are seen at time of observation

50. Which of the following is *never* a ceiling value?

   a. The vertical visibility into a fog layer that conceals the sky
   b. The vertical visibility into precipitation that obscures the sky
   c. The height of a thin overcast
   d. The height of a broken opaque layer of clouds

51. Thunder is heard from a thunderstorm that is west of the station, movement unknown, accompanied by frequent cloud to cloud lightning. It is reported in remarks as:

   a. T W MOVMT UNK FQT LTGCC
   b. T W FQT LTGCC
   c. CB W FQT LTG
   d. No remark necessary

52. At the time of an aviation observation, the anemometer cups, whose starting speed is 3 knots, are not moving. If smoke near the surface is observed to be drifting to the south, the wind is reported as:

   a. 0000
   b. E3602
   c. 1802E
   d. M

53. The essential difference between squalls and gusts is:

   a. The variation between peaks and lulls
   b. The intensity of precipitation with which they are associated
   c. Their peak speed
   d. The duration of the increased wind speed

54. Before the wind is characterized as gusty, variation between peaks and lulls must be at least:

   a. 25 knots
   b. 10 knots
   c. 15 knots
   d. 20 knots

55. The average surface wind speed is 18 knots and is rapidly varying between 10 and 25. This condition is reported as:

   a. 10G25
   b. 18G25
   c. 18G
   d. 18

56. Which of the following would be an *incorrect* entry in column 5 of MF1-10C?

   a. T+
   b. A+
   c. S+
   d. R+

57. You determine height of an overcast ceiling layer as 1500 feet using a ceiling balloon. Correct ceiling designator is:

   a. E
   b. W
   c. B
   d. M

58. The weather phenomena which are always written out in full in an aviation weather report are:

   a. Thunderstorms, heavy rain, or rain showers
   b. Blowing dust or sand
   c. Hail
   d. Tornadoes, waterspouts, and funnel clouds

59. A minus sign (−) to show light intensity is *not* used with:

   a. L
   b. IP
   c. T
   d. S

60. Ground fog cannot hide more than:

   a. 1/10 of the sky
   b. 3/10 of the sky
   c. 5/10 of the sky
   d. 4/10 of the sky

61. Haze, blowing sand, blowing snow, and smoke are coded as:

   a. Z, BS, GN, S
   b. H, BS, GN, K
   c. B, BN, BS, K
   d. H, BN, BS, K

62. When computations require the disposal of decimals and the decimal to be disposed of is five, the preceding digit will:

   a. Be increased by one
   b. Remain unchanged
   c. Remain unchanged if that digit is even
   d. Remain unchanged if that digit is odd

63. A severe thunderstorm with heavy rain showers and 3/4 inch hail is recorded as

   a. T+RW+A
   b. TRW+A
   c. T+ARW+
   d. T+RW+A+

64. Surface wind direction is reported in column 9 of MF1-10C with respect to:

   a. Magnetic north and 16 points of the compass
   b. Magnetic north and to the nearest 10 degrees
   c. True north and to the nearest 10 degrees
   d. True north and 16 points of the compass

65. Which of the following conditions is common only to intermittent type precipitation?

   a. Intensity increases and decreases gradually, and stops, and recommences at least once hourly
   b. Cumuliform clouds
   c. Abrupt beginning and ending
   d. Rapid variation of intensity

66. Which of the following would *not* restrict horizontal visibility?

   a. Drifting snow
   b. Blowing snow
   c. Dust
   d. Blowing dust

67. A temperature of 108°F is entered in column 7 (temperature) of MF1-10C as:

   a. 08
   b. 108
   c. 99+
   d. None of the above

68. Wind direction for aviation observations usually will be determined for an interval of:

   a. 1 hour
   b. 1 minute
   c. 5 minutes
   d. An instantaneous value

69. Which of the following is *not* entered in Col. 13?

    a. Distant cumulonimbus
    b. Dust devils
    c. Halo
    d. Distant precipitation

70. In an hourly aviation weather report, missing data pertaining to an element normally included in columns 3-12 of MF1-10C are indicated by:

    a. "X"
    b. The letter "M"
    c. A slant "/"
    d. The letter "O"

71. When surface based smoke completely obscures the sky and the observer uses a ceilometer to determine vertical visibility, ceiling is classified as:

    a. M
    b. W
    c. V
    d. E

72. Fog reduces the visibility in all directions to 1/4 mile and completely conceals the sky. The sky condition is reported as:

    a. OVC
    b. −X
    c. X
    d. CLR

73. The maximum allowable time interval between observation of elements reported in a special aviation observation and the time of the last entry of the observation on MF1-10C is:

    a. 15 minutes
    b. 5 minutes
    c. 30 minutes
    d. 20 minutes

74. A dry-bulb temperature of 0.5°F is recorded in column 7 of MF1-10C as:

    a. 05
    b. 00
    c. 0
    d. 1

75. A peak wind during the past hour of 27 knots from 300° occurred at 35 after the hour. At record observation time this condition is reported in Remarks as:

    a. PK WND 30/27 35
    b. PK WND 3027/35
    c. PK WND 2730/35
    d. No remark necessary

263

76. The beginning (1320 GMT) and ending (1345 GMT) of a thunderstorm moving from south to north over a station are reported in appropriate Special observations. The peak wind during the past hour was observed to be from 280° at 40 knots at 1340 GMT. These data will be reported in remarks of the 1400 GMT Record observation as:

   a. TE45 G40 MOVD N
   b. TB20E45 MOVD S-N PK WND 28/40 40
   c. TB20E45 MOVD N DSIPTD
   d. TB20E45 MOVD N PK WND 2840/40

77. Light rain showers are occurring, the average wind speed is 20 knots, and squalls accompanied by peak speeds of 50 knots are occurring. The squalls will be reported as:

   a. RWQ-
   b. RW-Q-
   c. 20Q50
   d. Q20G50

78. The ceiling and sky condition for an observation at a station with a field elevation of 1100 feet is measured at 1500 feet overcast. Prior to completion of this observation a pilot over the field reports the top of the overcast as 4000 feet MSL. In column 13 your coded PIREP would indicate the top of the overcast layer is:

   a. 5100 feet
   b. 4000 feet
   c. 6600 feet
   d. 2900 feet

79. A thunderstorm moving from south to north with occasional cloud to ground lightning is west of the station at the time of observation. Remarks concerning this storm are reported as:

   a. T MOVG S TO N OCNL LTGCG
   b. T W MOVG N
   c. T W MOVG N OCNL LTGCG
   d. CB W

80. Which of the following changes in sky ceiling does *not* require a special observation?

   a. 8 SCT to CLR
   b. E13 BKN to E6 BKN
   c. M9 OVC to M10 OVC
   d. CLR to clouds observed at 800 feet

81. Which of the following conditions always requires a special observation?

   a. A tornado is reported by the public to have occurred 5 hours ago.
   b. 1-minute wind speed suddenly increases from 6 to 26 knots.
   c. Sky condition changes from OVC to SCT
   d. Altimeter setting rises at a rate of .20 inch per hour.

Aviation Weather—Review Questions for FAA Written Test

82. In an aviation weather report, a dry-bulb temperature of −25.5°F is reported as:

   a. −25
   b. −26
   c. 75
   d. 74

83. In the past 10 minutes a pilot reported base of the overcast layer at 1500 feet over the field. Field elevation is 1000 feet. You are using this report in your observation for ceiling height. (The ceilometer is inoperative). The entry in column 3 would be:

   a. E5 OVC
   b. M15 OVC
   c. M5 OVC
   d. E25 OVC

84. To reduce station pressure to sea level, the temperature argument is:

   a. The addition of a constant correction
   b. The current temperature
   c. The average of current temperature and temperature 6 hours previously
   d. The average of current temperature and temperature 12 hours previously

85. Which of the following sky conditions is an *incorrect* report?

   a. W5 X
   b. −X 100 SCT
   c. −X M20 −OVC
   d. 5 −SCT 25 −SCT

86. Which of the following elements may be reported alone as a Special observation?

   a. Sky condition
   b. Tornadic activity
   c. Dewpoint
   d. Temperature

87. Rain completely obscures the sky. Vertical visibility is 800 feet. Scattered low stratus is visible at 300 feet. The entry in column 3 would be:

   a. M3 SCT W8 X
   b. 3 SCT W8 X
   c. W3 SCT 8 X
   d. 3 SCT M8 OVC

88. The dewpoint is always reported with respect to:

   a. Water at temperatures below 32°F. and ice above 32°F.
   b. Ice
   c. Water
   d. Ice at temperatures below 32°F. and water at temperatures above 32°F.

265

89. A smoke layer, based at the surface, completely hides 0.9 of the sky. Through the smoke, 0.1 of transparent cirrostratus estimated at 25,000 feet is observed overhead. The entry for sky and ceiling is:

   a. −X 250 OVC with no remark
   b. −X 250 −OVC with remark K9
   c. −X 250 −OVC with no remark
   d. −X E250 OVC with remark K9

90. The sky is covered by a combination of 0.6 fog (all opaque), 0.2 clouds (all transparent) at a measured 500 feet, and 0.2 clouds (all transparent) at 4000 feet. The correct sky and ceiling report including remarks if appropriate is:

   a. −X 5 −BKN 40 −OVC with a remark F6
   b. X M5 BKN 40 OVC no remarks
   c. −X M5 BKN 40 OVC with a remark F6
   d. −X M5 BKN 40 −OVC with a remark F6

91. In reading the mercury barometer, which of the following operations is performed first?

   a. Set the vernier
   b. Tap the barrel near the top of the mercury column
   c. Adjust the thumbscrew until the mercury just touches the ivory point
   d. Read the attached thermometer

92. The wind shifted at 1030EST from 180° at 15 knots to 270° at 40 knots with gusts to 50 knots. Indicate the correct remark concerning the wind shift.

   a. WSHFT 1030E FM 1815
   b. WSHFT 30
   c. WSHFT 30 TO 2740G50
   d. WSHFT 1530GMT

93. When it appears that the pen of the barograph is about to pass off the chart, the observer should:

   a. Reset the pen to a position corresponding to 1 inch of pressure difference and make a note of pressure and time
   b. Reset the pen by .20″ of pressure, and make a note of pressure and time
   c. Take mercurial barometer readings every 15 minutes while the pen is off the chart
   d. Note the time the pen passes below the chart

94. Maximum and minimum temperatures are recorded in columns 47 and 48:

   a. To the nearest tenth of a degree
   b. To the nearest whole degree
   c. To the nearest two degrees
   d. To the nearest two-tenths of a degree

95. Which of the following is one of the phenomena normally associated with the passage of a cold front?

    a. Gusty winds shifting in a clockwise manner
    b. Drop in pressure
    c. Rapid rise in dewpoint
    d. Decrease in the wind speed

96. The lowest temperature to be secured by evaporation is the:

    a. Dry-bulb temperature
    b. Wet-bulb temperature
    c. Dewpoint temperature
    d. Saturation point

97. The barograph correction should be entered on MF1-10C to the closest:

    a. .001 inch
    b. .01 inch
    c. .005 inch
    d. .05 inch

98. When using a psychrometer, the official dry-bulb reading is taken:

    a. By averaging several dry-bulb readings
    b. Before moistening the wet bulb
    c. At the time of the lowest wet-bulb reading
    d. At the time of the lowest dry-bulb reading

99. At the usual point of observation, the visibility, by sectors, is determined to be N=4, E=2, S=3, and W=1. Meanwhile, the transmissometer visibility is determined to be 2 miles. The correct entry in column 4 of MF1-10C is:

    a. 4
    b. 2 1/2
    c. 2
    d. 3

100. An altimeter setting of 28.96 would be reported as:

    a. 896
    b. 8.96
    c. LOW 896
    d. 896 LOW

ANSWERS TO REVIEW QUESTIONS

Question No.	Answer
1.	a.
2.	c.
3.	b.
4.	b.
5.	c.
6.	a.
7.	c.
8.	c.
9.	b.
10.	c.
11.	c.
12.	d.
13.	d.
14.	b.
15.	a.
16.	d.
17.	b.
18.	a.
19.	b.
20.	d.
21.	c.
22.	a.
23.	b.
24.	d.
25.	c.
26.	c.
27.	c.

Question No.	Answer
28.	d.
29.	b.
30.	a.
31.	a.
32.	c.
33.	a.
34.	d.
35.	a.
36.	d.
37.	b.
38.	b.
39.	c.
40.	c.
41.	c.
42.	a.
43.	b.
44.	d.
45.	a.
46.	d.
47.	c.
48.	b.
49.	c.
50.	c.
51.	b.
52.	b.
53.	d.
54.	b.
55.	b.
56.	b.

Question No.	Answer
57.	a.
58.	d.
59.	c.
60.	c.
61.	d.
62.	a.
63.	a.
64.	c.
65.	a.
66.	a.
67.	b.
68.	b.
69.	c.
70.	b.
71.	b.
72.	c.
73.	a.
74.	d.
75.	b.
76.	d.
77.	c.
78.	b.
79.	c.
80.	a.
81.	b.
82.	b.
83.	a.
84.	d.
85.	c.

Question No.	Answer
86.	b.
87.	b.
88.	c.
89.	d.
90.	c.
91.	d.
92.	b.
93.	a.
94.	b.
95.	a.
96.	b.
97.	c.
98.	c.
99.	d.
100.	a.

Part IV
NTSB Special Study—Nonfatal, Weather-Involved General Aviation Accidents

SYNOPSIS

The National Transportation Safety Board is concerned about the large number of weather-involved[1] general aviation accidents. This study is based on 7,856 such accidents, which have occurred from 1964 through 1974.

During the 11-year study period, "inadequate preflight planning preparation and/or planning" was the most frequently cited cause in which both pilots and weather were involved. Statistics reveal that most of the nonfatal, weather-involved general aviation accidents occurred during the landing regime, i.e., either during the landing roll or during leveloff and touchdown, when unfavorable wind conditions existed, and the weather was VFR. Unfavorable winds were cited 5 times more frequently as a cause or a factor than were low ceilings, and 16 times more frequently than was thunderstorm activity. Statistics also reveal that a pilot was 12 times more likely to encounter weather as predicted than to encounter weather worse than predicted.

As a result of its findings, the Safety Board urges general aviation pilots to attend the various safety seminars, clinics, and courses of instruction sponsored by both Government and industry. For familiarization purposes, there is no substitute for visiting National Weather Service and Federal Aviation Administration facilities to determine what data are available and the means by which they can be obtained. The Board urges all pilots to postpone any flight until a timely and thorough preflight weather briefing can be obtained and reiterates that if there is any doubt—DON'T GO.

OVERVIEW OF THE DATA

From 1964 through 1974, weather was the most frequently cited causal factor for nonfatal general aviation accidents. From 1964 through 1974, 7,856 nonfatal, weather-involved accidents resulted in injuries to 3,637 persons. (See Tables 1 and 2.) Serious

TABLE 1

GENERAL AVIATION ACCIDENTS

Year	Total	Nonfatal	Weather-Involved	Nonfatal Weather-Involved
1964	5,069	4,543	798	620
1965	5,196	4,658	669	457
1966	5,712	5,139	909	728
1967	6,115	5,512	1,112	912
1968	4,968[1]	4,276	1,067	820
1969	4,767	4,120	986	751
1970	4,718	4,071	1,014	780
1971	4,640	3,987	947	703
1972	4,256	3,561	983	688
1973	4,255	3,532	976	689
1974	4,343	3,694	1,010	708
TOTAL	54,039	47,093	10,471	7,856

---

[1] For purposes of this study, a weather-involved accident was considered to be one for which the Safety Board had determined that weather had been a cause or a contributing factor.

[1] The decrease in the total number of accidents was caused by a change in the definition of "substantial damage" included in the definition of an accident. The change was effective on January 1, 1968.

injuries resulted in 16.2 percent of these accidents. Although slightly more than half of the accidents resulted in no injuries, there was substantial damage to the aircraft. Although complete data on economic losses are not available, data are available for 1970 through 1972. Based on these data, hull damage alone cost $8,000,000 for that 3-year period.

TABLE 2

INJURIES
1964 THROUGH 1974

	Serious	Minor	None	Unknown	Total
Pilot	687	1,188	5,984	1	7,860
Copilot	18	25	106	–	149
Dual Student	11	43	231	–	285
Check Pilot	–	–	10	–	10
Flt. Eng.	–	1	1	–	2
Cabin Att.	1	1	6	–	8
Extra Crew	5	7	27	–	39
Passengers	553	1,097	5,629	–	7,285
TOTAL	1,275	2,362	11,994	1	15,638

Statistics show that from 1964 through 1974, there were 54,039 general aviation accidents. Of this total, 47,093 were nonfatal accidents, 16.7 percent of which were weather-involved accidents and 14.5 percent of which were nonfatal, weather-involved accidents. Nonfatal accidents comprised 75 percent of the total weather-involved accidents.

From 1965 to 1968, the percentage of nonfatal accidents which were weather-involved increased dramatically and that percentage has remained at a relatively high level. (See Figure 1.) On the other hand, the accident rate per 100,000 hours flown (all nonfatal accidents) has been downward over most of the period from 1964 through 1974.

About 50 percent of the nonfatal, weather-involved accidents have occurred during pleasure flying and more than 12 percent during noncommercial business flying; the remainder occurred during other kinds of flying. (See Table 3.)

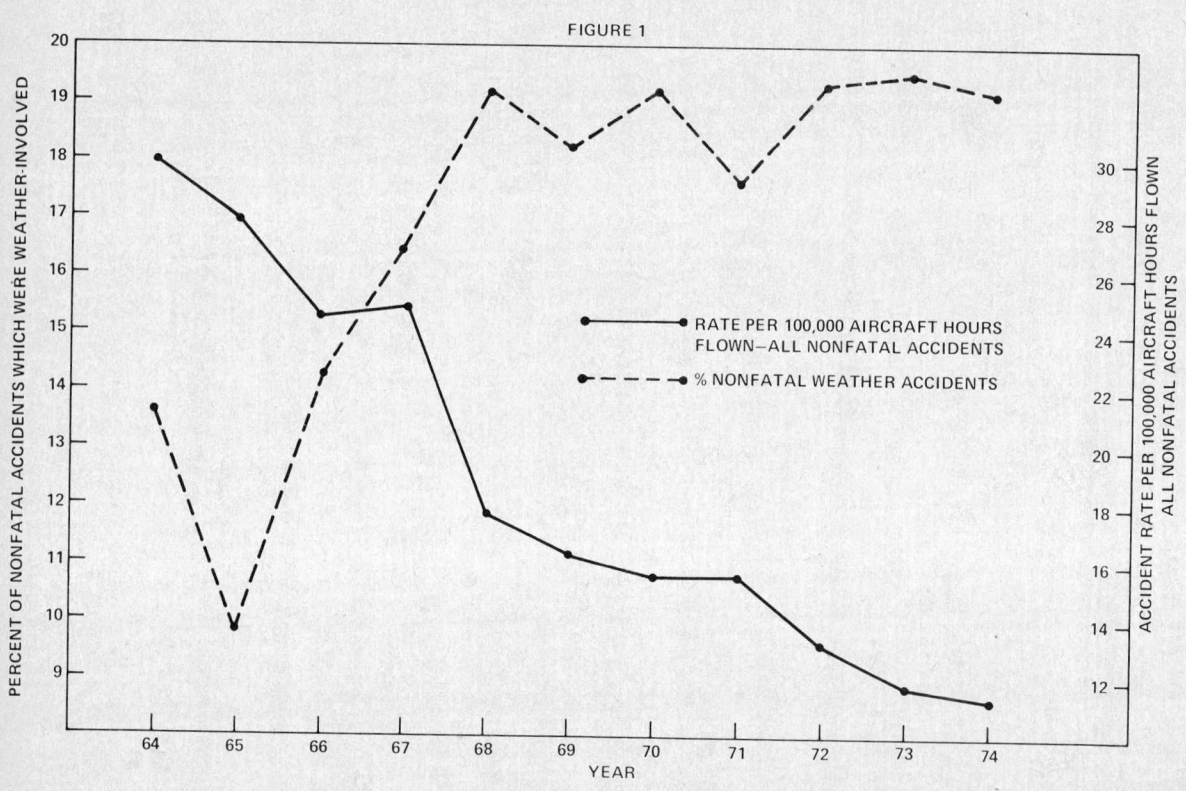

FIGURE 1

TABLE 3
KIND OF FLYING
1964-1974

Kind of Flying[1]	Accidents	Percent
Pleasure	3,925	49.94
Business (noncommercial)	994	12.65
Solo (Instructional)	491	6.26
Training (Instructional)	412	5.24
Dual (Instructional)	288	3.66
Crop Control Related	285	3.63
Air Taxi-Passenger Operations	239	3.04
TOTAL	6,634	84.42

[1] The foregoing represent the kind of flying being conducted in most of the accidents. The remaining accidents occurred during more than 30 other kinds of flying.

TABLE 4
CONDITIONS OF LIGHT Vs.
TYPE OF WEATHER CONDITIONS
1964 through 1974

WEATHER CONDITIONS

LIGHT CONDITIONS	None	VFR	IFR	Below Minima[1]	Unknown	Accidents	Percent
None	4[2]	–	–	–	–	4	.05
Dawn	–	33	15	3	–	51	.65
Daylight	–	6,426	478	38	29	6,967	88.69
Dusk (twilight)	–	176	60	1	2	239	3.04
Night (dark)	–	301	225	40	2	568	7.23
Night (moonlight-bright)	–	18	7	1	–	26	.33
Unknown/Not Reported	–	–	–	–	1	1	.01
Accidents	4	6,950	785	83	34	7,856	

[1] Landing and takeoff accidents only.
[2] Invalid data fields.

Almost 89 percent of the accidents examined in this study occurred during daylight hours and in visual flight rules (VFR) conditions; only about 7 percent occurred at night. (See Table 4.) By contrast, for fatal, weather-involved accidents from 1964 through 1972, 60 percent occurred during daylight hours, 36 percent occurred at night and only 40 percent occurred in VFR conditions.

The accidents examined in this study occurred during 56 phases of flight. The phase most frequently coded was the landing roll (fixed wing)–22.8 percent. The next six phases in descending order of frequency were: Level off/touchdown–17.9 percent, initial climb–11.6 percent, normal cruise–7.6 percent, final approach (VFR),–6.6 percent, takeoff run–5.8 percent, and taxi from landing–4.5 percent. (See Table 5.)

TABLE 5
PHASE OF FLIGHT
1964 through 1974

Phase[1]	Accidents	Percent
Landing Roll (fixed wing)	1,791	22.8
Leveloff/Touchdown	1,404	17.9
Initial Climb	912	11.6
Normal Cruise	595	7.6
Final Approach (VFR)	518	6.6
Takeoff Run	456	5.8
Taxi from Landing	353	4.5

[1] Top 7 of 56 phases coded.

PILOT DATA

General aviation pilots with low total flight times were frequently involved in weather accidents. (See Figure 2 and Table 6.) In about 84 percent of the accidents for which data were available, the pilots involved had less than 100 hours total flight time.

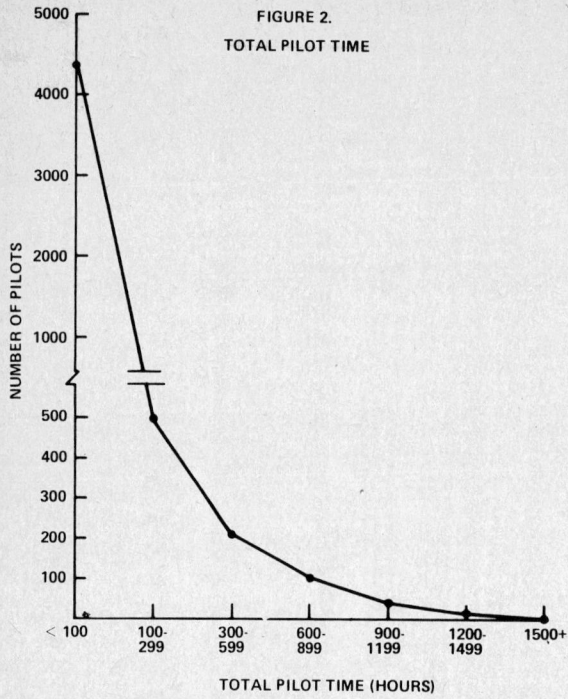

FIGURE 2. TOTAL PILOT TIME

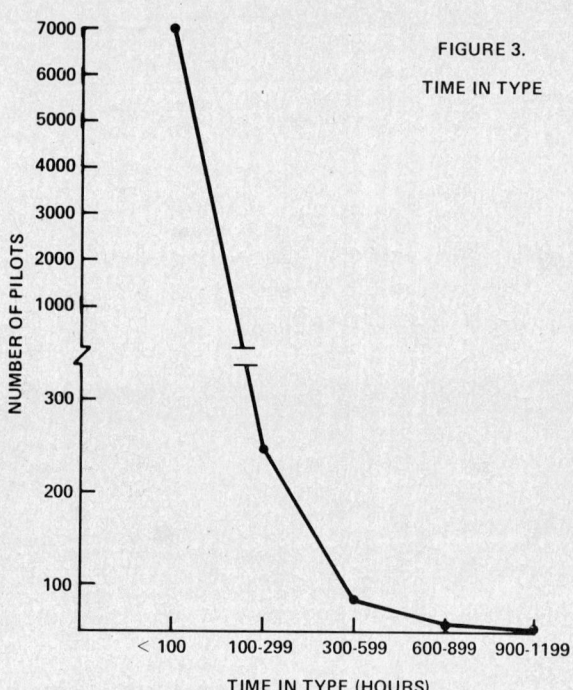

FIGURE 3. TIME IN TYPE

TABLE 7

TIME IN TYPE

HOURS	Student	Private	Commercial	ATP	Private/F.I.	Comm./F.I.	ATP/F.I.	Foreign	None/Exprd.	Unknown	TOTAL
<100	1,370	3,415	1,247	84	2	733	72	3	26	2	6,954
100-299	1	59	101	16	–	62	10	–	–	–	249
300-599	–	11	31	3	–	10	2	–	–	–	57
600-899	–	1	4	1	–	2	–	–	–	–	8
900-1199	–	–	4	1	–	–	–	–	–	–	5
TOTAL	1,371	3,486	1,387	105	2	807	84	3	26	2	7,273

TABLE 6

TOTAL PILOT TIME

HOURS	Student	Private	Commercial	ATP	Private/F.I.	Comm./F.I.	ATP/F.I.	Foreign	None/Exprd.	Unknown	TOTAL
<100	1,324	2,601	300		1	112	1	1	21	1	4,362
100-299	1	184	207	2		89	2			1	486
300-599		44	99	5		56	3		1		208
600-899		10	55	10		28	5	1			109
900-1199		1	26	4		10	1				42
1200-1499		1	10	1		7					19
1500-1799											
1800-2099			2			1	1				4
2100-2399			1			1					2
TOTAL	1,325	2,841	697	25	1	304	13	2	22	2	5,232

It is significant to note that in more than 95 percent of the cases examined, the pilots involved had less than 100 hours in the type of aircraft flown. (See Table 7 and Figure 3.) Nearly 67 percent of these pilots had 10 hours or less in type.

Information concerning pilot time during the 90-days before the accident was available in a majority of the accident cases examined. In 75 percent of the cases examined, the pilots involved had less than 50 hours of flight time during the 90 days before the accident. (See Table 8 and Figure 4.) More than 21 percent of those pilots had less than 10 hours of flight time during that 90-day period.

Pilot Certification

Since 1964, private pilots have been involved in more weather accidents than any other type of pilot and, as would be antici-

NTSB Special Study

TABLE 8
PILOT TIME LAST 90 DAYS

HOURS	Student	Private	Commercial	ATP	Private/F.I.	Comm./F.I.	ATP/F.I.	Foreign	None/Exprd.	Unknown	TOTAL
<50	927	2,228	504	18	2	162	21	2	18	1	3,883
50-99	41	322	196	16	–	132	10	–	2	–	719
100-299	1	65	248	41	–	163	21	–	3	–	542
300-599	–	–	8	–	–	9	–	–	–	–	17
TOTAL	969	2,615	956	75	2	466	52	2	23	1	5,161

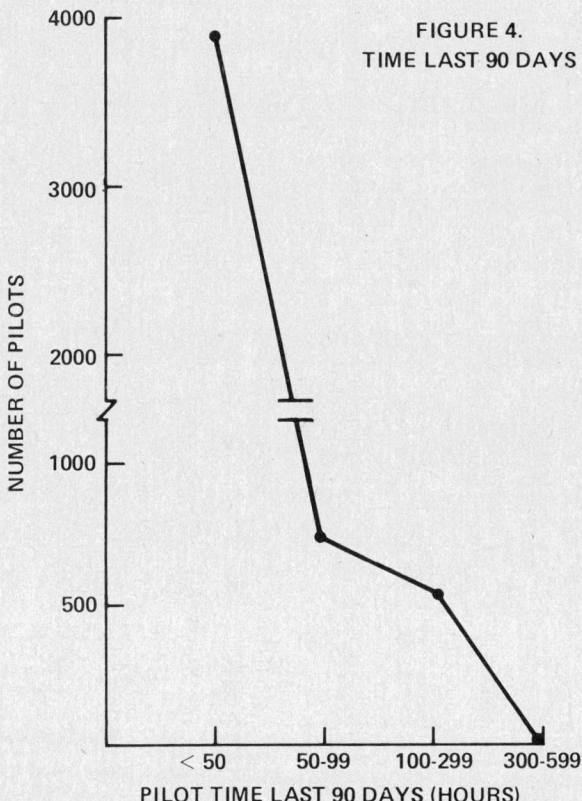

FIGURE 4. TIME LAST 90 DAYS

pated, most pilots involved in nonfatal, weather-involved accidents held private pilot certificates. (See Table 9 and 10.) Of the nonfatal, weather-involved accidents, 46.8 percent of the pilots held private certificates while 41.5 percent of the total pilot population had private certificates. As shown in Table 11, most pilots were in the single engine, land category. The Board's statistics also show that there were more than 40 pilots who either had no certificates or were flying with certificates which had expired.

Geographical Distribution of Accidents and Pilots Involved Compared with Total Active Pilots

Accident exposure to various geographical areas was determined by separating the nonfatal, weather-involved accidents into FAA regions and comparing the number of accidents in each region with the number of active pilots in that particular region. (See Figures 5 and 6; Tables 12 and 13.) The New England region had the best record with 252 accidents, and Alaska was next with 355 accidents. However, when those figures are compared with the numbers of active pilots in the respective regions, it can be seen that the record in Alaska is the worst for any FAA region and that the Alaskan record is about 6 times worse than that of the Southern region, which has the best record. The Great Lakes and the Western regions rank No. 1 and No. 2 both in numbers of accidents and in numbers of active pilots, but their records are equiva-

TABLE 9
ACTIVE AIRMAN CERTIFICATES HELD: JAN. 1, 1964 – DEC. 31, 1974

Category	1964	1965	1966	1967	1968	1969	1970	1971	1972	1973	1974	11-Year Average Number	Per. Cent
Pilot Total*	431,041	479,770	548,757	617,931	691,695	720,028	732,729	741,009	750,869	714,607	733,728	651,106	
Student	120,743	139,172	165,177	181,287	209,406	203,520	195,861	186,428	181,477	181,905	180,795	176,888	27.2
Private	175,574	196,393	222,427	253,312	281,728	299,491	303,779	312,656	321,418	298,921	305,848	270,141	41.5
Commercial	108,428	116,665	131,539	150,135	164,458	176,585	186,821	192,409	196,228	182,444	192,425	163,467	25.1
Airline transport	21,572	22,440	23,917	25,817	28,607	31,442	34,430	35,949	37,714	38,139	41,002	31,003	4.8

*Totals include other categories such as helicopter only and glider only.

TABLE 10

TYPE OF PILOT CERTIFICATE

Type of Certificate	Number	Percent
Student	1,353	17.2
Private	3,677	46.8
Commercial	1,601	20.4
ATP	144	1.8
Private/F.I.	2	0
Commercial/F.I.	923	11.7
ATP/F.I.	110	1.4
Foreign	3	0
None/Expired	37	.5
Unknown	6	.1
TOTAL	7,856	

TABLE 11

CATEGORY AND CLASS RATING

	Accidents	Percent
Single engine land	5,100	64.96
Multiengine land	17	.22
Single engine sea	14	.18
Single engine land and inst.	370	4.71
Multiengine land and inst.	31	.39
Single engine sea and inst.	1	.01
Single engine land and sea	180	2.29
Multiengine land and sea	1	.01
Single engine land and sea and inst.	31	.39
Multiengine land and sea and inst.	2	.03
Single-multiengine land	411	5.23
Single-multiengine sea	1	.01
Single-multiengine land and inst.	947	12.06
Single-multiengine sea and inst.	1	.01
Single-multiengine land and sea	57	.73
Single-multiengine land, sea and inst.	141	1.80
Rotorcraft	331	4.22
Rotorcraft and inst.	36	.46
Glider	110	1.40
Lighter than air	6	.08
None	49	.62
Unknown	19	.24
Accidents	7,856	

lent and rank high when compared with the other regions.

Pilot Age

The ages of pilots in nonfatal, weather-involved accidents were categorized into groups. (See Table 14.) The peak group was 31 to 35 years; however, both the 26 to 30 year group and the 36 to 40 group were near the peaks. By contrast, pilots involved most often in fatal, weather-involved accidents were between 41 and 46 years old. Updated statistics on pilot age were solicited from the FAA in order to determine the age group to which most pilots belong. (See Table 15.) These data indicate that at least since 1968 there have been more pilots in the younger age groups than in the 36 to 40 year group.

Pilot-Involvement as a Cause or Factor

During the 11 years studied, "Inadequate Preflight Preparation and/or Planning" was the most frequently cited cause in which both pilot and weather were involved (See Table 16.) The next most frequently cited causes by order of frequency were: "Failed to obtain/maintain flying speed," "improper leveloff," "failed to maintain directional control," and "selected unsuitable terrain."

Flight Plans

More than 80 percent of the pilots in the 7,856 nonfatal, weather-involved accidents did not file flight plans of any type, less than 14 percent filed VFR flight plans, and less than 5 percent filed instrument flight rules

NTSB Special Study

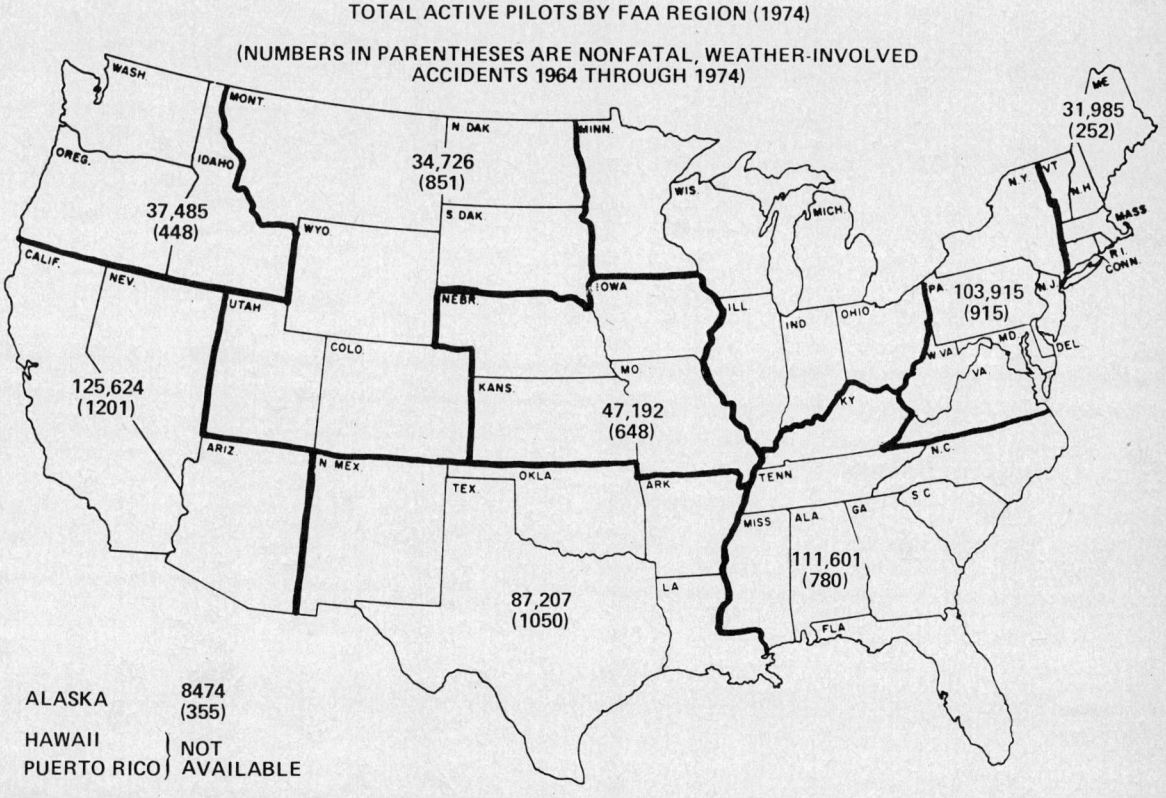

FIGURE 5
GEOGRAPHICAL DISTRIBUTION OF ACCIDENTS
NONFATAL, WEATHER-INVOLVED ACCIDENTS U.S.
GENERAL AVIATION
1964-1974

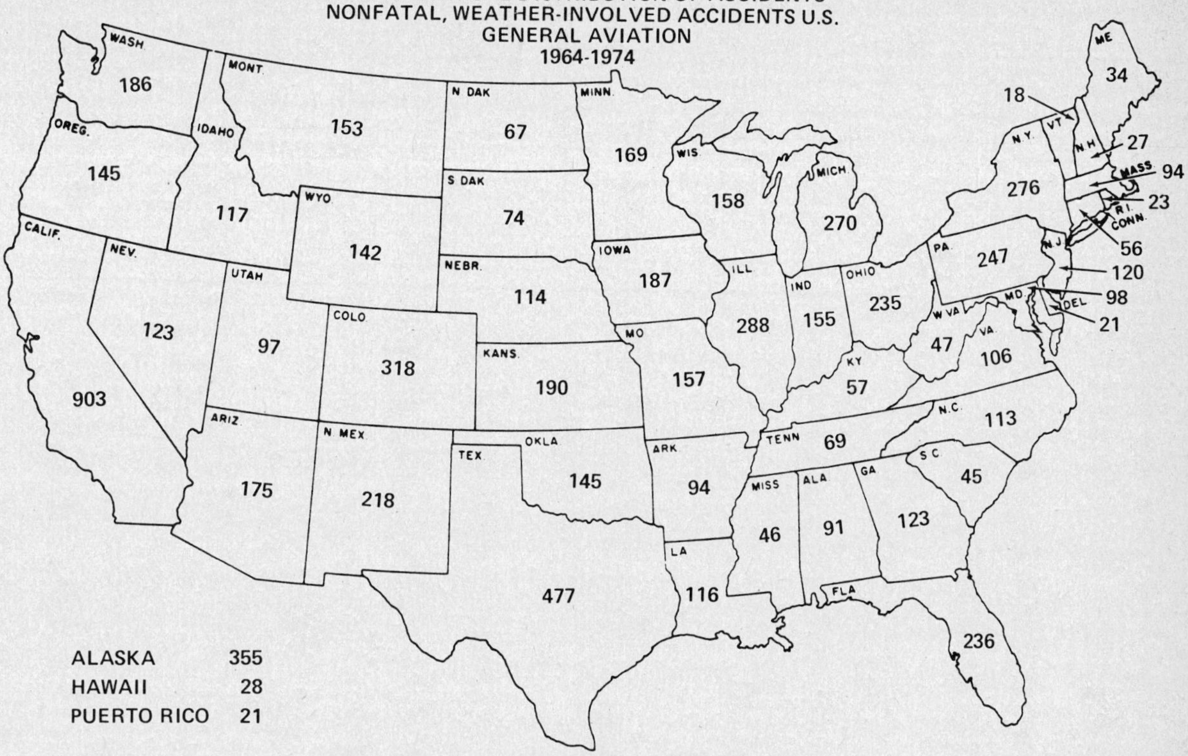

FIGURE 6
TOTAL ACTIVE PILOTS BY FAA REGION (1974)
(NUMBERS IN PARENTHESES ARE NONFATAL, WEATHER-INVOLVED ACCIDENTS 1964 THROUGH 1974)

TABLE 12

TOTAL ACTIVE PILOTS AND FLIGHT INSTRUCTORS[1] BY FAA REGION AND STATE: DECEMBER 31, 1974

FAA region and state	Total pilots	FAA region and state	Total pilots
Total	733,728[2]	Southern—total	111,601
		North Carolina	13,294
United States—total	724,366	South Carolina	6,655
		Georgia	16,163
New England—total	31,985	Florida	42,493
Maine	3,440	Mississippi	6,109
New Hampshire	3,251	Alabama	9,371
Rhode Island	1,609	Tennessee	11,216
Massachusetts	12,914	Kentucky	6,300
Connecticut	9,241		
Vermont	1,530	Southwest—total	87,207
		Louisiana	9,808
Eastern—total	103,915	Oklahoma	13,156
New York	31,080	Texas	52,006
Pennsylvania	23,428	New Mexico	5,880
Virginia	15,428	Arkansas	6,357
Maryland	10,576		
West Virginia	2,834	Rocky Mountain—total	34,726
Delaware	1,877	Colorado	15,659
New Jersey	17,835	Wyoming	2,208
District of Columbia	857	Utah	4,892
		Montana	4,776
Great Lakes—total	133,201	North Dakota	3,738
Illinois	32,963	South Dakota	3,453
Indiana	15,765		
Minnesota	17,172	Western—total	125,624
Michigan	24,562	California	107,702
Ohio	30,442	Arizona	13,268
Wisconsin	12,297	Nevada	4,654
Central—total	47,192	Northwest—total	37,485
Kansas	12,908	Washington	21,186
Iowa	11,009	Oregon	11,630
Missouri	16,000	Idaho	4,669
Nebraska	7,275		
		Alaskan region—total	8,474
		Pacific region—total	2,956
		Outside U.S.—total	9,362

NOTE: Puerto Rico and Virgin Island are included in Outside U.S. total.

[1] Not included in total.
[2] Includes Outside U.S.

NTSB Special Study

TABLE 13

PERCENTAGE OF TOTAL ACTIVE PILOTS WHO WERE INVOLVED IN NONFATAL WEATHER-INVOLVED ACCIDENTS (BY FAA REGION)

FAA Region	Pilots Involved in Nonfatal, Weather-Involved Accidents 1964-1974	Rank	Active Pilots (1974)	Rank	Pilots in Nonfatal Weather-Involved Accidents / Active Pilots Percent	Rank
New England	252	9	31,985	8	.79	7
Eastern	915	4	103,915	4	.88	6
Great Lakes	1,275	1	133,201	1	.96	5
Central	648	7	47,192	6	1.37	3
Southern	780	6	111,601	3	.70	8
Southwest	1,050	3	82,207	5	1.20	4
Rocky Mountain	851	5	34,726	8	2.45	2
Western	1,201	2	125,624	2	.96	5
Northwest	448	7	37,485	7	1.20	4
Alaska	355	8	8,474	9	4.19	1

TABLE 14

PILOT AGE

Age Group	Number of Accidents
<16	1
16-20	297
21-25	892
26-30	1,217
31-35	1,232
36-40	1,221
41-45	1,163
46-50	892
51-55	493
56-60	245
61-65	99
66-70	42
70+	18
Unknown	44
Total	7,856

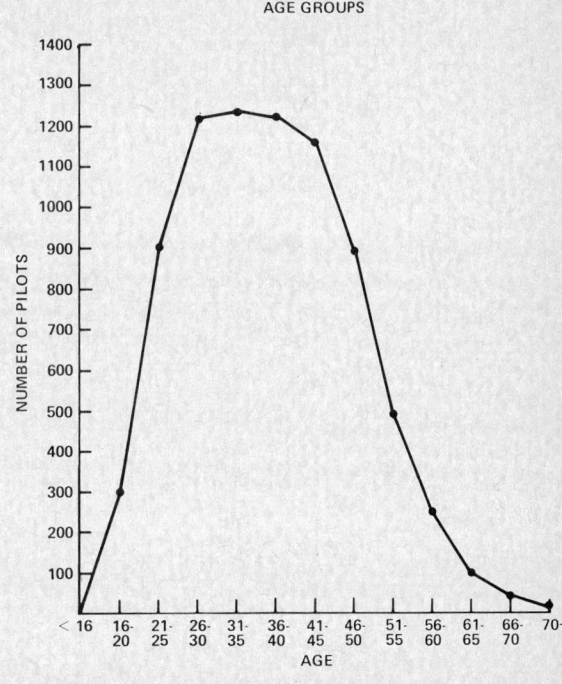

FIGURE 7. AGE GROUPS

TABLE 15

ACTIVE PILOT CERTIFICATES HELD BY AGE GROUP OF HOLDER

Age	1968 Number	1968 Percent	1970 Number	1970 Percent	1972 Number	1972 Percent	1974 Number	1974 Percent
14-15	—	—	—	—	—	—	173	<1
16-19	31,957	5	34,817	5	32,091	4	31,989	4
20-24	96,516	14	101,238	14	92,023	12	90,401	12
25-29	117,465	17	124,363	17	126,416	17	118,111	16
30-34	98,899	14	105,784	14	113,254	15	113,514	15
35-39	95,939	14	96,633	13	97,869	13	95,081	13
40-44	89,951	13	89,642	12	93,047	12	87,899	12
45-49	83,531	12	86,186	12	82,215	11	75,211	10
50-54	45,277	6	54,952	8	65,729	9	66,610	9
55-59	19,471	3	24,335	3	30,440	4	34,828	5
60+	12,689	2	14,779	2	17,785	3	19,911	3
Total	691,695		732,729		750,869		733,728	

TABLE 16

CAUSE/FACTOR TABLE NONFATAL WEATHER INVOLVED ACCIDENTS 1964 THROUGH 1974

Detailed Cause/Factor Pilot In-Command	Nonfatal Accidents Cause	Factor	Total
1. Inadequate preflight preparation and/or planning	3,949	418	4,367
2. Failed to obtain/maintain flying speed	3,900	4	3,904
3. Improper level off	3,712	8	3,720
4. Failed to maintain directional control	3,348	9	3,357
5. Selected unsuitable terrain	2,880	83	2,963

(IFR) flight plans. (See Table 17.) In more than 88 percent of the cases, the weather at the accident site was VFR. By comparison, from 1964 to 1974, for fatal, weather-involved accidents the weather at the accident site was VFR only in 40 percent of the accidents and IFR in 54 percent of the accidents.

WEATHER PHENOMENA AS A CAUSE/FACTOR

Unfavorable wind was the most frequently cited weather factor in nonfatal, weather-involved accidents. Low ceiling and updrafts or downdrafts were the next most frequently cited factors, followed by conditions conducive to carburetor icing, fog, rain, sudden windshifts and thunderstorm activity. (See Table 18.)

TABLE 17

TYPE OF WEATHER (AT ACCIDENT SITE) VS. TYPE OF FLIGHT PLAN

	None	VFR	IFR	Below Minima[1]	Unknown/ Not Reported	Accidents	Percent
None[2]	4	3	1	–	–	8	.10
None	–	5,828	429	32	27	6,313	80.36
VFR	–	960	130	5	4	1,098	13.98
IFR	–	114	194	44	–	352	4.48
Controlled VFR	–	2	4	1	–	7	.09
IFR (VFR on top)	–	–	3	–	–	3	.04
DVFR	–	5	1	–	–	6	.08
VFR Flt. Flwg.	–	14	5	–	–	19	.24
Special VFR	–	4	14	1	–	19	.24
Other	–	1	–	–	–	–	.01
Unknown/not reported	–	23	4	–	3	30	.38
Accidents	4	6,950	785	83	34	7,856	
Percent	.1	88.5	10.0	1.1	.4		

[1] Landing, and takeoff accidents only.
[2] Invalid data fields.

WEATHER FORECASTS

Information concerning weather forecasts was not available in many of the accidents examined because most of them occurred during VFR conditions and in the immediate vicinity of the airport. However, based on available data, a pilot was 12 times more likely to encounter weather as predicted than to encounter weather worse than predicted. (See Table 19.) As the Board indicated in its previous special study, experienced pilots generally are aware that forecasts cannot be considered completely accurate, but they also know that they cannot be ignored. Forecasts should be treated as the best professional advice available.

TABLE 18

WEATHER PHENOMENA CONSIDERED AS A CAUSE/FACTOR IN NONFATAL WEATHER-INVOLVED ACCIDENTS 1964 THROUGH 1974

Phenomenon	Cause	Factor	Total
Unfavorable wind	1,564	2,373	3,937
Low Ceiling	72	614	686
Downdraft/Updraft	190	435	625
Conditions conducive to carburetor icing	290	278	568
Fog	65	483	548
Rain	40	332	372
Sudden windshift	125	146	271
Thunderstorm activity	44	196	240

TABLE 19

ACCURACY OF WEATHER FORECASTS VS. TYPE OF WEATHER CONDITIONS AT ACCIDENT SITE NONFATAL, WEATHER-INVOLVED ACCIDENTS 1964 THROUGH 1974

	VFR	IFR	Below Minima[1]	Unknown	Accidents	Percent
Forecast substantially correct	1,786	388	60	4	2,237	41.56
Weather slightly better than forecast	1				1	.02
Weather considerably better than forecast	2				2	.04
Weather slightly worse than forecast	84	37	3		123	2.28
Weather considerably worse than forecast	31	22	2		55	1.02
Forecast completely erroneous	3	4			7	.13
Unknown/not reported	2,637	282	13	26	2,958	54.95
Accidents	4,542	733	78	30	5,383	
Percent	84.38	13.62	1.45	.55		

[1] Landing and takeoff accidents only.

WEATHER BRIEFINGS

Information concerning weather briefing was also not available in many accident cases examined, probably because most accidents occurred during VFR conditions and in the vicinity of airports. Nevertheless, for those cases in which the pilot was briefed, most briefings were received from Flight Service Station personnel by telephone. (See Table 20.) In those cases for which information was available, more than 45 percent of the pilots were not briefed. The Board's statistics also show that of the 2,704 pilots who were not briefed, 2,615 also did not file flight plans. Of the 2,698 pilots who were briefed on weather, 1,093 also filed flight plans of some type.

TIME OF YEAR

During April and May, more nonfatal, weather-involved accidents occurred than during any other time of the year. (See Table 21 and Figure 8.) Beginning in November, the accident trend began to rise and peaked in May. From May, the trend was downward until the low point was reached in November. The largest rise in accidents was from February to March and the largest drop in accidents was from August to September.

TABLE 20

SOURCE OF WEATHER BRIEFING NONFATAL WEATHER-INVOLVED ACCIDENTS 1964 THROUGH 1974

Source	Accidents	Percent
Pilot, self-help	54	.9
In person, NWS[1]	76	1.3
Phone, NWS	138	2.4
In person, FSS[2]	330	5.7
Phone, FSS	1,258	21.8
Radio, FSS	422	7.3
Recorded by phone	9	.2
Recorded L/MF radio	86	1.5
Pilot to forecaster	2	.0
Partial, phone NWS	2	.0
Partial, FSS in person	10	.2
Partial, FSS phone/radio	21	.4
No briefing received	2,704	46.8
Briefing received, method unknown	181	3.1
Company dispatch	1	.0
Other	108	1.9
Unknown, not reported	374	6.4
Total Accidents	5,775	

[1] National Weather Service
[2] Flight Service Station personnel

TABLE 21
MONTH OF OCCURRENCE

Month	1964	1965	1966	1967	1968	1969	1970	1971	1972	1973	1974	Accidents	Percent
January	49	27	29	75	43	53	60	57	63	39	77	572	7.28
February	75	35	37	96	62	60	54	61	60	39	52	631	8.03
March	74	47	57	81	88	76	63	64	72	65	72	759	9.66
April	73	43	52	107	82	75	71	60	68	93	81	803	10.24
May	72	52	91	86	88	75	79	68	67	69	72	819	10.42
June	49	44	59	67	81	75	69	64	69	64	67	708	9.01
July	34	41	97	89	74	71	84	63	63	57	67	739	9.41
August	49	54	60	77	90	72	64	74	63	60	69	732	9.31
September	53	32	49	63	59	51	72	46	46	65	41	576	7.34
October	34	31	69	57	56	52	53	47	49	43	37	528	6.72
November	25	27	59	53	37	50	50	50	26	39	27	443	5.64
December	33	24	70	62	60	41	62	49	42	56	47	546	6.95
Accidents	620	457	728	912	820	751	780	703	688	689	708	7856	
Percent	7.9	5.8	9.3	11.6	10.4	9.6	9.9	8.9	8.8	8.8	9.0		

STATISTICAL SUMMARY

Based only on the statistics presented, a pilot most likely to have been involved in a nonfatal, weather-involved general aviation accident:

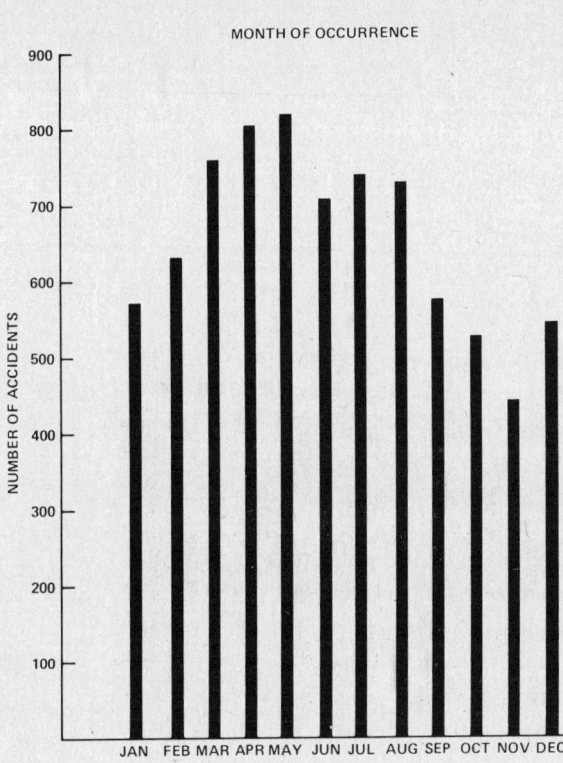

FIGURE 8.
MONTH OF OCCURRENCE

1. Received a preflight briefing from a Flight Service Station which utilized Weather Service Forecasts which were reasonably accurate.
2. Was proposing a pleasure flight.
3. Had less than 100 total flight-hours.
4. Had less than 100 hours in type of aircraft (and most probably had less than 10 hours in type.)
5. Had less than 50 hours in the 90 days before the accident (and probably had less than 10 hours.)
6. Had a private pilot certificate with a single-engine, land rating.
7. Had not filed a flight plan.
8. Was between the ages of 31 and 35.
9. Crashed in VFR conditions with unfavorable winds involved.
10. Was accompanied by at least one passenger.

ACTIONS BY GOVERNMENT AND INDUSTRY TO MINIMIZE WEATHER-INVOLVED GENERAL AVIATION ACCIDENTS

National Weather Service and Federal Aviation Administration

The growth of general aviation, particularly during the past 10 years, has caused many problems to surface, among them—the difficulty of providing required preflight weather information to pilots. There is little doubt that the best preflight briefing is one which is obtained, face to face, from a professional meteorologist who has used the latest current and forecast weather information, including charts and radar weather data. Unfortunately, because of the large increase in the pilot population, it is becoming physically impossible and economically impractical to provide individual, personal weather service.

The next most practical method for obtaining a preflight briefing is by telephone, and statistics indicate that the telephone is the most popular method. However, when weather is bad or marginal, it is difficult to contact the local Flight Service Station (FSS) or NWS office by phone because the lines are kept busy. To alleviate the problem, a few years ago the FAA and NWS initiated the program, "Pilot's Automatic Telephone Weather Answering Service (PATWAS)." By calling the PATWAS number, in 70 areas where the service is available, a pilot can receive a recorded weather briefing for areas within a 250-mile radius of his location. However, because of heavy demand, even the PATWAS lines have become saturated during periods of inclement weather. In order to overcome this difficulty, the NWS and the FAA are cooperating in a test at New York to improve the PATWAS. The FAA is providing the equipment, and the NWS is providing the space and the manpower to keep the tapes updated. The objective of the test is to determine if a sophisticated telephone recording system can provide an adequate briefing for pilots and reduce the demand for personal weather briefings. The New York City PATWAS is made up of 3 separate phone numbers for New York City and 3 for northern New Jersey. The PATWAS contains 34 incoming lines. In addition to providing route forecasts and a weather synopsis, the test PATWAS contains flight precautions, winds aloft reports, terminal forecasts, hourly weather observations, and notices to airmen (NOTAMS), which are updated and amended regularly. Normal PATWAS systems are time-limited, but the New York PATWAS system is flexible and the length of the tape is controlled by the forecaster.

Indications are that use of PATWAS in New York has increased 300 percent. If the test proves it to be successful, the New York PATWAS will become an integral part of a National PATWAS System.

As a result, in part, of a previous Safety Board recommendation, the FAA and the NWS have issued jointly, the updated booklet entitled, "Aviation Weather." The revised version has been published and is now available in a two-volume set; "Aviation Weather" (Advisory Circular 006A) and "Aviation Weather Services" (Advisory Circular 00-45). "Aviation Weather" is a general text on the principles of meteorology and their applications to aviation operations. It is used primarily by pilots and flight operations personnel, but also serves as an excellent introductory text on weather for nonaviation interests. "Aviation Weather Services" discusses services provided to the pilot by NWS and FAA facilities, the structure and interpretation of weather observations and forecasts, data communications, the use of analytic and prognostic charts, graphs, and conversion tables. The book is also used as a text in the NWS Pilot Weather Briefer's Course and will be updated periodically as new products, forecast techniques, and briefing services are developed. The two volumes are available at a nominal cost from the Superintendent of Documents, Government Printing Office, Washington, D.C. 20402.

After extensive coordination between Government agencies and the aviation community, a new and standardized format for pilot reports (PIREPS) has been developed and was implemented on a test basis. This is a coordinated program which includes the FAA, the military, and the air carriers. FAA Flight Service Stations in 15 States are involved in the test. Present plans call for implementation of the new format nationwide during the summer of 1976. The advantages of the new format include ease of reading, since the various parameters are in the same place in each report and the reports can be computer-processed to develop charts of phenomena such as icing and turbulence. These types of charts are planned for development in the near future and are expected to be of great benefit to briefers who will then have a pictorial guide to locate icing and turbulence areas.

The NWS has also implemented a program that summarizes the PIREPS, by States, into nine bulletins every 30 minutes. These PIREPS are disseminated on the Service-A weather teletype. The bulletins are then sent to the FAA's Weather Message Switching Center at Kansas City where they are retained for 2 hours and are available to those locations which have request-reply capability.

Since June 1975, the NWS has participated with the FAA and the Department of Defense (DOD) in a 1-year test at the Kansas City Air Route Traffic Control Center (ARTCC). Seven forecasters provide on-site, 24-hour weather information to controllers and keep them advised of weather that may impact the air traffic control system. They are also attempting to place more PIREPS into the system, alerting military bases at night of potentially hazardous weather, and briefing some military pilots. An interagency group is examining the possibility of extending the test at Kansas City and expanding it to other ARTCC's.

The FAA has recently awarded a multimillion dollar contract for automated FSS equipment which is designed to improve the speed and quality of preflight and in-flight pilot briefings.

The computer-based system, called Automated Weather and Notice to Airmen System (AWANS) is considered a key element in FAA's program to automate its FSS network. It will be installed in the Washington FSS for operational testing early next year.

The AWANS computer processes flight data automatically replacing the present manual system whereby FSS specialists must sort through lengthy teletypewriter printouts. Using a special keyboard, the specialists can call up a variety of vital flight and weather information from the computer for viewing on a cathode ray tube display. This enables them to brief pilots with increased speed and efficiency while providing more up-to-date flight information. A prototype of the AWANS system is undergoing a year long test at the FSS in Atlanta, Georgia.

AWANS will be installed at the Washington FSS after that facility is relocated from National Airport to Leesburg, Va., where it will be collocated with the FAA's ARTCC. Until AWANS is ready, FAA will use a less sophisticated, semi-automated display system at the relocated Washington FSS.

In its Special Study, "Fatal, Weather-Involved, General Aviation Accidents," the Safety Board recommended that the FAA, among other things, take priority action in order to adhere to the proposed 4-year implementation plan for the En Route Flight Advisory Service (EFAS) which had begun at four West Coast FSS's in 1972. Because of budgetary constraints and difficulties concerning actions involving FSS's in fiscal year 1975, the FAA was unable to adhere to the 4-year plan. However, at least eight FSS's in the eastern U.S. are expected to add the EFAS function in 1976. It is also anticipated that during 1976, 220 EFAS specialists will be given an additional 4-week intensive training course devised and instructed by NWS meteorologists. As envisioned, the EFAS program eventually will be composed of 44 FSS's coast-to-coast, specially equipped and manned to enable pilots in flight to get the latest weather information on a discrete radio frequency. The EFAS stations will be on the National Facsimile Circuit, they will have request-reply teletype capability, and NWS radar remote equipment wherever available. They will also have a telephone connection to the nearest NWS forecast office in order to make available professional meteorological consultation in problem situations.

With funding by the FAA, the NWS has developed a new computer-based, short-term technique which makes it possible to forecast thunderstorms 2 to 6 hours after an observation. The new technique was made available to forecasters on a nationwide basis at the end of March 1976. The second program is one which is designed to provide the general aviation pilot with much needed wind information at the runway threshold. The program is experimental and utilizes a pole and streamers to provide wind direction and speed indications to the pilot and is said to be an improvement over the windsock.

Partially in response to Safety Board recommendations, the NWS and the FAA have established a number of working groups to work on priority items to improve aviation weather services. Among these are groups working on:

(1) Faster and more complete dissemination of hazardous and potentially hazardous weather information to the National Airspace System,
(2) Improved pilot and air traffic specialist education,
(3) Improved aviation weather products and
(4) Better quality control of observations and services.

Aircraft Owners And Pilots Association (AOPA)

AOPA's Air Safety Foundation offers many training programs for airmen at various locations across the country. One of these programs is the Practical Aviation Weather Course. During 1975, the course was attended by more than 400 persons. Practical Aviation Weather is a 16-hour course. The course includes basic weather theory, types of weather-briefing information, weather flying, and preflight and in-flight decisionmaking. Finally, the course covers items such as in-flight weather situations, live weather briefings, and air route traffic control capabilities and limitations with regard to avoidance of hazardous weather situations.

CONCLUSIONS

Statistics reveal that most of the nonfatal, weather-involved, general aviation accidents occurred during the landing regime, i.e., either during the landing roll or during leveloff and touchdown, when unfavorable wind conditions existed and the weather was VFR. Unfavorable winds were cited 5 times more frequently as a cause or factor than were low ceilings, and 16 times more frequently than was thunderstorm activity.

Most of the pilots involved in the "unfavorable wind" accidents simply didn't compensate for the wind conditions or used poor judgment where they attempted to land. Although some of the pilots may not have been aware of the exact wind conditions, one pass over the intended runway would have revealed those conditions. On the other hand, the lack of appropriate wind measuring equipment on the ground or the misinterpretation of a windsock, for example, could have contributed to some of the accidents. Although a windsock provides valuable information concerning wind direction and some information on wind speed, the windsock is of little or no value for gust information.

The Safety Board believes that many of the accidents attributed to "unfavorable wind" could have been prevented by increased emphasis on the subject during pilot training and by the expedited development of a simple, economical wind-measuring system for use particularly at relatively small airports which are used primarily by general aviation aircraft.

The Safety Board continues to believe that the general aviation accident prevention efforts by both Government and industry have been progressive, but that even greater efforts are needed to reverse the trend of weather-involved accidents.

RECOMMENDATIONS

Based on the results of this study, the National Transportation Safety Board recommended that the Federal Aviation Administration:

"Expedite the development, for operational purposes, of a simple, economical wind measuring system for use particularly at relatively small airports which are used primarily by general aviation aircraft.

"... in coordination with the National Oceanic and Atmospheric Administration/National Weather Service:

"Through the FAA/NWS Working Group on Improving Pilot Education, place special emphasis on the hazards associated with unfavorable winds during the landing regime, by various means such as:

1. Discussions at safety seminars and clinics sponsored by the General Aviation Accident Prevention Program Specialists.
2. Changes in the Private Pilot's Test Guide (AC 61-32A).
3. Changes in the Private Pilot's Handbook of Aeronautical knowledge (AC 61-23A).

4. Changes in Pilot Exam-O-Grams.

5. Addition of appropriate questions in both written and oral pilot examinations and checks.

6. Assuring through FAA Inspectors that Pilot Schools certificated under 14 CFR 141, highlight the problem in their training syllabi specified in 14 CFR 141.55 (6)(b)(2)."

BY THE NATIONAL TRANSPORTATION SAFETY BOARD

/s/ WEBSTER B. TODD, JR.
Chairman

/s/ FRANCIS H. McADAMS
Member

/s/ PHILIP A. HOGUE
Member

/s/ ISABEL A. BURGESS
Member

/s/ WILLIAM R. HALEY
Member

May 27, 1976

# Part V
Glossary of Weather Terms

GLOSSARY OF WEATHER TERMS

A

absolute instability—A state of a layer within the atmosphere in which the vertical distribution of temperature is such that an air parcel, if given an upward or downward push, will move away from its initial level without further outside force being applied.

absolute temperature scale—*See* Kelvin Temperature Scale.

absolute vorticity—*See* vorticity.

adiabatic process—The process by which fixed relationships are maintained during changes in temperature, volume, and pressure in a body of air without heat being added or removed from the body.

advection—The horizontal transport of air or atmospheric properties. In meteorology, sometimes referred to as the horizontal component of *convection*.

advection fog—Fog resulting from the transport of warm, humid air over a cold surface.

air density—The mass density of the air in terms of weight per unit volume.

air mass—In meteorology, an extensive body of air within which the conditions of temperature and moisture in a horizontal plane are essentially uniform.

air mass classification—A system used to identify and to characterize the different *air masses* according to a basic scheme. The system most commonly used classifies air masses primarily according to the thermal properties of their *source regions:* "tropical" (T); "polar" (P); and "Arctic" or "Antarctic" (A). They are further classified according to moisture characteristics as "continental" (c) or "maritime" (m).

air parcel—*See* parcel.

albedo—The ratio of the amount of electromagnetic *radiation* reflected by a body to the amount incident upon it, commonly expressed in percentage; in meteorology, usually used in reference to *insolation* (solar radiation); i.e., the albedo of wet sand is 9, meaning that about 9% of the incident insolation is reflected; albedoes of other surfaces range upward to 80–85 for fresh snow cover; average albedo for the earth and its atmosphere has been calculated to range from 35 to 43.

altimeter—An instrument which determines the altitude of an object with respect to a fixed level. *See* pressure altimeter.

altimeter setting—The value to which the scale of a *pressure altimeter* is set so as to read true altitude at field elevation.

altimeter setting indicator—A precision *aneroid barometer* calibrated to indicate directly the altimeter setting.

altitude—Height expressed in units of distance above a reference plane, usually above mean sea level or above ground.

(1) **corrected altitude**—Indicated altitude of an aircraft altimeter corrected for the temperature of the column of air below the aircraft, the correction being based on the estimated departure of existing temperature from standard atmospheric temperature; an approximation of true altitude.

(2) **density altitude**—The altitude in the standard atmosphere at which the air has the same density as the air at the point in question. An aircraft will have the same performance characteristics as it would have in a standard atmosphere at this altitude.

(3) **indicated altitude**—The altitude above mean sea level indicated on a *pressure altimeter* set at current local *altimeter setting*.

(4) **pressure altitude**—The altitude in the standard atmosphere at which the pressure is the same as at the point in question. Since an altimeter operates solely on pressure, this is the uncorrected altitude indicated by an altimeter set at standard sea level pressure of 29.92 inches or 1013 millibars.

(5) **radar altitude**—The altitude of an aircraft determined by radar-type radio altimeter; thus the actual distance from the nearest terrain or water feature encompassed by the downward directed radar beam. For all practical purposes, it is the "actual" distance above a ground or inland water surface or the true altitude above an ocean surface.

(6) **true altitude**—The exact distance above mean sea level.

altocumulus—White or gray layers or patches of cloud, often with a waved appearance; cloud elements appear as rounded masses or rolls; composed mostly of liquid water droplets which may be supercooled; may contain ice crystals at subfreezing temperatures.

altocumulus castellanus—A species of middle cloud of which at least a fraction of its upper part presents some vertically developed, cumuliform protuberances (some of which are taller than they are wide, as castles) and which give the cloud a crenelated or turreted appearance; especially evident when seen from the side; elements usually have a common base arranged in lines. This cloud indicates instability and turbulence at the altitudes of occurrence.

anemometer—An instrument for measuring *wind speed*.

Glossary of Weather Terms

aneroid barometer—A *barometer* which operates on the principle of having changing atmospheric pressure bend a metallic surface which, in turn, moves a pointer across a scale graduated in units of pressure.

angel—In radar meteorology, an *echo* caused by physical phenomena not discernible to the eye; they have been observed when abnormally strong temperature and/or moisture *gradients* were known to exist; sometimes attributed to insects or birds flying in the radar beam.

anomalous propagation (sometimes called AP)—In radar meteorology, the greater than normal bending of the radar beam such that *echoes* are received from ground *targets* at distances greater than normal *ground clutter*.

anticyclone—An area of high atmospheric pressure which has a closed circulation that is anticyclonic, i.e., as viewed from above, the circulation is clockwise in the Northern Hemisphere, counterclockwise in the Southern Hemisphere, undefined at the Equator.

anvil cloud—Popular name given to the top portion of a *cumulonimbus* cloud having an anvil-like form.

APOB—A *sounding* made by an aircraft.

Arctic air—An air mass with characteristics developed mostly in winter over Arctic surfaces of ice and snow. Arctic air extends to great heights, and the surface temperatures are basically, but not always, lower than those of *polar air*.

Arctic front—The surface of discontinuity between very cold (Arctic) air flowing directly from the Arctic region and another less cold and, consequently, less dense air mass.

astronomical twilight—*See* twilight.

atmosphere—The mass of air surrounding the Earth.

atmospheric pressure (also called barometric pressure)—The pressure exerted by the atmosphere as a consequence of gravitational attraction exerted upon the "column" of air lying directly above the point in question.

atmospherics—Disturbing effects produced in radio receiving apparatus by atmospheric electrical phenomena such as an electrical storm. Static.

aurora—A luminous, radiant emission over middle and high latitudes confined to the thin air of high altitudes and centered over the earth's magnetic poles. Called "aurora borealis" (northern lights) or "aurora australis" according to its occurrence in the Northern or Southern Hemisphere, respectively.

attenuation—In radar meteorology, any process which reduces power density in radar signals.

(1) **precipitation attenuation**—Reduction of power density because of absorption or reflection of energy by precipitation.

(2) **range attenuation**—Reduction of radar power density because of distance from the antenna. It occurs in the outgoing beam at a rate proportional to 1/range^2. The return signal is also attenuated at the same rate.

B

backing—Shifting of the wind in a counterclockwise direction with respect to either space or time; opposite of *veering*. Commonly used by meteorologists to refer to a cyclonic shift (counterclockwise in the Northern Hemisphere and clockwise in the Southern Hemisphere).

backscatter—Pertaining to radar, the energy reflected or scattered by a *target*; an *echo*.

banner cloud (also called cloud banner)—A banner-like cloud streaming off from a mountain peak.

barograph—A continuous-recording *barometer*.

barometer—An instrument for measuring the pressure of the atmosphere; the two principle types are *mercurial* and *aneroid*.

barometric altimeter—*See* pressure altimeter.

barometric pressure—Same as *atmospheric pressure*.

barometric tendency—The change of barometric pressure within a specified period of time. In aviation weather observations, routinely determined periodically, usually for a 3-hour period.

beam resolution—*See* resolution.

Beaufort scale—A scale of wind speeds.

black blizzard—Same as *duststorm*.

blizzard—A severe weather condition characterized by low temperatures and strong winds bearing a great amount of snow, either falling or picked up from the ground.

blowing dust—A type of *lithometeor* composed of dust particles picked up locally from the surface and blown about in clouds or sheets.

blowing sand—A type of *lithometeor* composed of sand picked up locally from the surface and blown about in clouds or sheets.

blowing snow—A type of *hydrometeor* composed of snow picked up from the surface by the wind and carried to a height of 6 feet or more.

blowing spray—A type of *hydrometeor* composed of water particles picked up by the wind from the surface of a large body of water.

bright band—In radar meteorology, a narrow, intense *echo* on the *range-height indicator* scope resulting from water-covered ice particles of high reflectivity at the melting level.

Buys Ballot's law—If an observer in the Northern Hemisphere stands with his back to the wind, lower pressure is to his left.

C

calm—The absence of wind or of apparent motion of the air.

cap cloud (also called cloud cap)—A standing or stationary cap-like cloud crowning a mountain summit.

Glossary of Weather Terms

ceiling—In meteorology in the U.S., (1) the height above the surface of the base of the lowest layer of clouds or *obscuring phenomena* aloft that hides more than half of the sky, or (2) the *vertical visibility* into an *obscuration*. See summation principle.

ceiling balloon—A small balloon used to determine the height of a cloud base or the extent of vertical visibility.

ceiling light—An instrument which projects a vertical light beam onto the base of a cloud or into surface-based obscuring phenomena; used at night in conjunction with a *clinometer* to determine the height of the cloud base or as an aid in estimating the vertical visibility.

ceilometer—A cloud-height measuring system. It projects light on the cloud, detects the reflection by a photo-electric cell, and determines height by triangulation.

Celsius temperature scale (abbreviated C)—A temperature scale with zero degrees as the melting point of pure ice and 100 degrees as the boiling point of pure water at standard sea level atmospheric pressure.

Centigrade temperature scale—Same as *Celsius temperature scale*.

chaff—Pertaining to radar, (1) short, fine strips of metallic foil dropped from aircraft, usually by military forces, specifically for the purpose of jamming radar; (2) applied loosely to *echoes* resulting from chaff.

change of state—In meteorology, the transformation of water from one form, i.e., solid (ice), liquid, or gaseous (water vapor), to any other form. There are six possible transformations designated by the five terms following:
(1) **condensation**—The change of water vapor to liquid water.
(2) **evaporation**—The change of liquid water to water vapor.
(3) **freezing**—The change of liquid water to ice.
(4) **melting**—The change of ice to liquid water.
(5) **sublimation**—The change of (a) ice to water vapor or (b) water vapor to ice. See latent heat.

Chinook—A warm, dry *foehn* wind blowing down the eastern slopes of the Rocky Mountains over the adjacent plains in the U.S. and Canada.

cirriform—All species and varieties of *cirrus, cirrocumulus,* and *cirrostratus* clouds; descriptive of clouds composed mostly or entirely of small ice crystals, usually transparent and white; often producing *halo* phenomena not observed with other cloud forms. Average height ranges upward from 20,000 feet in middle latitudes.

cirrocumulus—A *cirriform* cloud appearing as a thin sheet of small white puffs resembling flakes or patches of cotton without shadows; sometimes confused with *altocumulus*.

cirrostratus—A *cirriform* cloud appearing as a whitish veil, usually fibrous, sometimes smooth; often produces *halo* phenomena; may totally cover the sky.

cirrus—A *cirriform* cloud in the form of thin, white feather-like clouds in patches or narrow bands; have a fibrous and/or silky sheen; large ice crystals often trail downward a considerable vertical distance in fibrous, slanted, or irregularly curved wisps called mares' tails.

civil twilight—See twilight.

clear air turbulence (abbreviated CAT)—Turbulence encountered in air where no clouds are present; more popularly applied to high level turbulence associated with *wind shear*.

clear icing (or clear ice)—Generally, the formation of a layer or mass of ice which is relatively transparent because of its homogeneous structure and small number and size of air spaces; used commonly as synonymous with *glaze*, particularly with respect to aircraft icing. Compare with *rime icing*. Factors which favor clear icing are large drop size, such as those found in *cumuliform* clouds, rapid accretion of supercooled water, and slow dissipation of *latent heat* of fusion.

climate—The statistical collective of the weather conditions of a point or area during a specified interval of time (usually several decades); may be expressed in a variety of ways.

climatology—The study of *climate*.

clinometer—An instrument used in weather observing for measuring angles of inclination; it is used in conjunction with a *ceiling light* to determine cloud height at night.

cloud bank—Generally, a fairly well-defined mass of cloud observed at a distance; it covers an appreciable portion of the horizon sky, but does not extend overhead.

cloudburst—In popular teminology, any sudden and heavy fall of *rain*, almost always of the *shower* type.

cloud cap—*See* cap cloud.

cloud detection radar—A vertically directed radar to detect cloud bases and tops.

cold front—Any non-occluded *front* which moves in such a way that colder air replaces warmer air.

condensation—*See* change of state.

condensation level—The height at which a rising *parcel* or layer of air would become saturated if lifted adiabatically.

condensation nuclei—Small particles in the air on which water vapor condenses or sublimates.

condensation trail (or contrail) (also called vapor trail)—A cloud-like streamer frequently observed to form behind aircraft flying in clear, cold, humid air.

conditionally unstable air—Unsaturated air that will become unstable on the condition it becomes saturated. See instability.

conduction—The transfer of heat by molecular action through a substance or from one substance in contact with another; transfer is always from warmer to colder temperature.

constant pressure chart—A chart of a constant pressure surface; may contain analyses of height, wind, temperature, humidity, and/or other elements.

Glossary of Weather Terms

continental polar air—*See* polar air.

continental tropical air—*See* tropical air.

contour—In meteorology, (1) a line of equal height on a constant pressure chart; analogous to contours on a relief map; (2) in radar meteorology, a line on a radar scope of equal *echo* intensity.

contouring circuit—On weather radar, a circuit which displays multiple contours of *echo* intensity simultaneously on the *plan position indicator* or *range-height indicator* scope. *See* contour (2).

contrail—Contraction for *condensation trail*.

convection—(1) In general, mass motions within a fluid resulting in transport and mixing of the properties of that fluid. (2) In meteorology, atmospheric motions that are predominantly vertical, resulting in vertical transport and mixing of atmospheric properties; distinguished from *advection*.

convective cloud—*See* cumuliform.

convective condensation level (abbreviated CCL)—The lowest level at which condensation will occur as a result of *convection* due to surface heating. When condensation occurs at this level, the layer between the surface and the CCL will be thoroughly mixed, temperature *lapse rate* will be dry adiabatic, and *mixing ratio* will be constant.

convective instability—The state of an unsaturated layer of air whose *lapse rates* of temperature and moisture are such that when lifted adiabatically until the layer becomes saturated, convection is spontaneous.

convergence—The condition that exists when the distribution of winds within a given area is such that there is a net horizontal inflow of air into the area. In convergence at lower levels, the removal of the resulting excess is accomplished by an upward movement of air; consequently, areas of low-level convergent winds are regions favorable to the occurrence of clouds and precipitation. Compare with *divergence*.

Coriolis force—A deflective force resulting from earth's rotation; it acts to the right of wind direction in the Northern Hemisphere and to the left in the Southern Hemisphere.

corona—A prismatically colored circle or arcs of a circle with the sun or moon at its center; coloration is from blue inside to red outside (opposite that of a *halo*); varies in size (much smaller) as opposed to the fixed diameter of the halo; characteristic of clouds composed of water droplets and valuable in differentiating between middle and cirriform clouds.

corposant—*See* St. Elmo's Fire.

corrected altitude (approximation of true altitude)—*See* altitude.

cumuliform—A term descriptive of all convective clouds exhibiting vertical development in contrast to the horizontally extended *stratiform* types.

cumulonimbus—A cumuliform cloud type; it is heavy and dense, with considerable vertical extent in the form of massive towers; often with tops in the shape of an *anvil* or massive plume; under the base of cumulonimbus, which often is very dark, there frequently exists *virga*, precipitation and low ragged clouds (*scud*), either merged with it or not; frequently accompanied by lightning, thunder, and sometimes hail; occasionally produces a tornado or a waterspout; the ultimate manifestation of the growth of a cumulus cloud, occasionally extending well into the stratosphere.

cumulonimbus mamma—A *cumulonimbus* cloud having hanging protuberances, like pouches, festoons, or udders, on the under side of the cloud; usually indicative of severe turbulence.

cumulus—A cloud in the form of individual detached domes or towers which are usually dense and well defined; develops vertically in the form of rising mounds of which the bulging upper part often resembles a cauliflower; the sunlit parts of these clouds are mostly brilliant white; their bases are relatively dark and nearly horizontal.

cumulus fractus—*See* fractus.

cyclogenesis—Any development or strengthening of cyclonic circulation in the atmosphere.

cyclone—(1) An area of low atmospheric pressure which has a closed circulation that is cyclonic, i.e., as viewed from above, the circulation is counterclockwise in the Northern Hemisphere, clockwise in the Southern Hemisphere, undefined at the Equator. Because cyclonic circulation and relatively low atmospheric pressure usually coexist, in common practice the terms cyclone and low are used interchangeably. Also, because cyclones often are accompanied by inclement (sometimes destructive) weather, they are frequently referred to simply as storms. (2) Frequently misused to denote a *tornado*. (3) In the Indian Ocean, a *tropical cyclone* of hurricane or typhoon force.

D

deepening—A decrease in the central pressure of a pressure system; usually applied to a *low* rather than to a *high*, although technically, it is acceptable in either sense.

density—(1) The ratio of the mass of any substance to the volume it occupies—weight per unit volume. (2) The ratio of any quantity to the volume or area it occupies, i.e., population per unit area, *power density*.

density altitude—*See* altitude.

depression—In meteorology, an area of low pressure; a *low* or *trough*. This is usually applied to a certain stage in the development of a *tropical cyclone*, to migratory lows and troughs, and to upper-level lows and troughs that are only weakly developed.

dew—Water condensed onto grass and other objects near the ground, the temperatures of which have fallen below the initial dew point temperature of the surface air, but is still above freezing. Compare with *frost*.

dew point (or dew-point temperature)—The temperature to which a sample of air must be cooled, while the *mixing ratio* and barometric pressure remain constant, in order to attain saturation with respect to water.

discontinuity—A zone with comparatively rapid transition of one or more meteorological elements.

disturbance—In meteorology, applied rather loosely: (1) any low pressure or cyclone, but usually one that is relatively small in size; (2) an area where weather, wind, pressure, etc., show signs of cyclonic development; (3) any deviation in flow or pressure that is associated with a disturbed state of the weather, i.e., cloudiness and precipitation; and (4) any individual circulatory system within the primary circulation of the atmosphere.

diurnal—Daily, especially pertaining to a cycle completed within a 24-hour period, and which recurs every 24 hours.

divergence—The condition that exists when the distribution of winds within a given area is such that there is a net horizontal flow of air outward from the region. In divergence at lower levels, the resulting deficit is compensated for by subsidence of air from aloft; consequently the air is heated and the relative humidity lowered making divergence a warming and drying process. Low-level divergent regions are areas unfavorable to the occurrence of clouds and precipitation. The opposite of *convergence*.

doldrums—The equatorial belt of calm or light and variable winds between the two tradewind belts. Compare *intertropical convergence zone*.

downdraft—A relative small scale downward current of air; often observed on the lee side of large objects restricting the smooth flow of the air or in precipitation areas in or near *cumuliform* clouds.

drifting snow—A type of *hydrometeor* composed of snow particles picked up from the surface, but carried to a height of less than 6 feet.

drizzle—A form of *precipitation*. Very small water drops that appear to float with the air currents while falling in an irregular path (unlike *rain*, which falls in a comparatively straight path, and unlike *fog* droplets which remain suspended in the air).

dropsonde—A *radiosonde* dropped by parachute from an aircraft to obtain *soundings* (measurements) of the atmosphere below.

dry adiabatic lapse rate—The rate of decrease of temperature with height when unsaturated air is lifted adiabatically (due to expansion as it is lifted to lower pressure). See adiabatic process.

dry bulb—A name given to an ordinary thermometer used to determine temperature of the air; also used as a contraction for *dry-bulb temperature*. Compare *wet bulb*.

dry-bulb temperature—The temperature of the air.

dust—A type of *lithometeor* composed of small earthen particles suspended in the atmosphere.

dust devil—A small, vigorous *whirlwind*, usually of short duration, rendered visible by dust, sand, and debris picked up from the ground.

duster—Same as *duststorm*.

duststorm (also called duster, black blizzard)—An unusual, frequently severe weather condition characterized by strong winds and dust-filled air over an extensive area.

D-value—Departure of true altitude from pressure altitude (*see* altitude); obtained by algebraically subtracting true altitude from pressure altitude; thus it may be plus or minus. On a constant pressure chart, the difference between actual height and *standard atmospheric* height of a constant pressure surface.

E

echo—In radar, (1) the energy reflected or scattered by a *target;* (2) the radar scope presentation of the return from a target.

eddy—A local irregularity of wind in a larger scale wind flow. Small scale eddies produce turbulent conditions.

estimated ceiling—A ceiling classification applied when the ceiling height has been estimated by the observer or has been determined by some other method; but, because of the specified limits of time, distance, or precipitation conditions, a more descriptive classification cannot be applied.

evaporation—*See* change of state.

extratropical low (sometimes called extratropical cyclone, extratropical storm)—Any *cyclone* that is not a *tropical cyclone*, usually referring to the migratory frontal cyclones of middle and high latitudes.

eye—The roughly circular area of calm or relatively light winds and comparatively fair weather at the center of a well-developed *tropical cyclone*. A *wall cloud* marks the outer boundary of the eye.

F

Fahrenheit temperature scale (abbreviated F)—A temperature scale with 32 degrees as the melting point of pure ice and 212 degrees as the boiling point of pure water at standard sea level atmospheric pressure (29.92 inches or 1013.2 millibars).

Fall wind—A cold wind blowing downslope. Fall wind differs from *foehn* in that the air is initially cold enough to remain relatively cold despite compressional heating during descent.

filling—An increase in the central pressure of a pressure system; opposite of *deepening;* more commonly applied to a low rather than a high.

first gust—The leading edge of the spreading downdraft, *plow wind*, from an approaching thunderstorm.

flow line—A *streamline*.

foehn—A warm, dry downslope wind; the warmness and dryness being due to adiabatic compression upon descent; characteristic of mountainous regions. See adiabatic process, Chinook, Santa Ana.

fog—A *hydrometeor* consisting of numerous minute water droplets and based at the surface; droplets are small enough to be suspended in the earth's atmosphere in-

Glossary of Weather Terms

definitely. (Unlike *drizzle*, it does not fall to the surface; differs from cloud only in that a cloud is not based at the surface; distinguished from haze by its wetness and gray color.)

fractus—Clouds in the form of irregular shreds, appearing as if torn; have a clearly ragged appearance; applies only to stratus and cumulus, i.e., *cumulus* fractus and *stratus* fractus.

freezing—*See* change of state.

freezing level—A level in the atmosphere at which the temperature is 0° C (32° F).

front—A surface, interface, or transition zone of discontinuity between two adjacent *air masses* of different densities; more simply the boundary between two different air masses. *See* frontal zone.

frontal zone—A *front* or zone with a marked increase of density gradient; used to denote that fronts are not truly a "surface" of discontinuity but rather a "zone" of rapid transition of meteorological elements.

frontogenesis—The initial formation of a *front* or *frontal zone*.

frontolysis—The dissipation of a *front*.

frost (also hoarfrost)—Ice crystal deposits formed by sublimation when temperature and dew point are below freezing.

funnel cloud—A *tornado* cloud or *vortex* cloud extending downward from the parent cloud but not reaching the ground.

G

glaze—A coating of ice, generally clear and smooth, formed by freezing of supercooled water on a surface. *See* clear icing.

gradient—In meteorology, a horizontal decrease in value per unit distance of a parameter in the direction of maximum decrease; most commonly used with pressure, temperature, and moisture.

ground clutter—Pertaining to radar, a cluster of *echoes*, generally at short range, reflected from ground *targets*.

ground fog—In the United States, a *fog* that conceals less than 0.6 of the sky and is not contiguous with the base of clouds.

gust—A sudden brief increase in wind; according to U.S. weather observing practice, gusts are reported when the variation in wind speed between peaks and lulls is at least 10 knots.

H

hail—A form of *precipitation* composed of balls or irregular lumps of ice, always produced by convective clouds which are nearly always *cumulonimbus*.

halo—A prismatically colored or whitish circle or arcs of a circle with the sun or moon at its center; coloration, if not white, is from red inside to blue outside (opposite that of a *corona*); fixed in size with an angular diameter of 22° (common) or 46° (rare); characteristic of clouds composed of ice crystals; valuable in differentiating between *cirriform* and forms of lower clouds.

haze—A type of *lithometeor* composed of fine dust or salt particles dispersed through a portion of the atmosphere; particles are so small they cannot be felt or individually seen with the naked eye (as compared with the larger particles of *dust*), but diminish the visibility; distinguished from *fog* by its bluish or yellowish tinge.

high—An area of high barometric pressure, with its attendant system of winds; an *anticyclone*. Also high pressure system.

hoar frost—*See* frost.

humidity—Water vapor content of the air; may be expressed as *specific humidity*, *relative humidity*, or *mixing ratio*.

hurricane—A *tropical cyclone* in the Western Hemisphere with winds in excess of 65 knots or 120 km/h.

hydrometeor—A general term for particles of liquid water or ice such as rain, fog, frost, etc., formed by modification of water vapor in the atmosphere; also water or ice particles lifted from the earth by the wind such as sea spray or blowing snow.

hygrograph—The record produced by a continuous-recording *hygrometer*.

hygrometer—An instrument for measuring the water vapor content of the air.

I

ice crystals—A type of *precipitation* composed of unbranched crystals in the form of needles, columns, or plates; usually having a very slight downward motion, may fall from a cloudless sky.

ice fog—A type of fog composed of minute suspended particles of ice; occurs at very low temperatures and may cause *halo* phenomena.

ice needles—A form of *ice crystals*.

ice pellets—Small, transparent or translucent, round or irregularly shaped pellets of ice. They may be (1) hard grains that rebound on striking a hard surface or (2) pellets of snow encased in ice.

icing—In general, any deposit of ice forming on an object. *See* clear icing, rime icing, glaze.

indefinite ceiling—A ceiling classification denoting *vertical visibility* into a surface based obscuration.

indicated altitude—*See* altitude.

insolation—Incoming solar *radiation* falling upon the earth and its atmosphere.

instability—A general term to indicate various states of the atmosphere in which spontaneous *convection* will occur when prescribed criteria are met; indicative of turbulence. *See* absolute instability, conditionally unstable air, convective instability.

intertropical convergence zone—The boundary zone between the trade wind system of the Northern and Southern Hemispheres; it is characterized in maritime climates by showery precipitation with cumulonimbus clouds sometimes extending to great heights.

inversion—An increase in temperature with height—a reversal of the normal decrease with height in the *troposphere*; may also be applied to other meteorological properties.

isobar—A line of equal or constant barometric pressure.

iso echo—In radar circuitry, a circuit that reverses signal strength above a specified intensity level, thus causing a void on the scope in the most intense portion of an echo when maximum intensity is greater than the specified level.

isoheight—On a weather chart, a line of equal height; same as *contour* (1).

isoline—A line of equal value of a variable quantity, i.e., an isoline of temperature is an *isotherm*, etc. See isobar, isotach, etc.

isoshear—A line of equal *wind shear*.

isotach—A line of equal or constant wind speed.

isotherm—A line of equal or constant temperature.

isothermal—Of equal or constant temperature, with respect to either space or time; more commonly, temperature with height; a zero *lapse rate*.

J

jet stream—A quasi-horizontal stream of winds 50 knots or more concentrated within a narrow band embedded in the westerlies in the high *troposphere*.

K

katabatic wind—Any wind blowing downslope. See fall wind, foehn.

Kelvin temperature scale (abbreviated K)—A temperature scale with zero degrees equal to the temperature at which all molecular motion ceases, i.e., absolute zero ($0°$ K $= -273°$ C); the Kelvin degree is identical to the Celsius degree; hence at standard sea level pressure, the melting point is $273°$ K and the boiling point $373°$ K.

knot—A unit of speed equal to one nautical mile per hour.

L

land breeze—A coastal breeze blowing from land to sea, caused by temperature difference when the sea surface is warmer than the adjacent land. Therefore, it usually blows at night and alternates with a *sea breeze*, which blows in the opposite direction by day.

lapse rate—The rate of decrease of an atmospheric variable with height; commonly refers to decrease of temperature with height.

latent heat—The amount of heat absorbed (converted to kinetic energy) during the processes of change of liquid water to water vapor, ice to water vapor, or ice to liquid water; or the amount released during the reverse processes. Four basic classifications are:

(1) **latent heat of condensation**—Heat released during change of water vapor to water.

(2) **latent heat of fusion**—Heat released during change of water to ice or the amount absorbed in change of ice to water.

(3) **latent heat of sublimation**—Heat released during change of water vapor to ice or the amount absorbed in the change of ice to water vapor.

(4) **latent heat of vaporization**—Heat absorbed in the change of water to water vapor; the negative of latent heat of condensation.

layer—In reference to sky cover, clouds or other obscuring phenomena whose bases are approximately at the same level. The layer may be continuous or composed of detached elements. The term "layer" does not imply that a clear space exists between the layers or that the clouds or *obscuring phenomena* composing them are of the same type.

lee wave—Any stationary wave disturbance caused by a barrier in a fluid flow. In the atmosphere when sufficient moisture is present, this wave will be evidenced by *lenticular clouds* to the lee of mountain barriers; also called *mountain wave* or *standing wave*.

lenticular cloud (or lenticularis)—A species of cloud whose elements have the form of more or less isolated, generally smooth lenses or almonds. These clouds appear most often in formations of orographic origin, the result of *lee waves*, in which case they remain nearly stationary with respect to the terrain (standing cloud), but they also occur in regions without marked orography.

level of free convection (abbreviated LFC)—The level at which a *parcel* of air lifted dry-adiabatically until saturated and moist-adiabatically thereafter would become warmer than its surroundings in a conditionally unstable atmosphere. See conditional instability and adiabatic process.

lifting condensation level (abbreviated LCL)—The level at which a *parcel* of unsaturated air lifted dry-adiabatically would become saturated. Compare *level of free convection* and *convective condensation level*.

lightning—Generally, any and all forms of visible electrical discharge produced by a *thunderstorm*.

lithometeor—The general term for dry particles suspended in the atmosphere such as dust, haze, smoke, and sand.

low—An area of low barometric pressure, with its attendant system of winds. Also called a barometric depression or *cyclone*.

M

mammato cumulus—Obsolete. See cumulonimbus mamma.

mare's tail—*See* cirrus.

maritime polar air (abbreviated mP)—*See* polar air.

maritime tropical air (abbreviated mT)—*See* tropical air.

Glossary of Weather Terms

maximum wind axis—On a constant pressure chart, a line denoting the axis of maximum wind speeds at that constant pressure surface.

mean sea level—The average height of the surface of the sea for all stages of tide; used as reference for elevations throughout the U.S.

measured ceiling—A ceiling classification applied when the ceiling value has been determined by instruments or the known heights of unobscured portions of objects, other than natural landmarks.

melting—*See* change of state.

mercurial barometer—A *barometer* in which pressure is determined by balancing air pressure against the weight of a column of mercury in an evacuated glass tube.

meteorological visibility—In U.S. observing practice, a main category of *visibility* which includes the subcategories of *prevailing visibility* and *runway visibility*. Meteorological visibility is a measure of horizontal visibility near the earth's surface, based on sighting of objects in the daytime or unfocused lights of moderate intensity at night. Compare *slant visibility*, *runway visual range*, *vertical visibility*. *See* surface visibility, tower visibility, and sector visibility.

meteorology—The science of the *atmosphere*.

microbarograph—An aneroid *barograph* designed to record atmospheric pressure changes of very small magnitudes.

millibar (abbreviated mb.)—An internationally used unit of pressure equal to 1,000 dynes per square centimeter. It is convenient for reporting *atmospheric pressure*.

mist—A popular expression for drizzle or heavy fog.

mixing ratio—The ratio by weight of the amount of water vapor in a volume of air to the amount of dry air; usually expressed as grams per kilogram (g/kg).

moist-adiabatic lapse rate—*See* saturated-adiabatic lapse rate.

moisture—An all-inclusive term denoting water in any or all of its three states.

monsoon—A wind that in summer blows from sea to a continental interior, bringing copious rain, and in winter blows from the interior to the sea, resulting in sustained dry weather.

mountain wave—A *standing wave* or *lee wave* to the lee of a mountain barrier.

N

nautical twilight—*See* twilight.

negative vorticity—*See* vorticity.

nimbostratus—A principal cloud type, gray colored, often dark, the appearance of which is rendered diffuse by more or less continuously falling rain or snow, which in most cases reaches the ground. It is thick enough throughout to blot out the sun.

noctilucent clouds—Clouds of unknown composition which occur at great heights, probably around 75 to 90 kilometers. They resemble thin *cirrus*, but usually with a bluish or silverish color, although sometimes orange to red, standing out against a dark night sky. Rarely observed.

normal—In meteorology, the value of an element averaged for a given location over a period of years and recognized as a standard.

numerical forecasting—*See* numerical weather prediction.

numerical weather prediction—Forecasting by digital computers solving mathematical equations; used extensively in weather services throughout the world.

O

obscuration—Denotes sky hidden by surface-based *obscuring phenomena* and *vertical visibility* restricted overhead.

obscuring phenomena—Any *hydrometeor* or *lithometeor* other than clouds; may be surface based or aloft.

occlusion—Same as *occluded front*.

occluded front (commonly called occlusion, also called frontal occlusion)—A composite of two fronts as a *cold front* overtakes a *warm front* or *quasi-stationary front*.

orographic—Of, pertaining to, or caused by mountains as in orographic clouds, orographic lift, or orographic precipitation.

ozone—An unstable form of oxygen; heaviest concentrations are in the stratosphere; corrosive to some metals; absorbs most ultraviolet solar radiation.

P

parcel—A small volume of air, small enough to contain uniform distribution of its meteorological properties, and large enough to remain relatively self-contained and respond to all meteorological processes. No specific dimensions have been defined, however, the order of magnitude of 1 cubic foot has been suggested.

partial obscuration—A designation of sky cover when part of the sky is hidden by surface based *obscuring phenomena*.

pilot balloon—A small free-lift balloon used to determine the speed and direction of winds in the upper air.

pilot balloon observation (commonly called PIBAL)—A method of winds-aloft observation by visually tracking a *pilot balloon*.

plan position indicator (PPI) scope—A radar indicator scope displaying range and azimuth of *targets* in polar coordinates.

plow wind—The spreading downdraft of a *thunderstorm;* a strong, straight-line wind in advance of the storm. *See* first gust.

polar air—An air mass with characteristics developed over high latitudes, especially within the subpolar highs. Continental polar air (cP) has cold surface temperatures, low moisture content, and, especially in its source regions, has great stability in the lower layers. It is shallow in comparison with *Arctic air*. Maritime polar (mP) initially possesses similar properties to those of continental polar air, but in passing over warmer water it becomes unstable with a higher moisture content. Compare *tropical air*.

polar front—The semipermanent, semicontinuous *front* separating air masses of tropical and polar origins.

positive vorticity—*See* vorticity.

power density—In radar meteorology the amount of radiated energy per unit cross sectional area in the radar beam.

precipitation—Any or all forms of water particles, whether liquid or solid, that fall from the atmosphere and reach the surface. It is a major class of *hydrometeor*, distinguished from cloud and *virga* in that it must reach the surface.

precipitation attenuation—*See* attenuation.

pressure—*See* atmospheric pressure.

pressure altimeter—An *aneroid barometer* with a scale graduated in altitude instead of pressure using *standard atmospheric* pressure-height relationships; shows indicated altitude (not necessarily true altitude); may be set to measure altitude (indicated) from any arbitrarily chosen level. *See* altimeter setting, altitude.

pressure altitude—*See* altitude.

pressure gradient—The rate of decrease of pressure per unit distance at a fixed time.

pressure jump—A sudden, significant increase in *station pressure*.

pressure tendency—*See* barometric tendency.

prevailing easterlies—The broad current or pattern of persistent easterly winds in the Tropics and in polar regions.

prevailing visibility—In the U.S., the greatest horizontal visibility which is equaled or exceeded throughout half of the horizon circle; it need not be a continuous half.

prevailing westerlies—The dominant west-to-east motion of the atmosphere, centered over middle latitudes of both hemispheres.

prevailing wind—Direction from which the wind blows most frequently.

prognostic chart (contracted PROG)—A chart of expected or forecast conditions.

pseudo-adiabatic lapse rate—*See* saturated-adiabatic lapse rate.

psychrometer—An instrument consisting of a *wet-bulb* and a *dry-bulb* thermometer for measuring wet-bulb and dry-bulb temperature; used to determine water vapor content of the air.

pulse—Pertaining to radar, a brief burst of electromagnetic radiation emitted by the radar; of very short time duration. *See* pulse length.

pulse length—Pertaining to radar, the dimension of a radar pulse; may be expressed as the time duration or the length in linear units. Linear dimension is equal to time duration multiplied by the speed of propagation (approximately the speed of light).

Q

quasi-stationary front (commonly called stationary front)—A *front* which is stationary or nearly so; conventionally, a front which is moving at a speed of less than 5 knots is generally considered to be quasi-stationary.

R

RADAR (contraction for radio detection and ranging)—An electronic instrument used for the detection and ranging of distant objects of such composition that they scatter or reflect radio energy. Since *hydrometeors* can scatter radio energy, *weather radars*, operating on certain frequency bands, can detect the presence of precipitation, clouds, or both.

radar altitude—*See* altitude.

radar beam—The focused energy radiated by radar similar to a flashlight or searchlight beam.

radar echo—*See* echo.

radarsonde observation—A *rawinsonde observation* in which winds are determined by radar tracking a balloon-borne target.

radiation—The emission of energy by a medium and transferred, either through free space or another medium, in the form of electromagnetic waves.

radiation fog—*Fog* characteristically resulting when radiational cooling of the earth's surface lowers the air temperature near the ground to or below its initial dew point on calm, clear nights.

radiosonde—A balloon-borne instrument for measuring pressure, temperature, and humidity aloft. Radiosonde observation—a *sounding* made by the instrument.

rain—A form of *precipitation;* drops are larger than *drizzle* and fall in relatively straight, although not necessarily vertical, paths as compared to drizzle which falls in irregular paths.

rain shower—*See* shower.

range attenuation—*See* attenuation.

range-height indicator (RHI) scope—A radar indicator scope displaying a vertical cross section of *targets* along a selected azimuth.

range resolution—*See* resolution.

RAOB—A *radiosonde* observation.

rawin—A *rawinsonde* observation.

rawinsonde observation—A combined winds aloft and radiosonde observation. Winds are determined by tracking the *radiosonde* by radio direction finder or radar.

refraction—In radar, bending of the *radar beam* by variations in atmospheric density, water vapor content, and temperature.
  (1) **normal refraction**—Refraction of the radar beam under normal atmospheric conditions; normal radius of curvature of the beam is about 4 times the radius of curvature of the Earth.
  (2) **superrefraction**—More than normal bending of the radar beam resulting from abnormal vertical gradients of temperature and/or water vapor.
  (3) **subrefraction**—Less than normal bending of the radar beam resulting from abnormal vertical gradients of temperature and/or water vapor.

relative humidity—The ratio of the existing amount of water vapor in the air at a given temperature to the maximum amount that could exist at that temperature; usually expressed in percent.

relative vorticity—*See* vorticity.

remote scope—In radar meteorology a "slave" scope remoted from weather *radar*.

resolution—Pertaining to radar, the ability of radar to show discrete *targets* separately, i.e., the better the resolution, the closer two targets can be to each other, and still be detected as separate targets.
  (1) **beam resolution**—The ability of radar to distinguish between targets at approximately the same range but at different azimuths.
  (2) **range resolution**—The ability of radar to distinguish between targets on the same azimuth but at different ranges.

ridge (also called ridge line)—In meteorology, an elongated area of relatively high atmospheric pressure; usually associated with and most clearly identified as an area of maximum anticyclonic curvature of the wind flow (*isobars*, *contours*, or *streamlines*).

rime icing (or rime ice)—The formation of a white or milky and opaque granular deposit of ice formed by the rapid freezing of supercooled water droplets as they impinge upon an exposed aircraft.

rocketsonde—A type of *radiosonde* launched by a rocket and making its measurements during a parachute descent; capable of obtaining *soundings* to a much greater height than possible by balloon or aircraft.

roll cloud (sometimes improperly called rotor cloud)—A dense and horizontal roll-shaped accessory cloud located on the lower leading edge of a *cumulonimbus* or less often, a rapidly developing *cumulus;* indicative of turbulence.

rotor cloud (sometimes improperly called *roll cloud*)—A turbulent cloud formation found in the lee of some large mountain barriers, the air in the cloud rotates around an axis parallel to the range; indicative of possible violent turbulence.

runway temperature—The temperature of the air just above a runway, ideally at engine and/or wing height, used in the determination of density *altitude;* useful at airports when critical values of density altitude prevail.

runway visibility—The *meteorological visibility* along an identified runway determined from a specified point on the runway; may be determined by a *transmissometer* or by an observer.

runway visual range—An instrumentally derived horizontal distance a pilot should see down the runway from the approach end; based on either the sighting of high intensity runway lights or on the visual contrast of other objects, whichever yields the greatest visual range.

S

St. Elmo's Fire (also called corposant)—A luminous brush discharge of electricity from protruding objects, such as masts and yardarms of ships, aircraft, lightning rods, steeples, etc., occurring in stormy weather.

Santa Ana—A hot, dry, *foehn* wind, generally from the northeast or east, occurring west of the Sierra Nevada Mountains especially in the pass and river valley near Santa Ana, California.

saturated adiabatic lapse rate—The rate of decrease of temperature with height as saturated air is lifted with no gain or loss of heat from outside sources; varies with temperature, being greatest at low temperatures. *See* adiabatic process and dry-adiabatic lapse rate.

saturation—The condition of the atmosphere when actual *water vapor* present is the maximum possible at existing temperature.

scud—Small detached masses of stratus *fractus* clouds below a layer of higher clouds, usually *nimbostratus*.

sea breeze—A coastal breeze blowing from sea to land, caused by the temperature difference when the land surface is warmer than the sea surface. Compare *land breeze*.

sea fog—A type of *advection fog* formed when air that has been lying over a warm surface is transported over a colder water surface.

sea level pressure—The *atmospheric pressure* at *mean sea level*, either directly measured by stations at sea level or empirically determined from the *station pressure* and temperature by stations not at sea level; used as a common reference for analyses of surface pressure patterns.

sea smoke—Same as *steam fog*.

sector visibility—*Meteorological visibility* within a specified sector of the horizon circle.

sensitivity time control—A radar circuit designed to correct for range *attenuation* so that echo intensity on the scope is proportional to reflectivity of the *target* regardless of range.

shear—*See* wind shear.

shower—*Precipitation* from a *cumuliform* cloud; characterized by the suddenness of beginning and ending, by the rapid change of intensity, and usually by rapid change in the appearance of the sky; showery precipitation may be in the form of rain, ice pellets, or snow.

slant visibility—For an airborne observer, the distance at which he can see and distinguish objects on the ground.

Glossary of Weather Terms

sleet—*See* ice pellets.

smog—A mixture of *smoke* and *fog*.

smoke—A restriction to visibility resulting from combustion.

snow—Precipitation composed of white or translucent ice crystals, chiefly in complex branched hexagonal form.

snow flurry—Popular term for snow *shower*, particularly of a very light and brief nature.

snow grains—*Precipitation* of very small, white opaque grains of ice, similar in structure to *snow* crystals. The grains are fairly flat or elongated, with diameters generally less than 0.04 inch (1 mm.).

snow pellets—*Precipitation* consisting of white, opaque approximately round (sometimes conical) ice particles having a snow-like structure, and about 0.08 to 0.2 inch in diameter; crisp and easily crushed, differing in this respect from *snow grains;* rebound from a hard surface and often break up.

snow shower—*See* shower.

solar radiation—The total electromagnetic *radiation* emitted by the sun. *See* insolation.

sounding—In meteorology, an upper-air observation; a *radiosonde* observation.

source region—An extensive area of the earth's surface characterized by relatively uniform surface conditions where large masses of air remain long enough to take on characteristic temperature and moisture properties imparted by that surface.

specific humidity—The ratio by weight of *water vapor* in a sample of air to the combined weight of water vapor and dry air. Compare *mixing ratio*.

squall—A sudden increase in wind speed by at least 15 knots to a peak of 20 knots or more and lasting for at least one minute. Essential difference between a *gust* and a squall is the duration of the peak speed.

squall line—Any nonfrontal line or narrow band of active *thunderstorms* (with or without *squalls*).

stability—A state of the atmosphere in which the vertical distribution of temperature is such that a *parcel* will resist displacement from its initial level. (*See also* instability.)

standard atmosphere—A hypothetical atmosphere based on climatological averages comprised of numerous physical constants of which the most important are:

(1) A surface *temperature* of 59° F (15° C) and a surface pressure of 29.92 inches of mercury (1013.2 millibars) at sea level;

(2) A *lapse rate* in the troposphere of 6.5° C per kilometer (approximately 2° C per 1,000 feet);

(3) A *tropopause* of 11 kilometers (approximately 36,000 feet) with a temperature of −56.5° C; and

(4) An *isothermal* lapse rate in the stratosphere to an altitude of 24 kilometers (approximately 80,000 feet).

standing cloud (standing lenticular altocumulus)—*See* lenticular cloud.

standing wave—A wave that remains stationary in a moving fluid. In aviation operations it is used most commonly to refer to a *lee wave* or *mountain wave*.

stationary front—Same as *quasi-stationary front*.

station pressure—The actual *atmospheric pressure* at the observing station.

steam fog—Fog formed when cold air moves over relatively warm water or wet ground.

storm detection radar—A weather radar designed to detect *hydrometeors* of precipitation size; used primarily to detect storms with large drops or hailstones as opposed to clouds and light precipitation of small drop size.

stratiform—Descriptive of clouds of extensive horizontal development, as contrasted to vertically developed *cumuliform* clouds; characteristic of stable air and, therefore, composed of small water droplets.

stratocumulus—A low cloud, predominantly *stratiform* in gray and/or whitish patches or layers, may or may not merge; elements are tessellated, rounded, or roll-shaped with relatively flat tops.

stratosphere—The atmospheric layer above the tropopause, average altitude of base and top, 7 and 22 miles respectively; characterized by a slight average increase of temperature from base to top and is very stable; also characterized by low moisture content and absence of clouds.

stratus—A low, gray cloud layer or sheet with a fairly uniform base; sometimes appears in ragged patches; seldom produces precipitation but may produce *drizzle* or *snow grains*. A *stratiform* cloud.

stratus fractus—*See* fractus.

streamline—In meteorology, a line whose tangent is the wind direction at any point along the line. A flowline.

sublimation—*See* change of state.

subrefraction—*See* refraction.

subsidence—A descending motion of air in the atmosphere over a rather broad area; usually associated with *divergence*.

summation principle—The principle states that the cover assigned to a layer is equal to the summation of the sky cover of the lowest layer plus the additional coverage at all successively higher layers up to and including the layer in question. Thus, no layer can be assigned a sky cover less than a lower layer, and no sky cover can be greater than 1.0 (10/10).

superadiabatic lapse rate—A *lapse rate* greater than the *dry-adiabatic lapse rate*. *See* absolute instability.

supercooled water—Liquid water at temperatures colder than freezing.

superrefraction—*See* refraction.

surface inversion—An *inversion* with its base at the surface, often caused by cooling of the air near the surface as a result of *terrestrial radiation*, especially at night.

surface visibility—Visibility observed from eye-level above the ground.

synoptic chart—A chart, such as the familiar weather map, which depicts the distribution of meteorological conditions over an area at a given time.

T

target—In radar, any of the many types of objects detected by radar.

temperature—In general, the degree of hotness or coldness as measured on some definite temperature scale by means of any of various types of thermometers.

temperature inversion—*See* inversion.

terrestrial radiation—The total infrared *radiation* emitted by the Earth and its atmosphere.

thermograph—A continuous-recording *thermometer*.

thermometer—An instrument for measuring *temperature*.

theodolite—An optical instrument which, in meteorology, is used principally to observe the motion of a *pilot balloon*.

thunderstorm—In general, a local storm invariably produced by a *cumulonimbus* cloud, and always accompanied by lightning and thunder.

tornado (sometimes called cyclone, twister)—A violently rotating column of air, pendant from a cumulonimbus cloud, and nearly always observable as "funnel-shaped." It is the most destructive of all small-scale atmospheric phenomena.

towering cumulus—A rapidly growing *cumulus* in which height exceeds width.

tower visibility—*Prevailing visibility* determined from the control tower.

trade winds—Prevailing, almost continuous winds blowing with an easterly component from the subtropical high pressure belts toward the *intertropical convergence zone;* northeast in the Northern Hemisphere, southeast in the Southern Hemisphere.

transmissometer—An instrument system which shows the transmissivity of light through the atmosphere. Transmissivity may be translated either automatically or manually into *visibility* and/or *runway visual range*.

tropical air—An air mass with characteristics developed over low latitudes. Maritime tropical air (mT), the principal type, is produced over the tropical and subtropical seas; very warm and humid. Continental tropical (cT) is produced over subtropical arid regions and is hot and very dry. Compare *polar air*.

tropical cyclone—A general term for a *cyclone* that originates over tropical oceans. By international agreement, tropical cyclones have been classified according to their intensity, as follows:

(1) **tropical depression**—winds up to 34 knots (64 km/h);

(2) **tropical storm**—winds of 35 to 64 knots (65 to 119 km/h);

(3) **hurricane or typhoon**—winds of 65 knots or higher (120 km/h).

tropical depression—*See* tropical cyclone.

tropical storm—*See* tropical cyclone.

tropopause—The transition zone between the *troposphere* and *stratosphere*, usually characterized by an abrupt change of *lapse rate*.

troposphere—That portion of the *atmosphere* from the earth's surface to the *tropopause;* that is, the lowest 10 to 20 kilometers of the atmosphere. The troposphere is characterized by decreasing temperature with height, and by appreciable water vapor.

trough (also called trough line)—In meteorology, an elongated area of relatively low atmospheric pressure; usually associated with and most clearly identified as an area of maximum cyclonic curvature of the wind flow (*isobars, contours,* or *streamlines*); compare with *ridge*.

true altitude—*See* altitude.

true wind direction—The direction, with respect to true north, from which the wind is blowing.

turbulence—In meteorology, any irregular or disturbed flow in the atmosphere.

twilight—The intervals of incomplete darkness following sunset and preceding sunrise. The time at which evening twilight ends or morning twilight begins is determined by arbitrary convention, and several kinds of twilight have been defined and used; most commonly civil, nautical, and astronomical twilight.

(1) **Civil Twilight**—The period of time before sunrise and after sunset when the sun is not more than 6° below the horizon.

(2) **Nautical Twilight**—The period of time before sunrise and after sunset when the sun is not more than 12° below the horizon.

(3) **Astronomical Twilight**—The period of time before sunrise and after sunset when the sun is not more than 18° below the horizon.

twister—In the United States, a colloquial term for *tornado*.

typhoon—A *tropical cyclone* in the Eastern Hemisphere with winds in excess of 65 knots (120 km/h).

U

undercast—A cloud *layer* of ten-tenths (1.0) coverage (to the nearest tenth) as viewed from an observation point above the layer.

unlimited ceiling—A clear sky or a sky cover that does not meet the criteria for a *ceiling*.

unstable—*See* instability.

updraft—A localized upward current of air.

upper front—A *front* aloft not extending to the earth's surface.

upslope fog—Fog formed when air flows upward over rising terrain and is, consequently, adiabatically cooled to or below its initial *dew point*.

V

vapor pressure—In meteorology, the pressure of water vapor in the atmosphere. Vapor pressure is that part of the total atmospheric pressure due to water vapor and is independent of the other atmospheric gases or vapors.

vapor trail—Same as *condensation trail*.

veering—Shifting of the wind in a clockwise direction with respect to either space or time; opposite of backing. Commonly used by meteorologists to refer to an anticyclonic shift (clockwise in the Northern Hemisphere and counterclockwise in the Southern Hemisphere).

vertical visibility—The distance one can see upward into a surface based *obscuration;* or the maximum height from which a pilot in flight can recognize the ground through a surface based obscuration.

virga—Water or ice particles falling from a cloud, usually in wisps or streaks, and evaporating before reaching the ground.

visibility—The greatest distance one can see and identify prominent objects.

visual range—*See* runway visual range.

vortex—In meteorology, any rotary flow in the atmosphere.

vorticity—Turning of the atmosphere. Vorticity may be imbedded in the total flow and not readily identified by a flow pattern.

   (a) **absolute vorticity**—the rotation of the Earth imparts vorticity to the atmosphere; absolute vorticity is the combined vorticity due to this rotation and vorticity due to circulation relative to the Earth (relative vorticity).

   (b) **negative vorticity**—vorticity caused by anticyclonic turning; it is associated with downward motion of the air.

   (c) **positive vorticity**—vorticity caused by cyclonic turning; it is associated with upward motion of the air.

   (d) **relative vorticity**—vorticity of the air relative to the Earth, disregarding the component of vorticity resulting from Earth's rotation.

W

wake turbulence—*Turbulence* found to the rear of a solid body in motion relative to a fluid. In aviation terminology, the turbulence caused by a moving aircraft.

wall cloud—The well-defined bank of vertically developed clouds having a wall-like appearance which form the outer boundary of the *eye* of a well-developed *tropical cyclone*.

warm front—Any non-occluded *front* which moves in such a way that warmer air replaces colder air.

warm sector—The area covered by warm air at the surface and bounded by the *warm front* and *cold front* of a *wave cyclone*.

water equivalent—The depth of water that would result from the melting of snow or ice.

waterspout—*See* tornado.

water vapor—Water in the invisible gaseous form.

wave cyclone—A *cyclone* which forms and moves along a front. The circulation about the cyclone center tends to produce a wavelike deformation of the front.

weather—The state of the *atmosphere*, mainly with respect to its effects on life and human activities; refers to instantaneous conditions or short term changes as opposed to *climate*.

weather radar—Radar specifically designed for observing weather. *See* cloud detection radar and storm detection radar.

weather vane—A *wind vane*.

wedge—Same as *ridge*.

wet bulb—Contraction of either *wet-bulb temperature* or *wet-bulb thermometer*.

wet-bulb temperature—The lowest *temperature* that can be obtained on a *wet-bulb thermometer* in any given sample of air, by evaporation of water (or ice) from the muslin wick; used in computing *dew point* and *relative humidity*.

wet-bulb thermometer—A thermometer with a muslin-covered bulb used to measure wet-bulb temperature.

whirlwind—A small, rotating column of air; may be visible as a dust devil.

willy-willy—A *tropical cyclone* of hurricane strength near Australia.

wind—Air in motion relative to the surface of the earth; generally used to denote horizontal movement.

wind direction—The direction **from** which wind is blowing.

wind speed—Rate of wind movement in distance per unit time.

wind vane—An instrument to indicate wind direction.

wind velocity—A vector term to include both *wind direction* and *wind speed*.

wind shear—The rate of change of *wind velocity* (direction and/or speed) per unit distance; conventionally expressed as vertical or horizontal wind shear.

X–Y–Z

zonal wind—A west wind; the westerly component of a wind. Conventionally used to describe large-scale flow that is neither cyclonic nor anticyclonic.